HIGH PERFORMANCE POLYMERS – POLYIMIDES BASED – FROM CHEMISTRY TO APPLICATIONS

Edited by **Marc Jean Médard Abadie**

High Performance Polymers – Polyimides Based – From Chemistry to Applications
http://dx.doi.org/10.5772/2834
Edited by Marc Jean Médard Abadie

Contributors

C. Aguilar-Lugo, A.L. Perez-Martinez, D. Guzman-Lucero, D. Likhatchev, L. Alexandrova, S. Diaham, M.-L. Locatelli, R. Khazaka, Evgenia Minko, Petr Sysel, Martin Spergl, Petra Slapakova, Anton Georgiev, Dean Dimov, Erinche Spassova, Jacob Assa, Peter Dineff, Gencho Danev, Shie-Chang Jeng, Shug-June Hwang, Takayuki Ishizaka, Hitoshi Kasai, Guangming Gong, Juntao Wu, Lutang Wang, Nian Fang, Zhaoming Huang, Andreea Irina Barzic, Iuliana Stoica, Camelia Hulubei, Aziz Paşahan, A.A. Périchaud, R.M. Iskakov, Andrey Kurbatov, T. Z. Akhmetov, O.Y. Prokohdko, Irina V. Razumovskaya, Sergey L. Bazhenov, P.Y. Apel, V. Yu. Voytekunas, M.J.M. Abadie

Published by InTech

Janeza Trdine 9, 51000 Rijeka, Croatia

Publishing Process Manager Marina Jozipovic
Technical Editor InTech Prepress, Novi Sad
Cover InTech Design team

Additional hard copies can be obtained from orders@intechopen.com

High Performance Polymers – Polyimides Based – From Chemistry to Applications,
Edited by Marc Jean Médard Abadie
p. cm.
ISBN 978-953-51-0899-3

For EU product safety concerns:
IN TECH d.o.o., Prolaz Marije Krucifikse Kozulić 3, 51000 Rijeka, Croatia,
info@intechopen.com or visit our website at intechopen.com.

Meet the editor

Professor Marc Jean Médard Abadie is Emeritus Professor at Université Montpellier 2, Sciences et Techniques du Languedoc. He was full Professor and head of Laboratory of Polymer Science and Advanced Organic Materials – LEMP/MAO. He is currently "Michael Fam" Visiting Professor at the School of Materials Science and Engineering MSE, Nanyang Technological University NTU, Singapore. Born in 1938, chemical engineer (1961) from Louis Pasteur University (Strasbourg – France), he successively got his BSc (1959), MSc (1960), Dr 3rd Cycle (1963) and Doctorate of Science (1972) in "Centre de Recherche sur les Macromolécules – Institute Charles Sadron" in Strasbourg with Professor Henry Benoit as President. He created the International Symposium on Polyimides and High Performance Polymers, called STEPI, which takes place every three years in Montpellier, France. Next STEPI 9 will be on June 3-5, 2013. His present activities concern high temperature polymers, high performance composites and nanocomposites, UV & EB coatings and biomaterials. He has published 11 books and 11 patents. He has advised nearly 95 MSc. and 52 PhD. students with whom he has published over 402 papers. Prof. Abadie has more than 40 years experience in Polymer Science with 10 years in the industry (IBM USA, MOD UK, SNPA/Total France).

Contents

Preface

The term "High Performance Polymers" covers a large number of organic materials. Some consider the thermal stability as a criterion of high performance and consider that polymers are able to sustain long-term service at 180°C and are termed as "High Temperature Polymers". Others referred to as "Heat Resistant Polymers" whose macromolecular backbone is constituted by alternate moieties of aromatic rings and heterocyclic units which is represented by Polyimides and derivatives.

Dr. Cyrus E. Sroog, at E.I. DuPont de Nemours and Company, Wilmington DE, was the precursor of high performance organic materials by developing a polymer based on the polycondensation of pyromelitic acid dianhydride and oxydianiline, universally known as the "PMDA-ODA" polyimide. This new concept of "polyheterocyclisation" was at the origin of an impressive series of new *high performance polymers.*

For the last three decades, increasing need in the high technology industries (space, micro and nano electronics, civil transportation, planes, automotive, membranes, fuel cells, etc.) has been the driving force for the development of new polymeric systems and materials combining both thermal mechanical and temperature resistance and also properties others such as lightweight, high corrosion resistance, good wear properties, dimensional stability, low flammability, separation properties, moisture resistance, insulating properties and ability to be transformed with conventional equipment.

In fact all these properties were unavailable in conventional materials but may be covered by aromatic and heterocyclic linear and thermosetting resins such as: acetylene terminated resins, bismaleimides, polyetherimides, polyamide-imides, polybenzimidazoles, polyimides, polyetherketones, polyphenylquinoxalines, polyphenylensulfides, polysulfones & derivatives, polystyrylpyridines, fluoropolymers, silicones, etc.

The feature of polyimides and other heterocyclic polymers are now well-established and used for long term temperature durability in the range of 250 – 350°C.

New structures coming from the chemical modifications of the heterocyclic backbones or from the creation of new architectures are under development for advanced technologies for the future. Some new chemical structures based on Polyimides and different applications are presented.

This book will review synthesis, mechanisms, ultimate properties, physico-chemical properties, processing and applications of such high performance materials needed in advanced technologies. It presents interdisciplinary papers on the state of knowledge of each topic under consideration through a combination of overviews and original unpublished research.

The volume contains eleven chapters by 41 authors from 11 countries. The book is divided into three sections as follow:

- Chemistry highlighting new chemical architectures – Chapters I to III
- Chemical and Physical Properties (Review) – Chapter IV
- Applications to :
  - Liquid Crystal Devices – Chapter V
  - Low-k – Chapter VI
  - Polyimides Fibers and Nano-fibers – Chapters VII and VIII
  - Optical Properties – Chapter IX
  - Sensors – Chapter X
  - Self-healing of Films – Chapter XI.

I would like to dedicate this book on "High Performance Polymers - Polyimides Based - From Chemistry to Applications" to Professor Alexander L'vovich Russanov known as Shura, and his team from Institute A. N. Nesmeyanov INEOS, Russian Academy of Science, Moscow. His contribution to polyheterocyclic polymers is immense by the originality of the chemistry developed and also by the novelty brought to the Polyimide family and derivatives. Shura has been for nearly thirty years my cheerful companion and during this period I was impressed not only by his creative chemistry but also by his sense of humor and his optimism. For sure, the scientific community will miss him, his competence and joviality.

Prof. Marc Jean Médard Abadie
POLYTECH' Montpellier - Université Montpellier; ICGM – AIME, France
Nanyang Technological University, School of Materials Science, Singapore

Section 1

Chemistry

Chapter 1

Polyimides Based on 4-4'-Diaminotriphenylmethane (DA-TPM)

C. Aguilar-Lugo, A.L. Perez-Martinez, D. Guzman-Lucero, D. Likhatchev and L. Alexandrova

Additional information is available at the end of the chapter

http://dx.doi.org/10.5772/53608

1. Introduction

Rigid-rod aromatic polyimides (PIs) constantly attract wider interest because of their unique combination of properties, such as the excellent thermo-oxidative stability and mechanical properties, good dielectric strength and dimensional stability.[1-4] Additionally in recent years, PIs have been considered as one of the best materials for gas separation membranes due to their reasonable permeability to CO_2 and high selectivity against CH_4.[5] However, synthesis and processing of these polymers are generally very difficult because of their limited solubility and infusibility.[6] Considerable efforts have been made to improve the solubility through the synthesis of new diamine or dianhydride monomers. The common strategy consists in the incorporation of bulky lateral substituents,[4,7-15] flexible alkyl side chains,[16,17] non-coplanar biphenylene moieties,[18] and kinked units,[19-22] into rigid polymer backbones.

4-4'-Diaminotriphenylmethane (DA-TPM) and its derivates, have attracted the attention of our research group as monomers for the synthesis of various rigid-rod polyamides (PAs) and PIs.[23,24] The pendant phenyl ring and practically free internal rotation of the triphenylmethane bridging group predicted from the theoretical calculations make DA-TPM an excellent candidate for synthesis of processable PAs and PIs without sacrificing the high thermal stability.[23] Earlier research carried out for the structurally similar N,N-diamine triphenylamine (DA-TPA) showed that the incorporation of a pendant phenyl group into the polymer backbone is a successful approach to increase solubility and processability of PIs.[15,20,25-26] The synthesis of DA-TPM developed in our group is simple and highly efficient using commercially available and cheap starting materials such as aniline and benzaldehyde. This is a big advantage in comparison to N,N-diamine triphenylamine, whose synthesis is much more complicated and required expensive reagents. Besides, the use of microwave irradiation instead of traditional heating allows reducing the reaction times as well as the amount of aniline employed, that facilitates the purification process. A wide variety of diamines derived from DA-TPM using different substituted anilines and benzaldehydes has been successfully synthesized.[27]

In spite of all these advantages, there was only one report on application of DA-TPM as a monomer for the PIs from the Koton group dated 1980,[28] where PIs based on DA-TPM and two anhydrides, pyromellitic and 4,4′-oxydiphtalic, had been obtained by traditional two-stage method using thermo-imidization. However the PIs synthesized yielded relatively low molecular weights and, as a result, their thermo-mechanical properties were not as good as those reported for other PIs based on traditional 4,4'-Diaminodiphenylmethane.

The two-stage process involving low temperature condensation with formation of pre-polymer, poly(amic acid), followed by the thermal cyclodehydration or imidization at 250-300 °C is the most frequently employed method for formation of PIs. This process has some inherent limitations for example, the generation of water, which would create voids and stresses in the final materials.[29] Additionally, high temperature leads to several undesirable side reactions, such as crosslinking or scissoring polymer chains that can result in brittle films.[1,2] The thermal imidization step may be substituted by the catalytic cyclodehydration at room temperature. Normally, a mixture of acetic anhydride with tertiary amines is applied for the chemical cyclization. Much milder reaction conditions in this process permitted to produce less damaged PIs and therefore the films of higher elasticity. Another method employed for producing soluble PIs is a one-step synthesis. It had been shown by various authors that this method may be the most effective for preparation of processable PIs of large molecular weights and linear structure.[30-33]

In this article we would like to report a comparative study of PIs based on DA-TPM and various dianhydrides, obtained by different methods: the two-step and one-step syntheses. The solubility, thermal, mechanical and preliminary gas transport properties of these materials have been studied.

2. Experimental

2.1. Materials

The reagents were purchased from Aldrich Co. and Chriskev. Dianhydrides were recrystallized from acetic anhydride (Ac_2O). Solvents: N-methyl-2-pyrrolidone (NMP), N,N-dimethylformamide (DMF) were dried and stored over molecular sieves. The nitrobenzene was distilled under vacuum prior to use. All other reagents and solvents were used as received.

2.2. Synthesis of DA-TPM

DA-TPM was prepared according to the literature.[23]

2.3. Polymerization procedures

The polymerizations were performed by two different routes; two-step and one-step polycondensations. The imidization in two-step synthesis was carried out thermally and chemically (Figure 1)

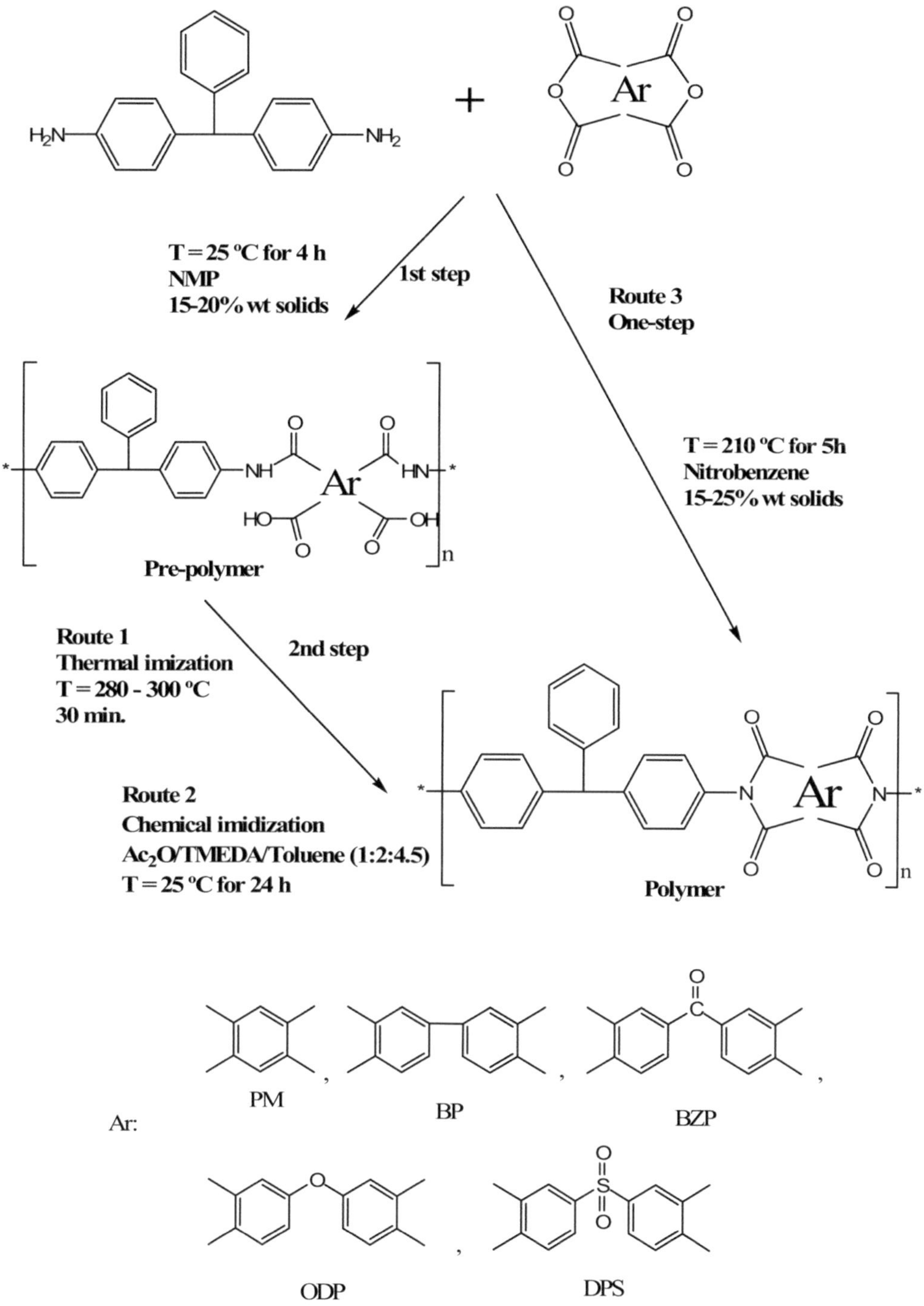

Figure 1. Synthesis of DA-TPM based PIs.

2.4. Synthesis of poly(amic acid)s

Poly(amic acid)s (PAA) were prepared by low-temperature solution polycondensation of DA-TPM and the corresponding dianhydride (25 wt % of solids). A stoichiometric amount of the solid dianhydride was added to the diamine solution in NMP at 0°C. The continuously stirred mixture was gradually heated to room temperature and allowed to stir for another 4-5 h.

2.5. Thermal imidization of PAA

The solutions of PAA were cast onto glass plates and dried at 60°C under vacuum for 5 h to produce solid transparent PAA films. The dry films were stripped off the glass plates, placed into metal frames, and heated at 270 °C or at 300 °C for ~ 0.5 h to produce the desirable PIs.

2.6. Chemical imidization of PAA

PPAs films were immersed in the imidization mixture of Ac_2O/TMEDA/Toluene (TMEDA=N,N,N',N'-tetramethylethylenediamine) (1:2:4.5) for 24 h at room temperature. Then the films were washed with distilled water and dried at 80 °C in vacuum till a constant weight (for ca. 6 h).

2.7. One-step high-temperature polycondensation

A solution of DA-TPM in nitrobenzene was placed into a three-neck round-bottom flask, equipped with a reflux condenser, under nitrogen atmosphere. A stoichiometric amount of the corresponding dianhydride was added to the DA-TPM solution (total 15-25 wt % solids). The reaction mixture was heated under intensive stirring and nitrogen flow at 210 °C for 5 h.. Films were cast from the reaction solutions at 50-60 °C onto glass plates and dried at 200 °C in vacuum till a constant weight (for ca. 12 h).

2.8. Measurements

Infrared (FT-IR) and UV-vis spectra were recorded with a Nicolet 510P FT-IR and a Shimadzu 3101PC UV spectrophotometer, respectively. Inherent viscosity (η_{inh}) was determined in 0.5 g /dL DMF solutions with an Ubbelohde viscometer at 25 °C. For BP-TPM, η_{inh} was determined in 0.5 g/dL nitrobenzene solution at 50 °C, because this polymer was insoluble in DMF at room temperature. A Du Pont, high resolution Thermogravimetric Analyzer, TGA 2950, was used for the thermal analysis at a heating rate of 5 °C/min. The glass transition temperature, T_g, was determined by a film-elongation technique using a Du Pont Thermo-mechanical Analyzer, Model TA 2940 (nitrogen atmosphere and 5 °C/min). Mechanical tests of polymer films (about 25 μm thickness) were performed by using an INSTRON Tester, Model 111, at a drawing rate 50 mm/min, on samples of 20 X 5 X 0.0025 mm size. Wide-angle x-ray diffractometry (WAXD) was performed on a Siemen's D-500 diffractometer, with $CuK_{\alpha 1}$ radiation of 1.5406 Å.

3. Results and discussion

The one-step polycondensation in nitrobenzene was less sensitive to the stoichiometry of reagents than the two-step synthesis.[1,34,35] FT-IR spectra of the obtained PIs showed intensive characteristic imide bands at 1773 cm^{-1} (imide C=O asymmetrical stretching), 1714 cm^{-1} (imide C=O symmetrical stretching), and 1380 cm^{-1} (imide CNC axial), confirming the complete imidization.

The solubility behavior of DA-TPM based PIs in common organic solvents is summarized in Table 1. The solubility was strongly correlated to the imidization technique. Remarkable differences in solubility were observed between the samples prepared by one step or two step syntheses with chemical imidization and those resulted from thermal imidization. Chemical imidization and one-stage methods yielded polymers readily soluble in NMP, DMF and pyridine (maximum concentration up to 10-20% by weight) at room temperature. In contrast, PIs obtained by thermal imidization exhibited poor or null solubility even at high temperature. The insoluble fraction increased with the temperature of imidization (270 or 300 °C), this behavior might be attributed to crosslinking occurred during the thermal process.[1-2,36-37] Thus, DA-TPM polymers displayed excellent solubility owing to the presence of the bulky pendent phenyl group in comparison to the analogue structures but obtained with conventional 4,4'-diaminodiphenylmethane.[38] Due to the bulkiness and free internal rotation in DA-TPM moieties, the chain packing of the polymer was disturbed, and consequently, the solvent molecules could easily penetrate between chains and dissolve the polymer. It should be noted that even PM-TPM showed good solubility in polar solvents although it was derived from dianhydride without any bridging groups and therefore had the most rigid structure. Only partial solubility at elevated temperature for the similar poly(triphenylaminepyromellitimide) has been reported.[15,25] Thus, PM-TPM is one of a few soluble poly(pyromellitimide)s. Besides, other soluble poly(pyromellitimide)s reported were synthesized using expensive 4,4'-hexafluroisopropylidene dianiline.[39]

Inherent viscosities, as molecular weight characteristics, and mechanical properties of the DA-TPM PIs are given in Table 2. Generally, the Young´s modulus (E_0) and tensile strength (σ_b) decrease, and the elasticity increases with increasing chain flexibility. The rigidity of dianhydride moiety decreases in the following order PM>BP>DPS>BZP>ODP and their mechanical properties changed as should be expected. As seen from the table, the molecular weights of PM-TPM and BZP-TPM, prepared by one-step, were practically the same or even slightly higher than those of their analogues obtained by chemical imidization of the corresponding PAAs. No big difference in mechanical properties between PI films prepared by one-step synthesis or by chemical imidization was noted, but it was not so for the PIs obtained through the thermal imidization. These PIs were not soluble and resulted in very brittle films in comparison to the same PIs synthesized by two other methods. The only suitable for analysis films formed by thermo-cyclization were casted from PAAs with flexible dianhidryde moieties, having –CO- or –O- groups between the phenyl rings; namely, BZP-TPM and ODP-TPM. Their mechanical properties were very poor; for example, the elongation at break (ε_b) was only 6 – 7 % whereas the elongations for the

Polymer	Route	DMF	NMP	THF	$C_2H_4Cl_2$	Py	NB*	*p*-chlorophenol*	*m*-cresol*
PM-TPM	Thermal	PS*	PS*	i	i	i	i	i	i
	Chemical	S	S	S	S	S	S	S	S
	One-step	S	S	-	-	S	S	S	S
BP-TPM	Thermal	PS*	PS*	i	i	i	i	i	i
	Chemical	S	S	S	S	S	S	S	S
	One-step	PS	PS	-	-	PS	S	S	S
BZP-TPM	Thermal	PS*	PS*	i	i	i	i	i	i
	Chemical	S	S	PS	PS	S	S	S	S
	One-step	S	S	-	-	S	S	S	S
ODP-TPM	Thermal	PS*	PS*	i	i	i	i	i	i
	Chemical	S	S	PS	PS	S	S	S	S
	One-step	S	S	-	-	S	S	S	S
DPS-TPM	Thermal	PS*	PS*	i	i	i	i	i	i
	Chemical	S	S	PS	PS	S	S	S	S
	One-step	S	S	-	-	S	S	S	S

Solubility at room temperature, S=totally soluble, PS=partially soluble, i=insoluble. DMF=N,N-dimethylformamide; NMP= N-methyl-2-pyrrolidone; Py=Pyridine; THF=Tetrahydrofuran; NB=Nitrobenzene.
* Solubility at high temperature (100-150°C)

Table 1. Solubility of DA-TPM Based PIs Obtained by Different Methods.

chemically cyclized or resulted from one-step method PIs were ten times higher (60 – 70 %). The same tendency was noted for other mechanical properties, Young´s modulus and tensile strength. Such results can be explained considering the reduction of the molecular weights and possible crosslinking occurred under the severe conditions of the thermal process. In spite of similar mechanical properties an important difference between one-step and chemically cyclized PIs was observed on the supramolecular structure level. It has been shown that one-step and chemical imidization processes led to remarkably different packing of polyimide molecules.[40] The WAXD diffraction patterns of PM-TPM films obtained by the chemical imidization of PAA and one-step high temperature polycondensation are presented in Figure 2. The significant difference in the positions, intensities and half-widths of the X-ray reflections suggested much better chain packing in the polymer prepared by one-step route. It is important to note that the PI films prepared by the one-step synthesis showed no changes of their initial properties after 6 months of storage at room temperature, even under long exposure to air and humidity, while PIs produced by chemical imidization rapidly lost their elasticity after several weeks under the same conditions. The higher stability of PIs from the one-step polycondensation may be attributed to their more regular structure. It is common even to find slight traces of imide isomeric unit, isoimide, or residual amic acid in PIs obtained by two-step synthesis with thermo- or chemo- cyclization.[1,2] These units should be considered as defective sites, because of their susceptibility to hydrolysis. This kind of defects is inevitable in the imidization process; it is particularly difficult to avoid formation of isoimide units because of their equilibrium with the imide structures. Isoimides are reactive and susceptible to nucleophilic attack, so the polymer chain may break and, as a consequence, the polymer molecular weights decrease. This

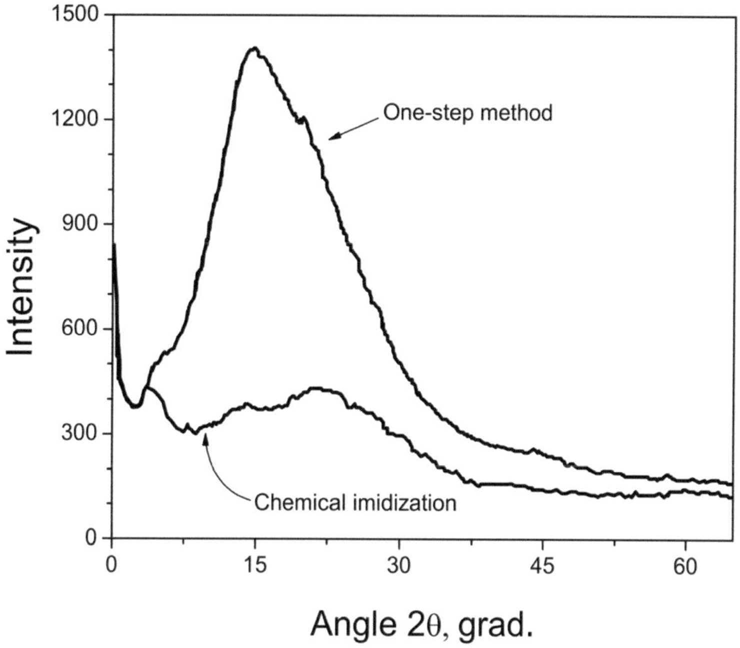

Figure 2. WAXD patterns of PM-TPM films obtained by the chemical imidization or by the one-step method.

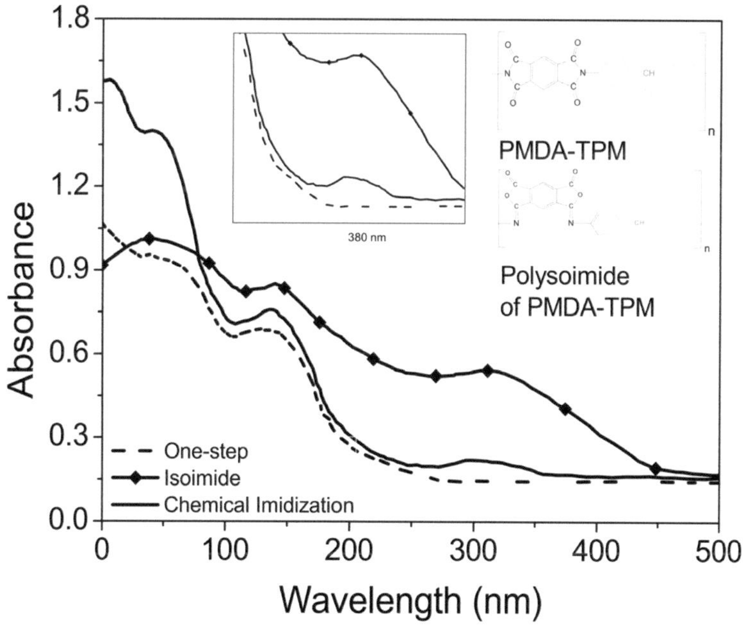

Figure 3. UV-vis spectra of PM-TPM obtained by chemical imidization (——) or by one-step high temperature polycondensation (— —), and the spectrum of the polyisoimide based on DA-TPM and PMDA (—♦—)

process is accelerated in wet media. Figure 3 shows the UV-vis spectra of two PM-TPMs obtained by one-step (dash line) and two-step (solid line) methods. Spectrum of polyisoimide is also given in Figure 3 for comparison. As can be seen, isoimide exhibits strong absorption at 380 nm. The absorption in this area is stronger for chemically cyclized PM-TPM (two-step method) than that for one-step PM-TPM which practically does not absorb at these wavelength. Therefore, the long wave tail in the UV-vis spectrum of PIs indicates the presence of isoimide units in the backbone.

Polymer	Route	η_{inh} dL/g	Young´s Modulus (E_0) GPa	Tensile Strength (σ_b) MPa	Elongation at Break (ε_b) %
PM-TPM	Thermal	-	*	*	*
	Chemical	1.1	1.8	158	28
	One-step	1.5	2.0	164	38
BP-TPM	Thermal	-	-	-	-
	Chemical	-	-	-	-
	One-step	0.6	1.8	150	42
BZP-TPM	Thermal	-	0.85	27.8	7
	Chemical	1.4	1.5	135	70
	One-step	1.2	1.5	138	75
ODP-TPM	Thermal	-	0.63	21.5	6
	Chemical	-	-	-	
	One-step	1.1	1.4	134	58
DPS-TPM	Thermal	-	-	-	-
	Chemical	-	-	-	-
	One-step	0.5	1.6	107	20

** Brittle*

Table 2. Mechanical Properties of DA-TPM Based Pis.

Glass transition temperatures, T_g, and temperatures for 5 and 10% weight loss of PIs obtained by all methods are listed in Table 3. The TMA analysis showed that PIs obtained by chemical imidization and one-step polycondensation exhibit well-distinguished T_gs in the range of 260 - 320 °C, depending on the chain rigidity. T_gs for the films formed by thermal imidization were not so well defined and difficult to detect. Flexible linkages, such as –O– in ODP-TPM, tend to lower T_g. The T_g values are close to those reported for the flexible chain polyimides based in 4,4′-diaminodiphenylmethane.[38] All synthesized polymers demonstrated excellent thermal stability. According to TGA data (Table 3), thermal decomposition of DA-TPM based PIs started above 400 °C, no important difference was observed for PIs obtained by chemical imidization and one-step method. The difference in the weight loss values for the different PIs depended on the dianhydride moiety. The highest thermal and thermo-oxidative resistance, among the synthesized polymers, was observed for PM-TPM, which shows a 5 and 10% weight loss in an inert atmosphere at 540 °C and 560 °C respectively.

Polymer	Route	Glass Transition Temperature [°C]	Weight Loss Temperature by TGA [°C]			
			Air		Nitrogen	
			5%	10%	5%	10%
PM-TPM	Thermal	-	470	500	-	-
	Chemical	365	505	529	538	564
	One-step	321	502	524	530	556
BP-TPM	Thermal	-	-	-	-	-
	Chemical	-	-	-	-	-
	One-step	315	498	516	525	541
BZP-TPM	Thermal	-	-	-	-	-
	Chemical	288	457	482	506	531
	One-step	282	463	488	509	536
ODP-TPM	Thermal	-	460	510	-	-
	Chemical	-	-	-	-	-
	One-step	269	459	476	502	538
DPS-TPM	Thermal	-		-	-	-
	Chemical	-	-	-	-	-
	One-step	305	431	457	492	519

Table 3. Thermal Properties of DA-TPM Based PIs.

Preliminary gas transport properties for some of the PIs and ideal separation factors for selected gas pairs are summarized in Table 4. The polymer permeability coefficients decreased in the following order $P(H_2) \approx P(He) > P(CO_2) > P(O_2) > P(N_2) > P(CH_4)$. This tendency is very similar to the behavior reported for the most glassy polymer membranes indicating a relationship between the permeability and the kinetic diameter of the tested gases.[5] Gas permeability typically increases with increasing free volume of the polymer which is determined in a great extent by the chemical structure. The presence of bulky pendant groups enhances interchain spacing and reduces the packing efficiency of the polymer chains and, thus, free volume and gas permeability increase.[41,42] Results of structure/property optimization studies for polymers suitable for such separation suggest that polymers with high selectivity exhibit low permeability and vice versa.[5] Aromatic PIs are one of the best candidates for gas separation membranes, particularly for the natural gas purification, due to their high CO_2/CH_4 selectivity. However, the low permeability is the principal obstacle for their wide industrial applications.[43] The goal is to improve permeability of PIs with the minimum loss in the selectivity. PIs containing DA-TPM exhibit much better gas separation characteristics combined with higher permeability coefficients than the similar polymers but synthesized with other non-fluorinated diamines.[5] Such good membrane characteristic of our PIs may be related to the pendant phenyl group, which creates a larger free volume and the possibility of molecular rotations. The anhydride bridging groups with low rotational barriers, such as –O-, facilitate chain motions and results in higher CO_2 permeability while the incorporation of the bulky linkage groups, like $-SO_2-$, lowers the gas permeability. This tendency could be seen for DA-TPM PIs, the order of $P(CO_2)$ is the following ODP-TPM>DPS-TPM>BZP-TPM. The polarity may also affect chain-to-chain interactions and subsequently modify the chain rigidity and packing efficiency. The greater polarity of C=O group in BZP-TPM explains the lower permeability and higher selectivity of this polymer in comparison to ODP-TPM and DPS-TPM.

Polymer	Permeability (Barrers*)						Ideal separation factors	
	H_2	He	O_2	N_2	CH_4	CO_2	$\alpha O_2/N_2$	$\alpha CO_2/CH_4$
BZP-TPM	8.69	7.86	0.46	0.17	0.086	2.94	2.7	34.2
DPS-TPM	10.95	10.75	0.90	0.16	0.14	4.53	5.6	32.4
ODP-TPM	9.42	9.36	0.73	0.14	0.13	3.66	5.2	28.2

* Barrer =$1x10^{-10}$ cm^3(STP)cm/(cm^2 s cmHg)

Table 4. Permeability Coefficients and Ideal Separation Factors Measured for Pure Gases at 35 ^{0}C and 10atm Upstream Pressure.

4. Conclusions

DA-TPM was found to be a suitable monomer for synthesis of processable PIs with good mechanical and thermal properties. Influence of synthetic method on the polymer properties has been studied. The PIs obtained by one-step high-temperature polycondensation and by two-step method with chemical imidization demonstrated better solubility and mechanical properties than PI films synthesized by thermo-imidization. However, PI films prepared by one-step method conserved their properties for much longer time upon exposure to air and humidity than the chemically imidized films. This is probably because of the formation of the isoimide defect units during the chemical imidization and differences in the supramolecular structures. The high solubility can be attributed to the effect of pendant phenyl ring and the free internal rotation in DA-TPM. Preliminary studies demonstrated that PIs based on DA-TPM exhibit also very promising gas transport properties.

Author details

C. Aguilar-Lugo, A.L. Perez-Martinez, D. Likhatchev and L. Alexandrova
Instituto de Investigaciones en Materiales, Universidad Nacional Autonoma de Mexico, Circuito Exterior s/n, Ciudad Universitaria, Mexico D.F., Mexico

D. Guzman-Lucero
Programa de Ingeniería Molecular, Instituto Mexicano del Petróleo. Eje Central Lázaro Cárdenas No. 152, México DF., Mexico

5. References

[1] Sroog, C.E. *Prog. Polym. Sci.* 1991, *16*, 561.

[2] Bessonov, M.I.; Koton, M. M.; Kudryavtsev, V.V.; Laius, L. A. (Eds.) *Polyimides, Thermally Stable Polymers,* Consultants Bureau: New York, 1987.

[3] Wilson, D.; Stenzenberger, H. D.; Hergenrother, P. M. *Polyimides,* Blackie: New York, 1990.

[4] Spiliopoulos, I. K.; Mikroyannidis, J. A.; Tsivgoulis, G.M. *Macromolecules* 1998, *31*, 522.

[5] Ayala, D.; Lozano, A. E.; de Abajo, J.; Garcia-Perez, C.; de la Campa, J. G.; Peinemann, K.V.; Freeman, B. D.; Prabhakar, R. *J. Membr. Sci.* 2003, *215*, 61.

[6] Ballauff, M. *Angew. Chem., Int. Ed.* 1989, *28*, 253.

[7] Liaw, D.J.; Liaw, B.Y. *Macromol. Symp.* 1997, *122*, 343.

[8] Jeong, H.J.; Oishi, Y.; Kakimoto, M.A.; Imai, Y. *J. Polym. Sci., Part A: Polym. Chem* 1990, *28*, 3193.

[9] Liaw, D.J.; Liaw, B.Y.; Li, L.J.; Sillion, B.; Mercier, R.; Thiria, R.; Sekiguchi, H. *Chem. Mater.* 1998, *10*, 734.

[10] Sun, X.; Yang, Y.K.; Lu, F. *Macromolecules* 1998, *31*, 4291.

[11] Akutsu, F.; Inoki, M.; Araki, K.; Kasashima, Y.; Naruchi, K.; Miura, M. *Polym. J.* 1997, *29*, 529.

[12] Lozano, A.E.; de Abajo, J.; de la Campa, J. G.; Preston, J. *J. Polym. Sci., Part A: Polym. Chem.* 1995, *33*, 1987.

[13] Park, K.H.; Tani, T.; Kakimoto, M.A.; Imai, Y. *J. Polym. Sci., Part A: Polym. Chem.* 1998, *36*, 1767.

[14] Kasashima, Y.; Kumada, H.; Yamamoto, K.; Akutsu, F.; Naruchi, K.; Miura, M. *Polymer* 1995, *36*, 645.

[15] Liaw, D.J.; Hsu, P.N.; Chen, W.H.; Lin, S.L. *Macromolecules* 2002, *35*, 4669.

[16] Ballauff, M.; Schmidt, G.F. *Macromol. Chem. Rapid Commun.* 1987, *8*, 93.

[17] Steuer, M.; Horth, M.; Ballauff, M. *J. Polym. Sci., Part A: Polym. Chem.* 1993, *31*, 1609.

[18] Kaneda, T.; Katsura, T.; Nakagawa, K.; Makino, H.; Horio, M. *J. Appl. Polym. Sci.* 1986, *32*, 3151.

[19] Liaw, D.J.; Liaw, B.Y.; Hsu, P.N.; Hwang, C.Y. *Chem. Mater.* 2001, *13*, 1811.

[20] Liaw, D.J.; Liaw, B.Y.; Yang, C.M. *Macromolecules* 1999, *32*, 7248.

[21] Liaw, D.J.; Liaw, B.Y. *Macromol. Chem. Phys.* 1998, *199*, 1473.

[22] Glatz, F.P.; Mulhaupt, R. *Polym. Bull.* 1993, *31*, 137.

[23] Likhatchev, D.; Alexandrova, L.; Tlenkopatchev, M.; Vilar, R.; Vera-Graziano, R. *J. Appl. Polym. Sci.* 1995, *57*, 37.

[24] Likhatchev, D.; Alexandrova, L.; Tlenkopatchev, M.; Martinez-Richa, A.; Vera-Graziano, R. *J. Appl. Polym. Sci.* 1996,61, 815.

[25] Vasilenko, N.A.; Akhmet'eva, Ye.D.; Sviridov, Ye.B.; Berendyayav, V.I.; Rogozhkina, Ye.D.; Alkayeva, O.F.; Koshelev, K.K; Izyumnikov, A.L.;Kotov, B.V. *Polym. Sci. USSR* 1991, *33*, 1439.

[26] Oishi, Y.; Ishida, M.; Kakimoto, M.; Imai, Y.; Kurosaki, T. *J. Polym. Sci., Part A: Polym. Chem.* 1992, *30*, 1027.

[27] Guzmán-Lucero, D.; Guzmán, J.; Likhatchev, D.; Martínez-Palou, R. *Tetrahedron Lett.* 2005, *46*, 1119.

[28] Koton, M.M.; Romanova, M.S.; Laius, L.A.; Sazanov, Yu.N.; Fjodorova, G.N. *Z. Prikl. Khim.* 1980, *53*, 1591.

[29] Hergenrother, P.M. *High Perform. Polym.* 2003, *15*, 3.

[30] Kuznetsov, A.A.; Yablokova, M.; Buzin, P.V.; Tsegelskaya, A.Y. *High Perform. Polym.* 2004, *16*, 89.

[31] Imai, Y.; Maldar, N.N.; Kakimoto, M.A. *J. Polym. Sci., Part A: Polym. Chem.* 1984, *22*, 2189.
[32] Giesa, R.; Keller, U.; Eiselt, P.; Schmidt, H.W. *J. Polym. Sci., Part A: Polym. Chem.* 1993, *31*, 141.
[33] Kaneda, T.; Katsura, T.; Nakagawa, K.; Makino, H. *J. Polym. Sci., Part A: Polym. Chem.* 1986, *32*, 3133.
[34] Dine-Hart, R.A.; Wright, W.W. *J. Appl. Polym. Sci.* 1967, *11*, 609.
[35] S. V. Vinogradova, S.V.; Slonimskii, G.L.; Vygodskii, Ya.S.; Askadskii, A.A.; Mzhel'skii, A.I.; Churochkina, N.A.; Korshak, V.V. *Polym. Sci. USSR* 1969, *11*, 3098.
[36] Saini, A.K.; Carlin, C.M.; Patterson, H.H. *J . Polym. Sci., Part A: Polym. Chem.* 1993, *31*, 2751.
[37] Snyder, R.W.; Thomson, B.; Bartges, B.; Czerniawski, D.; Painter, P.C. *Macromolecules* 1989, *22*, 4166.
[38] St. Clair, T.L. *Polyimides* Wilson, D.; Stenzenberger, H.D.; Hergenrother, P.M. (Eds.) Blackie, London, 1990.
[39] Likhatchev, D.; Gutierrez-Wing, C.; Kardash, I.; Vera-Graziano; R. *J. Appl. Polym. Sci.* 1996, *59*, 725.
[40] Likhatchev, D.; Chvalum, S. *Advances in Polyimides and Low Dielectric Polymers* Sachdev, H.S.; Khojasteh, M.M; Feger, C. (Eds.) SPE, Inc., New York, 1999, 167.
[41] Xiao, Y.; Low, B.T.; Hosseini, S.S.; Chung, T.S.; Paul, D.R. *Prog. Polym. Sci.* 2009, *34*, 561.
[42] Coleman, M.R.; Koros, W.J. *J. Membr. Sci.* 1990, *50*, 285.
[43] Scholes, C.A.; Stevens, G.W.; Kentish, S.E. *Fuel* 2012, *96*, 15.

Chapter 2

BPDA-PDA Polyimide: Synthesis, Characterizations, Aging and Semiconductor Device Passivation

S. Diaham, M.-L. Locatelli and R. Khazaka

Additional information is available at the end of the chapter

http://dx.doi.org/10.5772/53994

1. Introduction

Polyimides (PIs) are advanced polymeric materials well-known for their excellent thermal, electrical, mechanical and chemical properties [1]. PIs are particularly attractive in the microelectronics industry due to their high thermal stability (T_d), high glass transition temperature (T_g), low dielectric constant, high resistivity, high breakdown field, inertness to solvent, radiation resistance, easy processability, etc [2,3]. Recently, the emergence of novel wide bandgap semiconductor (SiC, GaN or Diamond) devices aiming to operate between 200 °C and 400 °C make PIs as one of the most potential organic materials for the surface secondary passivation [4]. In such a high temperature range of operation with large thermal cycling constraints imposed by both the devices and the ambient temperature, the thermo-mechanical properties of passivation PIs appear as fundamental to ensure long lifetime of the materials and reliable behaviour of the devices. For instance, Table I shows some electrical, thermal and mechanical properties of wide band gap semiconductors compared to those of silicon (Si).

	E_g (eV)	ε_r	E_{br} (MV/cm)	λ_{th} (W/m/K)	CTE (ppm/°C)
Si	1.1	11.8	0.2	150	3
GaN	2.3-3.3	9-11	1.3-3.3	80-130	5.6
SiC	2.2-3.3	10	1.2-2.4	450	4
Diamond	5.4	5.5	5-10	2000	1

E_g: forbidden energy band gap; ε_r: dielectric constant; E_{br}: breakdown field; λ_{th}: thermal conductivity; CTE: coefficient of thermal expansion.

Table 1. Main physical properties of wide band gap semiconductor materials at 300 K.

It is possible to observe that the semiconductor materials own a low coefficient of thermal expansion (CTE) below 6 ppm/°C. Thus, among the criteria for the PI passivation material choice, the CTE, T_g and T_d temperatures are of prior importance. Table II shows some electrical, thermal and mechanical properties of the main aromatic PIs.

The classical poly(4,4′-oxydiphenylene pyromellitimide) (PMDA-ODA) appears as not well adapted for a severe thermal cycling operation due to the strong mismatch between its CTE (30-40 ppm/°C) and the one of semiconductor materials (<6 ppm/°C). This mismatch induces strong mechanical stresses in PMDA/ODA films coated on Si wafer (29 MPa). The poly(4-4′-oxydiphenylene biphenyltetracarboximide) (BPDA-ODA), the poly(4-4′-oxydiphenylene benzophenonetetracarboximide) (BTDA-ODA), the poly(*p*-phenylene benzophenonetetracarboximide) (BTDA-PDA) and the poly(*p*-phenylene oxydiphthalimide) (ODPA-PDA) have a chemical packing which leads to similar issues on Si wafer due to large CTEs mismatching (≥28 ppm/°C). Moreover, PMDA-ODA, BPDA-ODA, BTDA-ODA and ODPA-PDA exhibit the lowest degradation temperature values due to the degradation of the C–O–C ether bond present in the PI monomer unit [1,15,16].

	T_g (°C)	T_d (°C)	ε_r	E' (GPa)	TS (MPa)	Film stress on Si wafer (MPa)	CTE (ppm/°C)
PMDA-ODA	>380	586	3.4	3.0	170	29	30-40
PMDA-PDA	>450	610	3.0	12.2	296	-10 (compression)	2
BPDA-ODA	325	531	3.3	3.0	230	34	28
BPDA-PDA	>330	595	3.1	10.2	597	5	3-7
BTDA-ODA	280	554	3.0	3.0	150	40	40
BTDA-PDA	>370		2.98	7.1	248	30	35
ODPA-PDA	>370		3.0	8.1	263	42	35

T_g: glass transition temperature; T_d: degradation temperature defined at 10% wt. loss; ε_r: dielectric constant; E': Young's modulus; TS: tensile strength; CTE: coefficient of thermal expansion between 50 °C and 300 °C. Film stress is given for film below 20 μm of thickness [5-14].

Table 2. Thermal, electrical and mechanical properties of the main aromatic PIs.

On the contrary, PIs synthesized from pyromellitic dianhydride (PMDA) or 3,3′,4,4′-biphenyltetracarboxilic dianhydride (BPDA) with *p*-phenylene diamine (PDA) in order to form PMDA-PDA and BPDA-PDA, respectively, present a higher thermal stability (T_d=610 °C and 595 °C, respectively) than PIs owning C–O–C ether bonds. Moreover, both appear as better candidates for the wide band gap semiconductor passivation due to their low CTEs (2 ppm/°C and 3-7 ppm/°C, respectively). For instance, PMDA-PDA and BPDA-PDA show internal stresses of -10 MPa (compression) and 5 MPa when they are coated on Si wafers (3 ppm/°C). Thus, BPDA-PDA seems to be, as given by the main thermo-mechanical properties, the most compatible PI for SiC and GaN semiconductor passivation while PMDA-PDA should be preferred for diamond passivation.

In this chapter, a particular attention is focused on the electrical properties of unaged BPDA-PDA and their evolution during a thermal aging on Si wafers in both oxidative and inert atmospheres. A comparative aging study with higher CTE's PIs (PMDA-ODA and BPDA-ODA) is carried out in order to highlight the longer lifetime of BPDA-PDA. Prior to this, a paragraph dealing with the optimization of the thermal imidization of BPDA-PDA is reported through a simultaneous analysis of the infrared spectra and the electrical properties evolutions as a function of the imidization curing temperature. Finally, an application of BPDA-PDA to the passivation of SiC semiconductor devices will be presented through the PI on-wafer etching process and the electrical characterization of bipolar diodes at high temperature and high voltage.

2. Synthesis and optimization of the imidization of BPDA-PDA polyimide

The final physical properties of PIs and their integrity during aging depend strongly on the control and on the optimization of the imidization reaction (i.e. the curing process) [17,18]. This process step appears as crucial for industrial applications. Unfortunately, it is quite difficult to predict *a priori* the imidization temperature optimum which leads to the best electrical properties. Literature presents a large range of imidization temperature from 200 °C to 425 °C without always indicating if it corresponds to an optimum [6,18-23]. Moreover, these works present mainly the optimization of the imidization reaction from a chemical point of view only based on *qualitative* infrared measurements. Even if the dielectric properties are strongly linked to the PIs chemical structure, it would be more adequate to optimize the imidization reaction taking into account both the chemical structure and the dielectric properties simultaneously. Indeed, dielectric characterizations can be more sensitive than infrared measurements regarding the determination of the imidization temperature optimum. In the section 2, the results are extracted from [24].

2.1. Material, sample preparation and curing process

BPDA-PDA PI was purchased as a polyamic acid (PAA) solution. It was obtained through the two-steps synthesis method from its precursor monomers [25]. The PAA solution was obtained by dissolving the precursor monomers in an organic polar solvent N-methyl-2-pyrrolidone (NMP). Two different vicosity types of the PAA solution were used for controlling the thickness. To convert PAA into PI, the solution was heated up to remove NMP and to induce the imidization through the evaporation of water molecules. Figure 1 shows the synthesis steps of BPDA-PDA.

The PAA solution was spin-coated on both square stainless steel substrates (16 cm^2) and highly doped 2″ Si N^{++} wafers ($<3\times10^{-3}$ Ω cm). PAA was first spread at 500 rpm for 10 seconds followed by a spin-cast at different rotation speeds between 2000 rpm and 4000 rpm for 30 seconds. Two successive curing steps followed the coatings. After a soft-bake (SB) at a low temperature (T_{SB}) of 150 °C for 3 minutes on a hot-plate in air, coatings were hard-cured (HC) at a higher temperature (T_{HC}) in a regulated oven under nitrogen atmosphere and during a time t_{HC}. In the following, the T_{HC} temperature represents the imidization

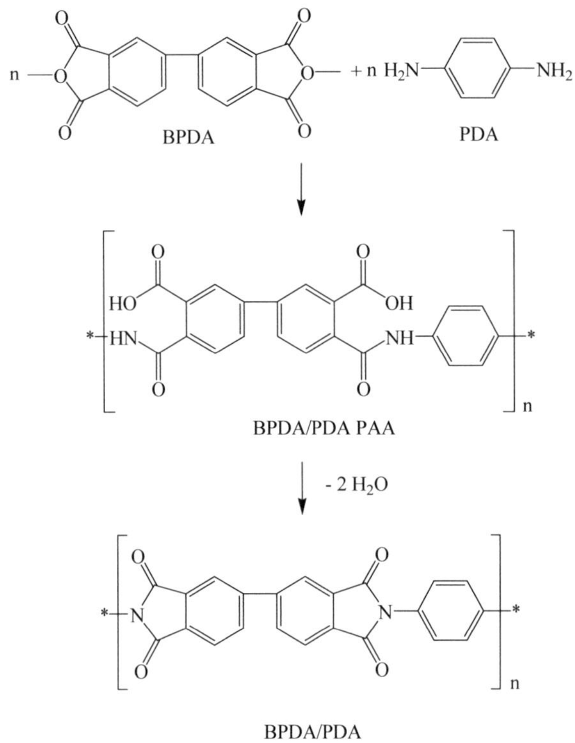

Figure 1. Synthesis steps of the BPDA-PDA polyimide.

temperature. The final film thicknesses have been measured using a KLA Tencor Alpha-Step IQ profilometer in a range from 1.5 to 20 μm depending on the spin-coating parameters, the viscosity of PAA and T_{HC}. In order to perform the electrical measurements, an upper gold metallization was evaporated after imidization onto the PI film surface under vacuum (10^{-4} Pa). This metal layer was then patterned using successively a photolithography step through a selective mask and a humid gold etching to form different circular electrodes from 300 μm to 5 mm in diameter.

The imidization cure is necessary to drive off solvent (boiling point of 202 °C for NMP), and to achieve the conversion of the PAA into PI by the formation of the imide rings. PAA coatings were hard-cured at T_{HC} in the range from 175 to 450 °C under nitrogen for a time t_{HC} of 60 minutes. The heating and cooling rates for all the samples were 2.5 and 4 °C min^{-1}, respectively.

2.2. Optimization of the imidization reaction

2.2.1. Fourier transform infrared spectroscopy (FTIR)

In order to detect the chemical bond changes during the imidization of PAA into PI, assignments of the absorption bands in FTIR spectra are necessary to identify the amide and imide peaks. The characteristic IR absorption peaks were assigned thanks to previous works

[17,26-36]. Usually, PAA spectra are compound of the N–H stretch bonds at 2900–3200 cm^{-1}, the C=O carbonyl stretch from carboxylic acid at 1710–1720 cm^{-1}, the symmetric carboxylate stretch bonds at 1330–1415 cm^{-1}, the C=O carbonyl stretch of the amide I mode around 1665 cm^{-1}, the 1540–1565 cm^{-1} amide II mode and the 1240–1270 cm^{-1} band due to the C–O–C ether aromatic stretch (if present in the monomer).

After the conversion reaction, the absence of the absorption bands near 1550 cm^{-1} (amide II) and 1665 cm^{-1} (amide I) indicates that PAA has been converted into PI. Simultaneously, this is confirmed by the occurrence of the C=O stretch (imide I) peaks at 1770–1780 cm^{-1} (symmetric) and 1720–1740 cm^{-1} (asymmetric), the typical C–N stretch (imide II) peak around 1380 cm^{-1}, the C–H bend (imide III) and C=O bend (imide IV) absorption bands respectively in the ranges of 1070–1140 cm^{-1} and of 720–740 cm^{-1}. The presence in PI films of a large absorption band between 2900 and 3100 cm^{-1} is associated to the C–H stretch bonds. Finally, the measurements may highlight the occurrence of a shoulder on the asymmetric C=O stretch bonds at 1710 cm^{-1} which corresponds to the out-of-plane optical response of the imide I conformation [37].

Figure 2 shows FTIR spectra of both PAA coatings after the SB at 150 °C and PI films after a HC at 250 °C. Spectra have been normalized to the classical C=C absorption band appearing at 1518 cm^{-1}. The spectrum performed after the SB shows the typical absorption bands of PAA coatings. The large absorption band observed between 2300 and 3400 cm^{-1} corresponds to the N–H stretch vibration modes, the C–H stretch bonds and the O–H stretch bonds present in both the PAA and NMP solvent. The FTIR spectrum of PI films already shows the typical completion of the imidization reaction with the presence of the four absorption bands from the imide rings. They occur at 1775/1734 cm^{-1} (imide I), 1371 cm^{-1} (imide II), 1124/1080 cm^{-1} (imide III) and 737 cm^{-1} (imide IV). Moreover, it is possible to observe the large absorption band induced by the C–H stretch vibration modes between 2600 and 3100 cm^{-1}. At 1415 cm^{-1}, a shoulder appears near the C–N stretch peak. This absorption band could be attributed to symmetric stretch of carboxylate ion COO^-. The carboxylic acid groups present in PAA appear through the O–H stretching bonds at 3400 cm^{-1} but free

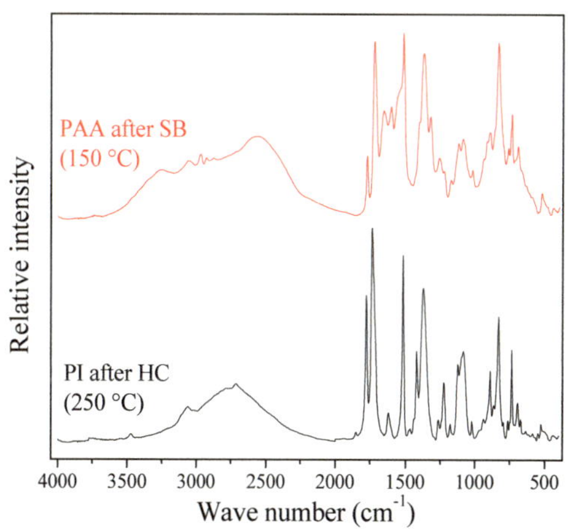

Figure 2. FTIR spectra of BPDA-PDA PAA coatings and PI films (thickness: 1.5 μm). Taken from [24].

carboxylic acid groups can be deprotonated by the weak amine base [36]. Consequently, COO^- carboxylate ions are usually also present in PIs exhibiting two peaks around 1606 and 1415 cm^{-1}. This could explain the release of mobile H^+ protons (from COOH) responsible of electrical conduction in PI [38].

FTIR measurements have been performed for different imidization temperatures T_{HC} to determine the completion of the imidization reaction of the PI films. This analysis rests on the absorption peak magnitude changes in the functional groups or in the characteristic linkages during the reaction. Figure 3a shows the changes in FTIR spectra of BPDA-PDA for different imidization temperatures. When T_{HC} increases, a general increase in all the absorption peaks is observed. This suggests that residual PAA monomers continue to be converted into PI. This evolution is stabilized after exposure to temperature above 350 °C.

To study the imidization kinetics of PI films, the peak of aromatic ring (C=C) stretching around 1500 cm^{-1} is chosen as a reference and the peak height method is adopted to calculate the amount of the appearing imide groups formed. The degree of imidization (DOI) is thus defined by comparing the intensity of an imide absorption peak normalized to the intensity of the C=C reference band and is given by [27]:

$$DOI_{T_{HC}} = \frac{(A/A^*)_{T_{HC}}}{(A/A^*)_i} \quad (1)$$

where A^* is the peak height of the C=C reference band at 1518 cm^{-1} and A is the imide peak height (i.e. 1775, 1732, 1420, 1371, 1080 and 737 cm^{-1}). Subscripts i and T_{HC} indicate the reaction at the initial and a given imidization temperature, respectively.

Figure 3b shows the extent of imidization of the main bonds of BPDA-PDA versus the imidization temperature. Most of the imidization reaction takes place rapidly with a

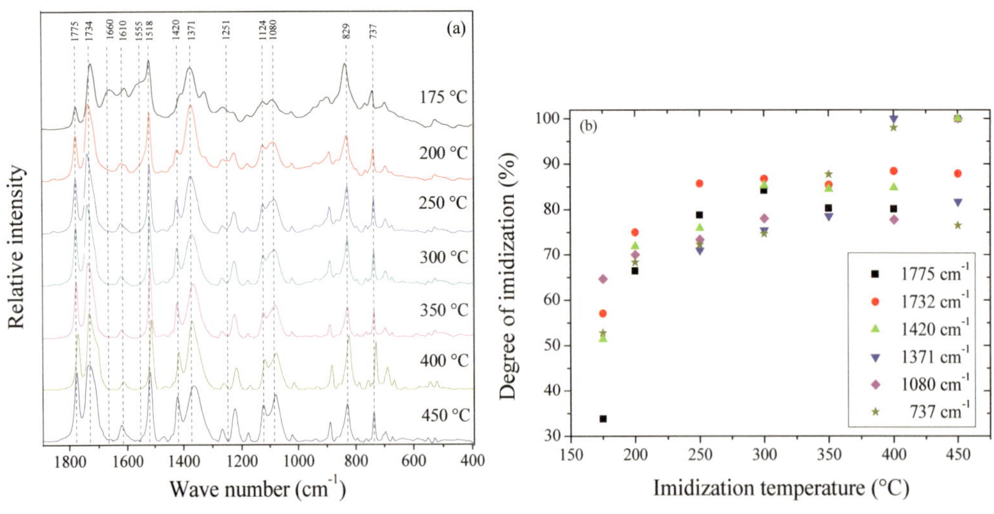

Figure 3. (a) FTIR spectra of BPDA-PDA for different T_{HC} (thickness: 1.5 μm). (b) Degree of imidization for the main absorption bands versus the imidization temperature. Taken from [24].

conversion rate as high as 70-85% at 250 °C and still continues slightly up to 400 °C as shown through the increase in the magnitude of the imide bands. However, it is difficult to detect the optimum imidization temperature (i.e. the highest magnitude) to not exceed in order to preserve PI from degradation. For instance at 450 °C, the imide II and IV absorption bands decrease of 20% and 10% respectively showing the initiation of a desimidization of the structure. Therefore, the use of complementary electrical measurements as a probe of the imidization advancement can allow obtaining a higher accuracy regarding the optimal temperature of the curing.

2.2.2. Electrical properties

As for the DOI, the electrical properties strongly depend on the imidization temperature. Changes in the electrical conductivity, dielectric properties or in the dielectric breakdown field of the PI films can be used to determine precisely the optimal imidization temperature. Larger the DOI is, better the electrical properties are expected due to a lower impurities amount in the PI films.

Current-Field (*j-F*) measurements up to 1 MV/cm show that, as soon as the low field range (10 kV/cm), the minimum of the dc conductivity (σ_{DC}, evaluated from the *j-F* curves) is obtained for an imidization cure of 400 °C (as seen in Figure 4a). Whereas the σ_{DC} values are poorly dispersed between 10^{-17} and 10^{-16} Ω^{-1} cm^{-1} up to 200 kV/cm whatever T_{HC}, the measurements show a strong divergence at high fields from 400 kV/cm. In this field range, the dc conductivity of BPDA-PDA increases much more for imidization curings at 350 °C and 450 °C (as seen in Figure 4a). Thus, the highest insulation quality (i.e. the lowest σ_{DC}) has been obtained for the films imidized at 400 °C. For T_{HC}=400 °C, the charges density and/or their mobility seem to be strongly reduced with this optimal imidization temperature. Usually, the electrical conduction in PI films is related to the motion of H^+ protons coming from unreacted PAA [38]. This is in likelihood agreement with the evolution of the COO^- band intensity which reaches its lowest magnitude at 400 °C before it increases again

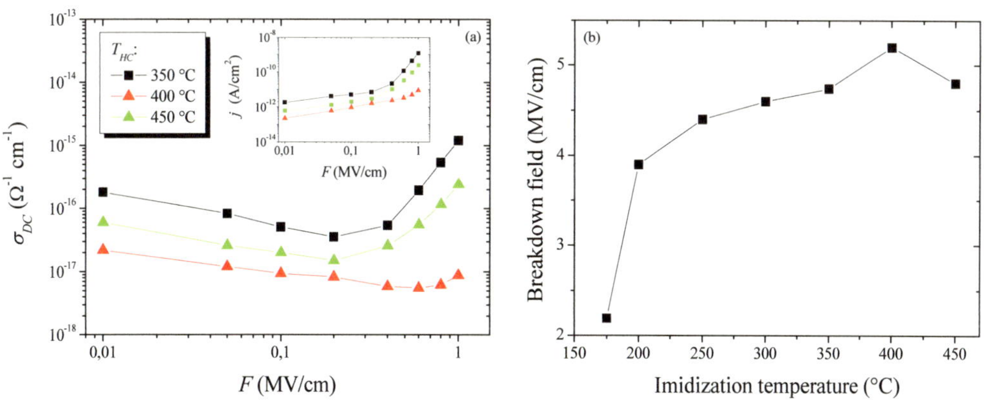

Figure 4. (a) Volume conductivity versus electric field of BPDA-PDA for different T_{HC} (thickness: 1.5 µm). (b) Breakdown field versus the imidization temperature (mean of 20 tested samples). Measurements performed at room temperature. Taken from [24].

at 450 °C (as seen in Figure 3b). This observation is emphasized by the changes in the dielectric breakdown field (E_{BR}) as a function of the imidization temperature (as seen in Figure 4b). Indeed, after a continuous increase in E_{BR} up to T_{HC}=400 °C, a sudden decrease in its magnitude when T_{HC} is raised up to 450 °C is observed. The degradation of the electrical properties above T_{HC}=400 °C appears as a consequence of desimidization (i.e. the decrease in the imide bands) of BPDA-PDA structure leading to the release of free mobile charges in the bulk.

3. Thickness influence on the structural and dielectric properties of BPDA-PDA polyimide

3.1. Influence on the chemical structure

The influence of the thickness of PI films on the chemical structure is rarely investigated. Figure 5 shows FTIR spectra of BPDA-PDA imidized at 400 °C for different film thicknessses. As represented by the downward arrows, one can observe that the quantity and the intensity of the bands corresponding to the amide bonds increase when increasing the film thickness. Hence, for higher film thicknesses, the conversion rate of PAA into PI is strongly affected either due to a bad diffusion of the temperature within the medium of the coating bulk during the curing process (presence of unreacted PAA) or due to a higher difficulty to remove by-products such as solvent and water molecules inherent in the imidization reaction. Unfortunately, this issue cannot be solved by higher temperatures or longer curings because in this range the desimidization of PI starts. Consequently, all these remaining impurities can act as ionizable centers supplying free mobile charges.

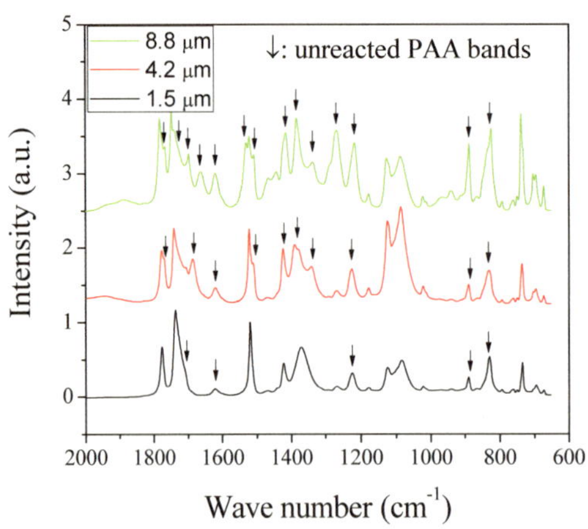

Figure 5. FTIR spectra of 400 °C-imidized BPDA-PDA for different film thicknesses.

3.2. Influence on the electrical properties

Wheras such a phenomenon can be negligeable for low temperature applications (< 150 °C) because of the low mobility of free charges, this can be more influent at high temperature

(>200 °C). Figure 6a shows the temperature dependence of the dc conductivity of BPDA-PDA between 200 °C and 350 °C for thicknesses from 1.5 µm up to 20 µm. At 200 °C, the dc conductivity is one order of magnitude higher for the thickest films compared to the one of the thinnest films. In comparison at 300 °C, the dc conductivity is two orders of magnitude higher for the thickest films than for the thinnest ones. The fact that the low field dc conductivity is thickness-dependent, particularly in the high temperature range, is directly related to the presence into BPDA-PDA of unreacted PAA impurities for which temperature supplies sufficient energy to the free charges to become mobile.

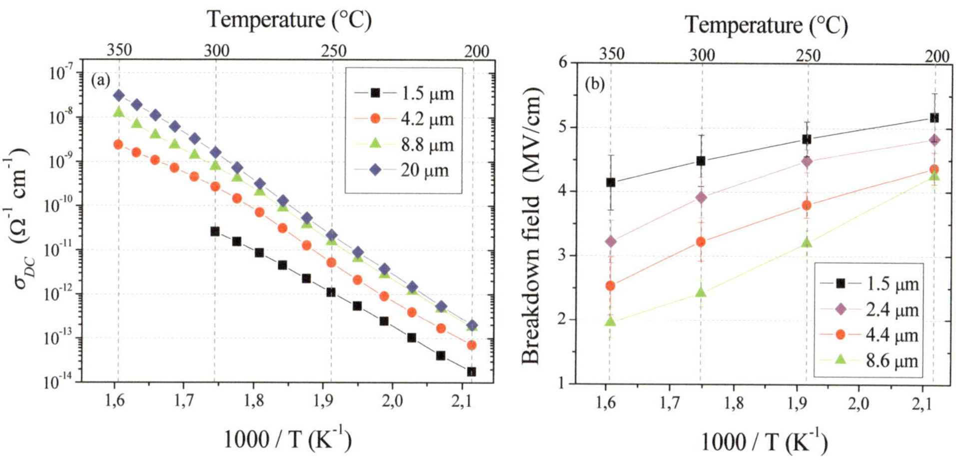

Figure 6. Temperature dependence of the dc conductivity (a) and breakdown field (b) of BPDA-PDA for different film thicknesses.

Figure 6b shows the temperature dependence of the dielectric strength of BPDA-PDA between 200 °C and 350 °C for thicknesses from 1.5 µm up to 8.6 µm. Same findings can be done in this high electric field region. The larger presence of PAA impurities in the thickest films leads to substantially decrease the dielectric breakdown field of 15% at 200 °C (compared to thinnest films) and of 55% at 350 °C. In these thick films, the earlier breakdown event could be explained by a prematured Joule effect occuring when a higher conduction current magnitude happens across the film during the voltage raising. Thus, the breakdown channel appears for lower applied electric fields.

4. Thermal aging of BPDA-PDA polyimide

The effect of long time aging of polyimide at high temperature (>200 °C) and in oxidative environment on the mechanical properties [39], weight loss [40,41], and chemical properties [42,43], was widely investigated for thick polyimide matrix composites (1 mm thick) used in high temperature aerospace applications. It was found that while thermal degradation occurred throughout the material, the oxidative degradation occurs mainly within a thin surface layer where oxygen diffuses into the material. Few papers discussed the effect of thermal aging on the electrical properties of PIs and this is always for thick and freestanding films [44,45]. Consequently, an overall understanding of the thermo-oxidative aging mechanisms (for PI thickness <20 μm) and their effects on the electrical properties are still lacking.

All the measurements presented below were performed on PI films deposited on highly doped N^{++} cleaned silicon wafers (resistivity < 3×10^{-3} Ω cm) and aged at high temperature. The time at the beginning of the aging is noted as t_0. At scheduled times, a part of the specimens were removed from the oven and subjected to non-destructive and destructive tests. The thicknesses noted in the text and labels refer to the initial ones measured at t_0.

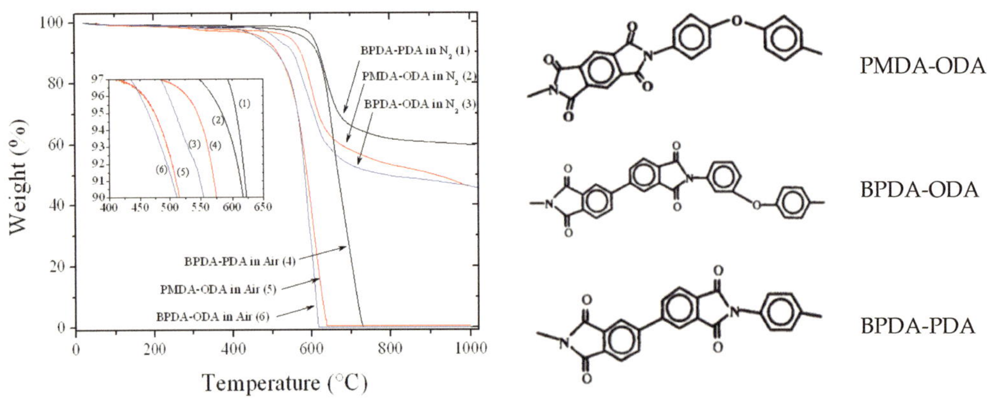

Figure 7. Dynamical TGA of PMDA-ODA, BPDA-ODA and BPDA-PDA in air and nitrogen.

4.1. Thermal stability of PIs

For PIs, it has been shown that the increase in the number of benzene rings contributes to an increase in the degradation temperature [1]. However, the degradation temperature can be also affected by the presence of low thermo-stable bonds in the macromolecular structure. For instance, even if BPDA-PDA and PMDA-ODA (Kapton-type) own the same number of benzene rings (i.e. three in elementary monomer backbone), the absence of the C—O—C ether group in the case of BPDA-PDA allows increasing T_d (defined at 10% wt. loss) of 48 °C in nitrogen and 100 °C in air in comparison to T_d of PMDA-ODA (see Figure 7). Moreover, if the diamine ODA is replaced by PDA in BPDA-based PIs, T_d increases of 68 °C in nitrogen and 105 °C in air. Indeed, this is due to the lower thermal stability of the ether bonds inducing earlier degradations than the rest of the structure [1,44].

4.2. Thermal aging in inert atmosphere

Figure 8 shows the evolution of the FTIR spectrum of BPDA-PDA before and after an aging at 300 °C in nitrogen during 1000 h and the evolution of the film thickness and the related breakdown field during this aging. One can notice that at this temperature in inert atmosphere, no change in the vibration bonds is remarkable even after a long period of aging. This is in agreement with a good stability of both the film thickness and the high field dielectric properties.

Moreover, a similar observation were done for aging at higher temperature. Indeed, up to T_g at 360 °C in nitrogen, both a stability of the chemical structure and the breakdown field were observed during 1000 h. This concludes that BPDA-PDA does not evolved up to 360 °C in inert atmosphere.

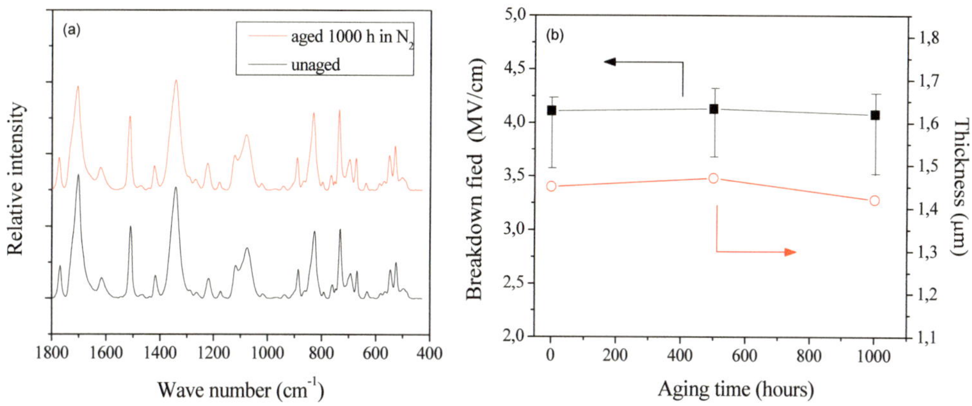

Figure 8. (a) FTIR spectra before and after aging in N_2 at 300 °C and (b) breakdown field and thickness during aging for 1.5 μm-thick BPDA-PDA films. The breakdown field is measured at 300 °C.

4.3. Thermal aging below T_g in oxidative atmosphere

The aging effects up to 5000 h at 300 °C in air on different properties of BPDA-PDA were measured on three different initial thicknesses varying from 1.5 μm to 8 μm. For comparison, a same aging was performed on BPDA-ODA, with a glass transition T_g of 330 °C, up to 1000 h.

The chemical structure variation was measured by FTIR and the spectra of the 4.2 μm-thick film is presented in Figure 9a. A quasi-stabilisation of almost all the peaks can be revealed during 5000 h specially the imide ones (see Figure 9b). However, an increase in the peak localized at 1212 cm^{-1} related to the asymmetric vibration of the C-O-C band can be observed. This can indicate the occurrence of additional oxidation of the unreacted polyamic acid, which was not completely imidized during the curing cycle.

On the contrary, for BPDA-ODA films (see Figure 9c and 9d), the same aging at 300 °C during 1000 h shows as soon as the first 200 h a strong decrease in all the main vibration bonds. Consequently, such an aging affects the chemical integrity of the chemical backbone and the physical properties would be modified.

If we look at the film thicknesses, the BPDA-PDA films do not show any thickness variation during 5000 h of aging, indicating that neither densification nor degradation occurred. On the other hand, the BPDA-ODA films loose more than 50% of their initial thickness after 1000 h of aging, reflecting the strong degradation in this case.

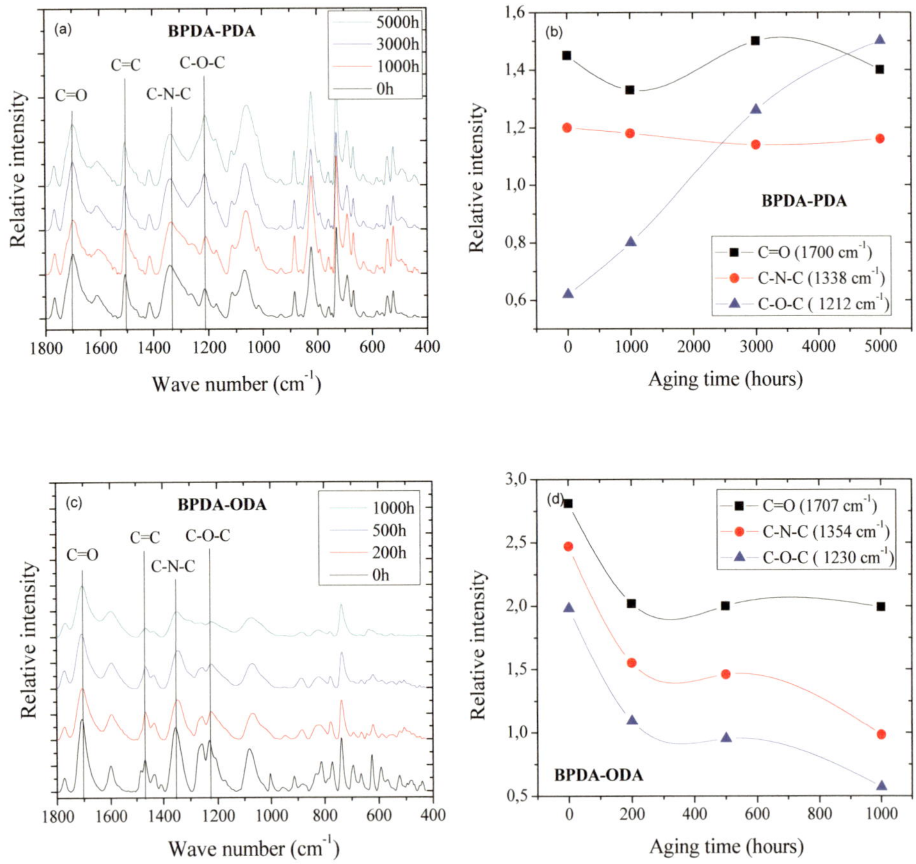

Figure 9. FTIR spectra during the aging in air at 300 °C for 4.2 μm-thick BPDA-PDA and 13.7 μm-thick BPDA-ODA films.

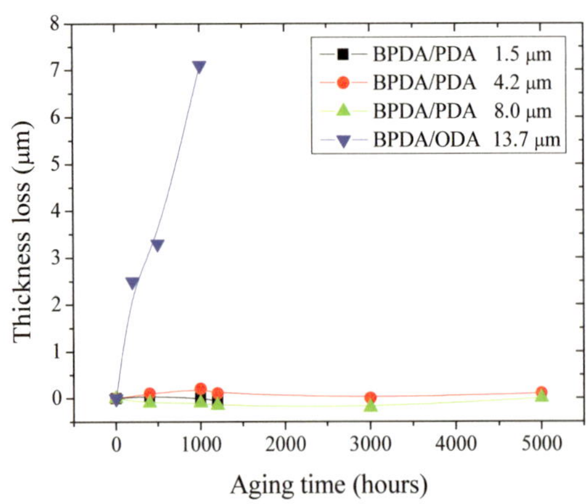

Figure 10. Thickness loss during the aging at 300 °C in air for three initial thicknesses of BPDA-PDA and a 13.7 μm initial thickness of BPDA-ODA.

The effect of the aging under air atmosphere on the breakdown field and low field dielectric properties measured at the aging temperatures, for the BPDA-PDA and BPDA-ODA films, are now presented and discussed. The breakdown field, performed by polarizing positively the gold electrode, for different initial BPDA-PDA thicknesses and one BPDA-ODA thickness are presented in Figure 11. Whereas a stabilization of the breakdown field during the 5000 h aging is observed for the BPDA-PDA films, a continuous decrease is observed for the BPDA-ODA films. This invariance of the breakdown field of BPDA-PDA is in good agreement with the good stability of FTIR spectra during aging. On the contrary, the strong decrease in the breakdown field of BPDA-ODA after only 1000 h highlights the progressive and fast degradation of the imide bonds in this kind of PIs.

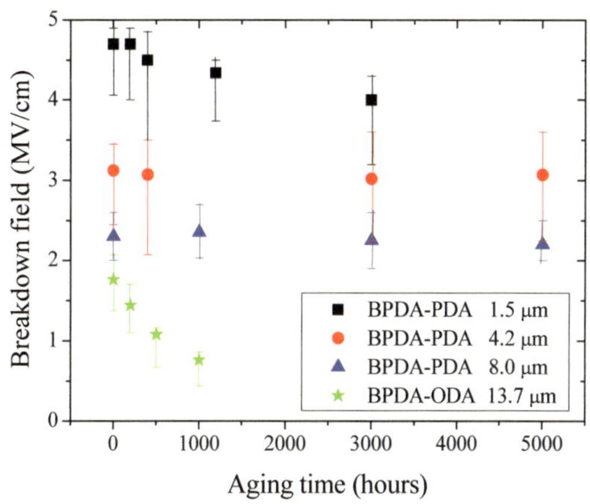

Figure 11. Breakdown field during the aging at 300 °C in air for three different thicknesses of BPDA-PDA and the 13.7 μm-thick films of BPDA-ODA. The breakdown field is measured at 300 °C.

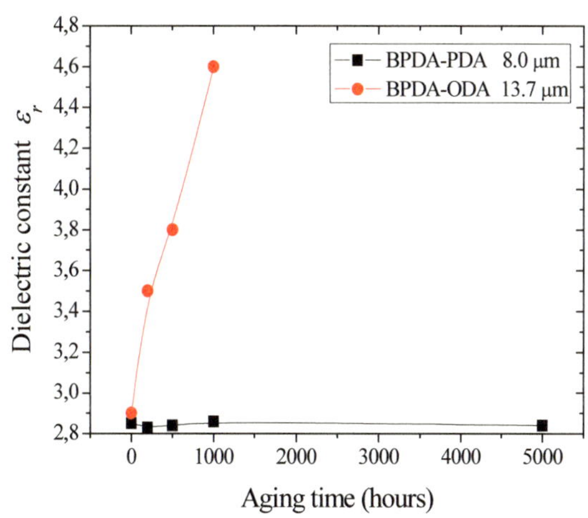

Figure 12. Dielectric constant measured at 300 °C and at 1 kHz during the aging for BPDA-PDA and BPDA-ODA films.

The dielectric constant ε_r measured at 300 °C and at a frequency of 1 kHz during the aging is illustrated in Figure 12. Here also, while an invariance in the ε_r values can be observed for the BPDA-PDA films, a huge increase in ε_r is observed during the aging for the BPDA-ODA films. This increase could be attributed to the chain breakage, already observed on the FTIR spectra, leading to the formation of polar groups in the dielectric bulk. Different authors observed such behavior during the aging of PIs films at high temperature in air [45, 46]. The degradation of BPDA-ODA films is also reflected on the DC conductivity values (not presented here) that increase from 2×10^{-12} to 4×10^{-11} Ω^{-1} cm^{-1} after 1000 h of aging.

4.4. Thermal aging above T_g in oxidative atmosphere

In order to check the stability of BPDA-PDA films at higher temperatures, aging has been performed at different temperatures above T_g. Figure 13 shows a comparison of the evolution of FTIR spectra of BPDA-PDA and PMDA-ODA during an accelerated aging at 400 °C during 200 minutes in air. It can easily be observed that BPDA-PDA remain stable after aging while PMDA-ODA is strongly degraded and even if the latter owns a comparable thermal stability and a higher T_g (see Table II).

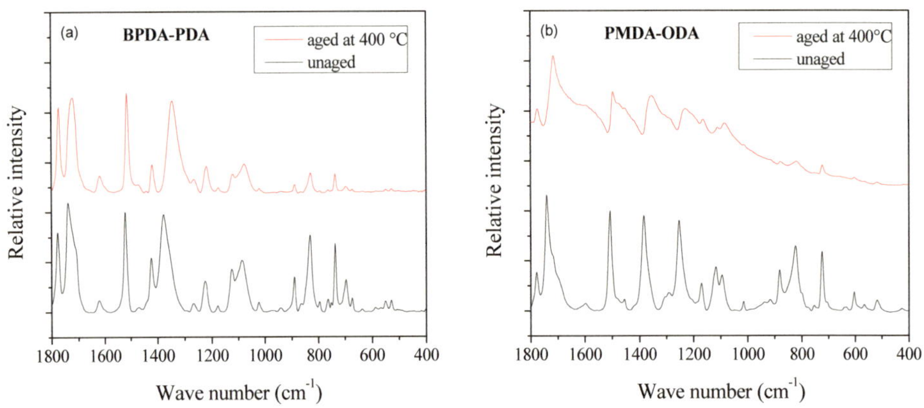

Figure 13. FTIR spectra before and after aging at 400 °C in air for 200 minutes for BPDA-PDA and PMDA-ODA films.

BPDA-PDA films 4.2 μm-thick were aged in air at 360 °C during 800 h. The FTIR spectrum does not show any variation in the imide bonds during this aging period (not shown here) indicating that the bulk of the material is not affected by the aging. Here also, the measured relative permittivity at 1 kHz remains constant with a value of 2.8 during all the aging period (not shown here), indicating that no additional polar groups are formed during the aging. In contrast the film thickness and the breakdown field measured at 300 °C, present a slight continuous decrease during the aging as illustrated in Figure 14. After 800 h a breakdown field decrease of about 50 % while a thickness reduction of 14% can be measured. During this aging, an increase in the surface roughness of BPDA-PDA and the formation of craters were observed (not shown here). They can cause local field intensification and assist the dielectric breakdown mechanism. Such a surface state degradation can explain the breakdown field decrease during the aging. So, it is believed that the observed degradations are related to the oxygen presence since no variation occurs during aging at the same temperature but in inert gas. Thus, BPDA-PDA is attacked at the outer layer face exposed to oxygen.

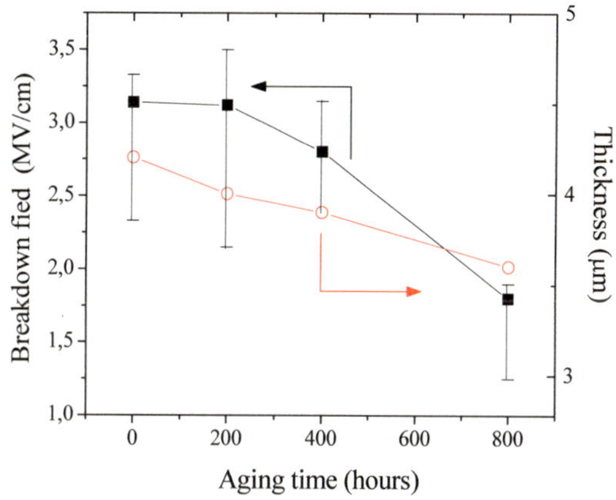

Figure 14. Breakdown field and thickness for 4.2 µm-thick BPDA-PDA films during aging at 360 °C in air. The breakdown field is measured at 300 °C.

All these results lead to show that BPDA-PDA is a reliable kind of PIs in order to passivate wide band gap semiconductor devices up to 360 °C during extremely long duration in inert atmosphere without any remarkable properties degradation.

5. Secondary passivation of SiC bipolar diodes with BPDA-PDA polyimide

The secondary passivation is the last fabrication step, at the wafer level, of a semiconductor device. It aims to reinforce the die protection against mechanical aggressions, chemical contamination, and surface electric flashover under blocking voltage operation. Note that after the wafer dicing and die packaging, a complementary insulating environment (moulding case, silicone gel, …), may be required over the device depending on the maximal voltage rating, in order to avoid electrical arcing in air during operation. Coming back at the wafer level, the component secondary passivation fabrication step consists in two main phases: the first one corresponds to the wafer PI coating, the second one corresponds to the PI film etching at the component metal electrode areas, in order to allow ulterior electrical contacting.

The feasibility of using BPDA-PDA polyimide for the passivation of high voltage silicon carbide bipolar diodes has been studied [47]. As BPDA-PDA material is not photosensitive and not removable using wet etching, its local etching requires applying a plasma process through a previously deposited metal mask. The following sections will first present the aimed diode structure and the used BPDA-PDA coating and ectching processes. Then, the component electrical characterizations will be reported. The results of this paragraph are extracted from [47].

5.1. Diodes fabrication, passivation process and plasma etching

4H-SiC PiN diodes were realized on a 2''-SiC wafer in a 60 μm-thick N^- epilayer (N_D=8.4×10^{14} cm^{-3}), with a Al-implanted P^+ emitter (N_E=10^{19} cm^{-3}), and a Al-implanted junction terminaison extention (JTE) periphery (with L_{JTE}≥125 μm, an implanted-Al dose of 1.1×10^{13} cm^{-2}, assuming a 80% electrical activation [48]). The common post-implantation annealing was performed at 1650 °C for 30 minutes in Ar. The theoretical breakdown voltage of these PiN JTE diodes, derived from ionization integral calculations realized by numerical 2D-simulations, taking into account the Konstantinov's ionization coefficients [49], is equal to 7.8 kV for the 0.9×10^{13} cm^{-2} JTE doping dose assumed. The primary passivation layer is a 40 nm-thick thermal SiO_2 oxide grown by dry oxidation at 1150 °C for 2 h followed by a post oxidation annealing at 1150 °C for 1 h in N_2. Then, after the SiO_2 layer selective opening, a first Ni/Al thin metallization was deposited, patterned and annealed to realize the low resistive ohmic contact to the diode P^+ emitter areas (of 250 μm to 1 mm in diameter). Then a 2 μm-thick Al metallisation was evaporated and patterned as well to obtain thick anode plots. Note that some electrical tests at that stage were performed on the wafer outside the clean room.

Then, the second passivation layer has been realized with the spin-coated BPDA-PDA through a multi-layer (3 layers) coating process onto the wafer. In order to reinforce the adhesion with the primary passivation layer, an adhesion promottor (containing silane groups) was first spin-cast onto the wafer at 3000 rpm. A first layer of BPDA-PDA PAA solution was spin-coated at 500 rpm followed by a rotation speed at 3000 rpm. The wafer was consecutively baked at 100 °C for 1 minute and 175 °C for 3 minutes on a hot plate. A second and a third BPDA-PDA PAA layers were coated with the same process in order to increase the passivation layer thickness. The wafer was finally hard cured *(apart from the clean room)* in a regulated oven under nitrogen atmosphere at 200 °C for 15 minutes (to drive off NMP solvent) followed by a slight raising slope of 2 °C min^{-1} up to 400 °C for 1h. After the imidization, the wafer was cooled down with a low slope of -4 °C min^{-1}. The resulting BPDA-PDA thickness onto the wafer was 4 μm.

Before the etching, a mask metallic layer of 300 nm was evaporated into vacuum onto the BPDA-PDA passivation layer. The metallic layer was opened (metal etch) just above the SiC diode anodes using the photolithography technique. Plasma etching of the unmasked BPDA-PDA areas was performed into a plasma reactor containing a 100% O_2 atmosphere at pressure of 1 mTorr. The injected micro-wave electrical power was 600 W. An auto-polarization voltage of -84 V was applied between the plasma electrodes. The incident RF power was 70 W while the reflected power was negligeable (around 3 W). The wafer was placed onto the ground electrode of the reactor which was cooled with water to maintain the temperature of the wafer around 10 °C. The etching was performed during 20 minutes by short steps of 5 minutes long in order to avoid an increase in the temperature of the wafer and until the total removing of exposed BPDA-PDA areas. Figure 15 shows optical and SEM images of the upper electrode of a SiC diode after the plasma etching of BPDA-PDA.

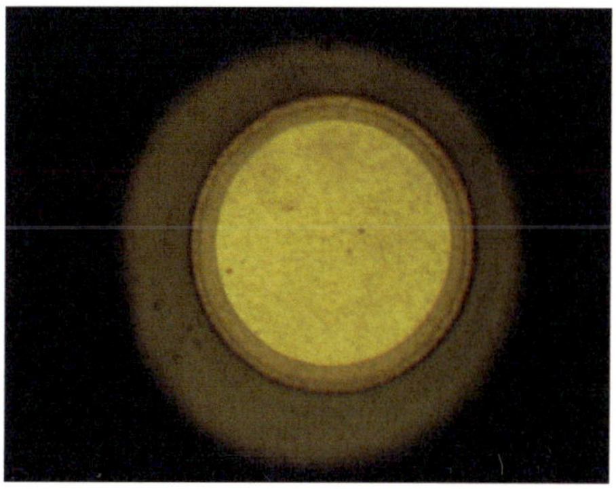

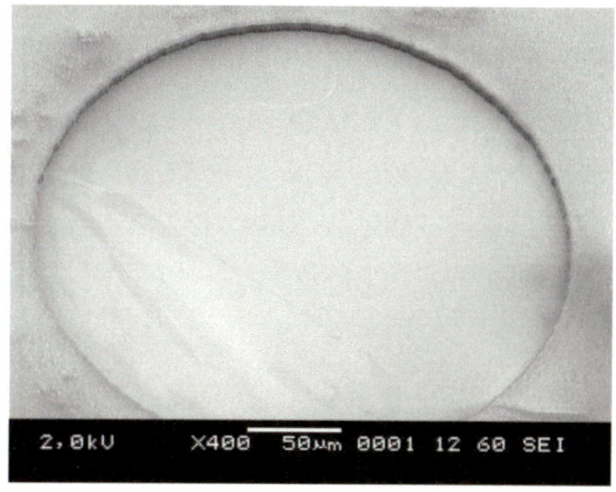

Figure 15. (a) Optical microscope image and (b) SEM image of the topside view of a SiC diode after the etching of BPDA-PDA secondary passivation.

5.2. Electrical characterizations of BPDA-PDA passivated SiC bipolar diodes

The diode forward and reverse current *vs.* voltage I(V) characteristics were measured using a Keithley 2410 source meter, at low levels (less than 1 A/cm² forward current density and -600 V reverse voltage) to avoid stressing the devices. These preliminary measurements were performed at different temperatures (the wafer sample being attached to a heating chuck), and in air ambient at low pressure (10^{-3} mbar). Then the diodes exhibiting the lowest reverse currents (less than 10^{-6} A up to -600 V) were selected for high reverse voltage measurements, performed in a perfluoropolyether oil (PFPE Galden® fluid, with a high dielectric strength around 16 kV/mm) in order to avoid arcing in air ambient. Figure 16 presents an example of diode forward I(V) characteristics measured at different temperatures before and after the secondary passivation. It is observed that the latter did not affect the functionality

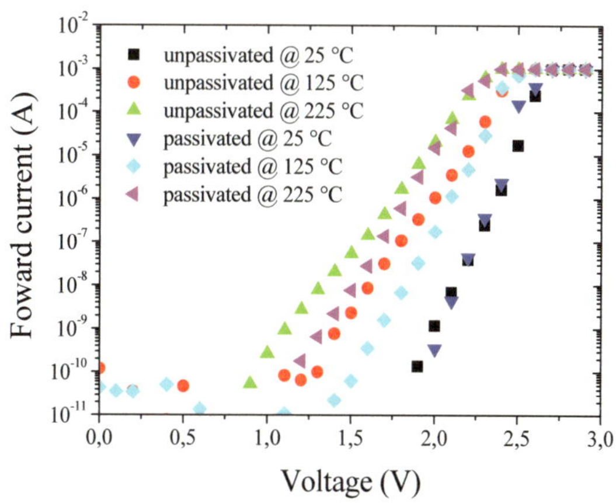

Figure 16. Forward current vs. voltage characteristics at three different temperatures before and after the BPDA-PDA passivation realization. The current was limited at 1 mA. The diode P+-emitter diameter and JTE length are 400 µm and 500 µm, respectively. Characterizations were performed in vacuum.

of the component. On the contrary, an improvement of the diode ideality factors at low forward current levels resulted compared to the non coated diode ones, especially at higher temperatures. A positive effect could be noticed on the reverse I(V) curves up to -600 V applied voltage as well (not shown here).

The typical high voltage reverse characteristics measured at ambient temperature for the selected best diodes are shown on Figure 17a. The breakdown event always occurred suddenly, at a voltage value between 5 kV and 6 kV, and leaving a visible mark on the sample located relatively far from the tested component (at a distance longer than 1 mm) as presented on Figure 17b.

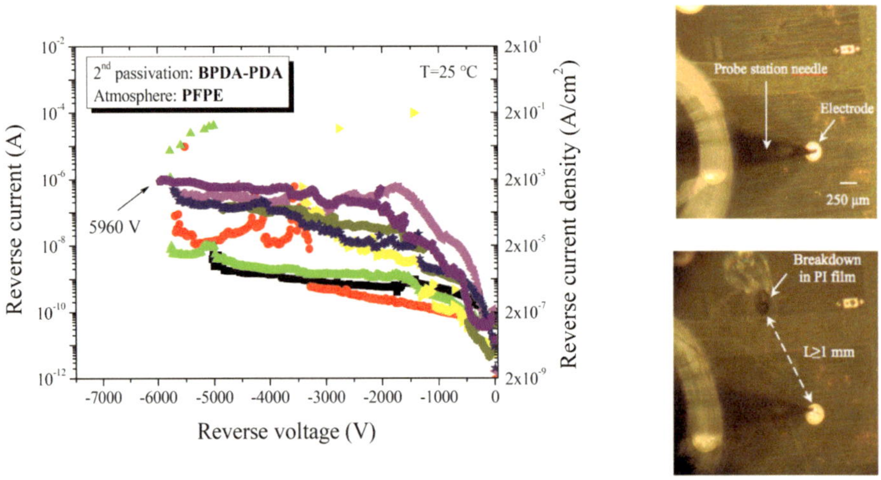

Figure 17. (a) Typical reverse characteristics of low leakage diodes passivated with BPDA-PDA (measurements performed in PFPE environment at 25 °C). (b) Views of the same probed diode in PFPE oil, before (top image) and after (bottom image) high voltage measurements.

The tested diodes were not destroyed after the first high voltage measurement, exhibiting approximately the same breakdown voltage values when polarized again several times. Considering the reverse voltage values achieved and the distance observed between the resulting crater and the device, the second passivation layer is certainly at the origin of the failure (the PFPE environment being able to withstand around three times higher voltages), due to the presence of local defects. For another diode, probably situated in a better quality area, a maximal breakdown voltage was measured with a value of 7.3 kV, with a I(V) characteristics presenting a current knee before breakdown, as can be seen on Figure 18a. Such a breakdown voltage value is very close to the theoretical maximal value assuming an avalanche breakdown in the SiC tested structure. Such a SiC avalanche mechanism in this particular case was further supported by the observation of the post-breakdown degraded zone, situated at the diode itself, the latter being totally destroyed as exhibited on Figure 18b.

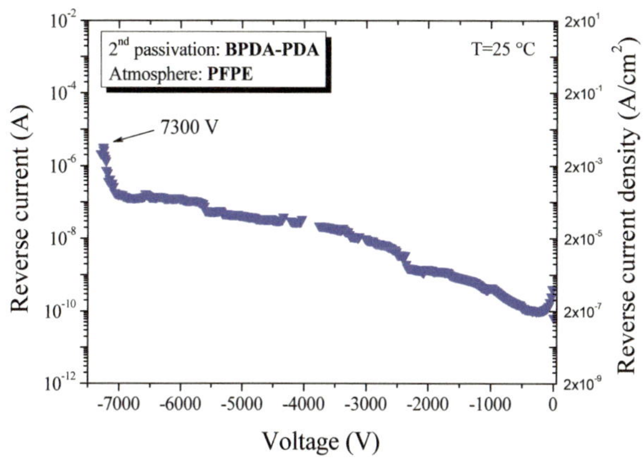

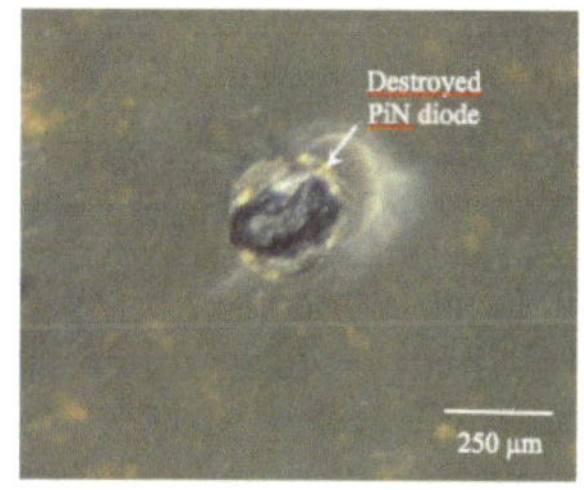

Figure 18. (a) Reverse characteristic of a BPDA/PDA passivated diode with the maximal breakdown voltage obtained (performed in PFPE environment at 25 °C). (b) View of the destroyed diode in PFPE oil after the high voltage measurement.

So the feasibility and the potentiality of the BPDA-PDA films for high voltage SiC device secondary passivation could be experimentally demonstrated, positively affecting the current voltage characteristics and allowing high breakdown voltage typical values with a maximum value close to the theoretical limit to be reached. Moreover, a significant improvement in the BPDA-PDA protection efficiency should result from a fabrication entirely performed in clean room conditions and from the use of thicker PI layers.

6. Conclusion

This chapter deals with polyimide materials (PIs), having in mind the emerging high temperature semiconductor devices currently demanded for high temperature and high power electronics. Among several PIs already well known for their best thermal properties, very good dielectric characteristics, chemical and radiation resistance, and easy processability, this chapter focuses on BPDA-PDA polyimide, evaluating its superiority for semiconductor insulating coating in the temperature range up to 400 °C.

It is shown that the BPDA-PDA's CTE, which is the closest to the semiconductors (as SiC and GaN) ones, is not the only advantage of this material with regards to the targeted application. In fact, though exhibiting comparable T_g and T_d to PMDA-ODA and PPDA-ODA ones in particular, BPDA-PDA on silicon substrate demonstrates a higher properties stability under thermo-oxidative ambient aging than its counterparts, thanks to its chemical nature exempt of C-O-C ether bonds. Indeed, the presented results highlight the insulation long term reliability at 300 °C in air, and at 360 °C in nitrogen ambient, of BPDA-PDA films on semiconductor (no chemical and no electrical degradations having been evidenced up to 5000 h of aging).

The presence of impurities (source of free charges) within the PI films playing a major role in the degradation of their dielectric characteristics above 200 °C, the highest degree of imidization has to be looked for, as considered in this chapter. An imidization cure (400 °C-temperature, 1 h-duration) is found optimal for maximizing both the low field resistivity and the dielectric strength, in correlation with FTIR spectrometry analysis. Because of its impact on the intrinsic free charges density as well, the film thickness parameter is also taken into account. Its strong influence on the high temperature dielectric properties is underlined, which can not be solved by a higher temperature or longer curing (leading to desimidization). As an example, at 350 °C, the mean dielectric strength of a 8.6 µm-thick film is measured two fold lower than that of a 1.5 µm-thick layer; however it is remaining as high as 2 MV/cm, so comparable to SiC critical field.

Going up to the application, the chapter finally describes an experiment demonstrating the feasibility of the secondary passivation of 7.8 kV SiC bipolar diodes, using BPDA-PDA. The PI coating and etching processes are detailed, resulting in a 4 µm-thick PI layer. The electrical characterization results arise that the applied final fabrication step positively affected the high temperature forward I(V) curves of the diodes. In reverse bias, the typical breakdown voltage, of around 70% of the theoretical maximum value, could be attributed to the presence of local defects throughout the PI coating. So, such a first experiment already attests the potentiality of BPDA-PDA for high voltage secondary passivation, knowing that one can expect an even higher protection efficiency using clean room elaboration conditions, and thicker PI layers.

Author details

S. Diaham, M.-L. Locatelli and R. Khazaka
University of Toulouse – UPS – INPT – LAPLACE Laboratory– CNRS, Toulouse, France

7. References

[1] Sroog C E, Endrey A L, Abramo S V, Berr C E, Edwards W M, Olivier K L (1965). J. Polym. Sci. A Polym. Chem. 3: 1373.
[2] Hougham G, Tesero G, Shaw J (1994). Macromolecules 27: 3642-3649.
[3] Gosh M K, Mittal K L (1996). Polyimides, Fundamentals and Applications. New-York: Mercel-Dekker.
[4] Wayne Johnson R, Wang C, Liu Y, Scofield J D (2007). IEEE Trans. Elec. Pack. Manuf. 30: 182-193.
[5] Ree M, Swanson S, Volksen W (1994). Polymer. 34: 1423-1430.
[6] Ree M, Kim K, Woo S H, Chang H (1997). J. Appl. Phys. 81: 698-708.
[7] Chung H S, Lee C K, Joe Y I, Han H S (1998). J. Kor. Inst. Chem. Eng. (Hwahak Konghak). 36: 329-335 (in Korean language).
[8] Chung H, Lee C, Han H (2001). Polymer. 42: 319-328.
[9] Cho K, Lee D, Lee M S, Park C E (1997). Polymer. 38: 1615-1623.
[10] Poon T W, Silverman B D, Saraf R F, Rossi A R, Ho P S (1992). Phys. Rev. B. 46: 11456-11462.

[11] Numata S, Oohara S, Fujisaki K, Imaizumi J, Kinjo N (1986). J. Appl. Polym. Sci. 31: 101-110.
[12] Liou H C, Ho P S, Stierman R (1999). Thin Solid Films. 339: 68-73.
[13] Ree M, Shin T J, Lee S W (2001). Kor. Polym. J. 9: 1-19.
[14] Maier G (2001). Prog. Polym. Sci. 26: 3-65.
[15] Tsukiji M, Bitoh W, Enomoto J (1990). Proc. Inter. Symp. Elec. Insul. 88.
[16] Hsaio S H, Chen Y J (2002). Eur. Polym. J. 38: 815.
[17] Pramoda K P, Liu S, Chung T S (2002). Macromol. Mater. Eng. 287: 931.
[18] Ree M, Chu C W, Goldberg M J (1994). J. Appl. Phys. 75: 1410.
[19] Shin T J, Ree M (2007). J. Phys. Chem. B. 111: 13894.
[20] Lee S A, Yamashita T, Horie K, Kozawa T (1997). J. Phys. Chem. B. 101: 4520.
[21] Sung J, Kim D, Whang C N, Oh-e M, Yokoyama H (2004). J. Phys. Chem. B. 108: 10991.
[22] Factor B J, Russell T P, Toney M F (1991). Phys. Rev. Lett. 66: 1181.
[23] Sasaki T, Moriuchi H, Yano S, Yokota R (2005). Polymer. 46: 6968.
[24] Diaham S, Locatelli M L, Lebey T, Malec D (2011). Thin Solid Films. 519: 1851-1856.
[25] Sroog C E (1991). Prog. Polym. Sci. 16: 561.
[26] Chen K M, Wang T H, King J S, Hung A (1993). J. Appl. Polym. Sci. 48: 291.
[27] Hsu T C J, Liu Z L (1992). J. Appl. Polym. Sci. 46: 1821.
[28] Ishida H, Wellinghoff S T, Baer E, Koenig J L (1980). Macromolecules. 13: 826.
[29] Thomson B, Park Y, Painter P C, Snyder R W (1989). Macromolecules. 22: 4159.
[30] Synder R W, Thompson B, Bartges B, Czerniawski D, Painter P C (1989). Macromolecules. 22: 4166.
[31] Pryde C A (1993). J. Polym. Sci. A: Polym. Chem. 31: 1045.
[32] Karamancheva I, Stefov V, Soptrajanov B, Danev G, Spasova E, Assa J (1999). Vibr. Spectr. 19: 369.
[33] Spassova E (2003). Vacuum. 70: 551.
[34] Deligöz H, Yalcinyuva T, Özgümüs S, Yildirim S (2006). Eur. Polym. J. 42: 1370.
[35] Saeed M B, Zhan M S (2006). Eur. Polym. J. 42: 1844.
[36] Anthamatten M, Letts S A, Day K, Cook R C, Gies A P, Hamilton T P, Nonidez W K (2004). J. Polym. Sci. A: Polym. Chem. 42: 5999.
[37] Hietpas G D, Allara D L (1998). J. Polym. Sci. B Polym. Phys. 36: 1247.
[38] Ito Y, Hikita M, Kimura T, Mizutani T (1990). Jpn. J. Appl. Phys. 29: 1128.
[39] Ruggles-Wrenn M B, Broeckert J L (2009). J. Appl. Polym. Sci. 111: 228-236.
[40] Schoeppner G A, Tandon G P, Ripberger E R (2007). Composites: Part A. 38: 890-904.
[41] Tandon G P, Pochiraju K V, Schoeppner G A (2008). Mater. Sci. Eng. A. 498: 150-161.
[42] Meador M A B, Johnston C J, Cavano P J, Frimer A A (1997). Macromolecules. 30: 3215-3223.
[43] Meador M A B, Johnston C J, Frimer A A, Gilinsky-Sharon P (1999). Macromolecules. 32: 5532-5538.
[44] Tsukiji M, Bitoh W, Enomoto J (1990). Proc. IEEE Int. Symp Elec. Insul. pp. 88-91.
[45] Li L, Bowler N, Hondred P R, Kessler M R (2011). J. Phys. Chem. Solids. 72: 875-881.
[46] Dine-Hart R A, Parker D B V, Wright W W (1971). Br. Polym. J. 3: 222-236.

[47] Diaham S, Locatelli M L, Lebey T, Raynaud C, Lazar M, Vang H, Planson D (2009). Mater. Sci. Forum. 615-617: 695-698.
[48] Lazar M, Raynaud C, Planson D, Chante J-P, Locatelli M-L, Ottaviani L, Godignon P (2003). J. Appl. Phys. 94: 2992-2999.
[49] Konstantinov A O, Wahah Q, Nordell N, Lindefelt U (1997). Appl. Phys. Lett. 71: 90-92.

Chapter 3

Hyperbranched Polyimides Prepared from 4,4′,4′′-Triaminotriphenylmethane and Mixed Matrix Materials Based on Them

Evgenia Minko, Petr Sysel, Martin Spergl and Petra Slapakova

Additional information is available at the end of the chapter

http://dx.doi.org/10.5772/2834

1. Introduction

Polymers are macromolecules built up by the linking together, under formation of chemical bonds, of large numbers of much smaller molecules (monomers). Due to their structure they can be classified as linear, branched, or crosslinked polymers [1]. The dendritic topology (dendrimers, hyperbranched polymers) has recently been recognized as a new class of macromolecular architecture. Dendritic polymers are expected to play a key role as enabling building blocks for nanotechnology during the 21st century [2].

Hyperbranched polymers are highly branched macromolecules. The highly branched structure and a large number of terminal functional groups, resulting, e.g., in better solubility and lower viscosity, are their important structural features which distinguish them from linear polymers. They do not have the perfectly branched architectures but they are thought to have similar physical properties to dendrimers and for some cases they can be used to replace them. (Dendrimers consist of highly branched, monodisperse spherical macromolecules containing a control core surrounded by repeating units, all enclosed by a terminal group shell. They are prepared using multi-step syntheses [2]). Hyperbranched polymers often can be simply prepared by a direct one-step polymerization of multifunctional monomers using a single-monomer (starting compound is generally ABx monomer, most frequently AB_2 (Figure 1)) or double-monomer (A_2+B_3) methodology [3]. (A and B represent two kinds of functional groups which can react with each other but cannot undergo self-reaction.) Many kinds of hyperbranched polymers, e.g., polyesters, polyethers, poly(ether ketone)s, poly(ether sulfone)s, polyamides, have been investigated as novel dendritic macromolecules so far [3]. An increasing attention has been also devototed to hyperbranched polyimides [4] due to a potentially possible connection of the known advantages of linear or crosslinked polyimides [5] with those of hyperbranched polymers [2,3].

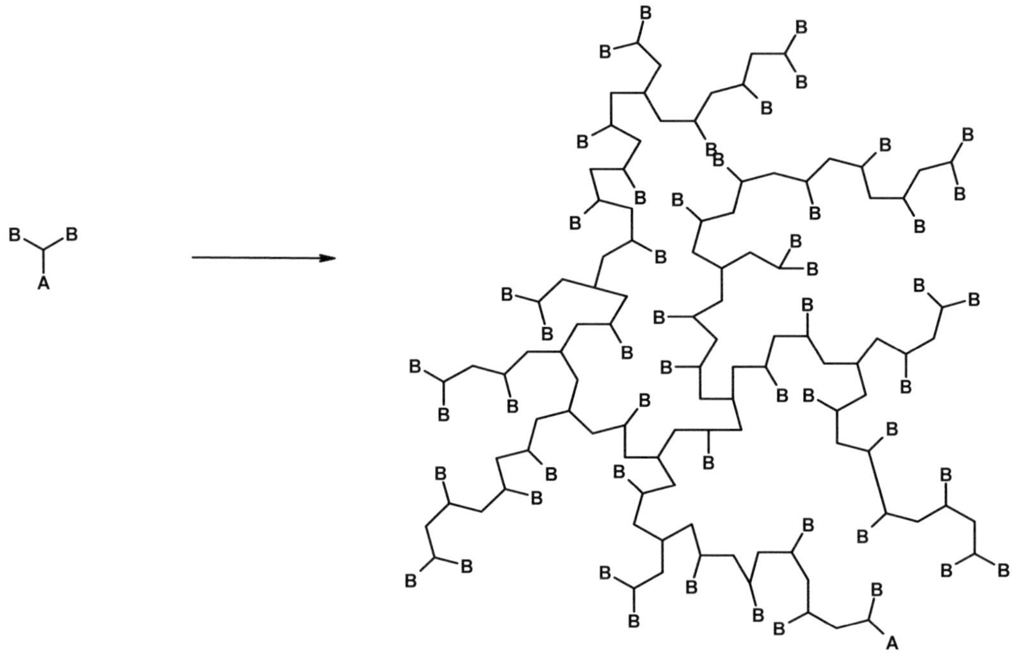

Figure 1. Schematic structure of the hyperbranched polyimide precursor from AB_2 monomer

Polyimides (PI) exhibit very good chemical, mechanical and dielectric stability at temperatures from –150 to 250 °C. These rigid polymers with a high glass transition temperature are mostly used in (micro)electronics, aircraft industry, space exploration and as polymeric separation membranes. Linear polyimides (LPI) are traditionally prepared by the two-step polymerization. The polyimideprecursor, polyamic acid (PAA) (the most often a solution in N-methyl-2-pyrrolidone), is prepared from an aromatic dianhydride and an aromatic diamine. This precursor is transformed into a polyimide using thermal or chemical imidization (Figure 2) [5].

It is difficult to prepare hyperbranched polyimides (HBPI) from AB_x monomers due to the high reactivities of A and B groups. Therefore, they are often prepared by the combination of bifunctional (A_2) and trifunctional (B_3) monomers. The ratio of anhydride and amine component determines the kind and the ratio of endgroups. Nevertheless, their synthesis from trifunctional monomers often a rather complex structure requires special and controlled reaction conditions (namely low monomer concentrations, strictly controlled slow addition rates and molar ratios of monomers) to avoid gel formation [4] because it is known that direct polycondensation of A_2 and B_x (x is higher than 2) monomers generally results in three-dimensional product [1]. As such triamines were used, e.g., tris(4-aminophenyl)amine or 1,3,5-tris(4-aminophenoxy)benzene [4].

In the paper [6] it was shown that AB_2 monomer provides HBPI with a higher degree of branching (degree of branching is defined as the ratio of (branching + terminal units)/(branching + terminal + linear units)) and a less level of entanglements in comparison with the use of A_2 and B_3 monomers (Figure 3) although their chemical structures are very close. A different chain architecture also influenced the viscosity behaviour, glass transition temperature and thermooxidative stability of the products.

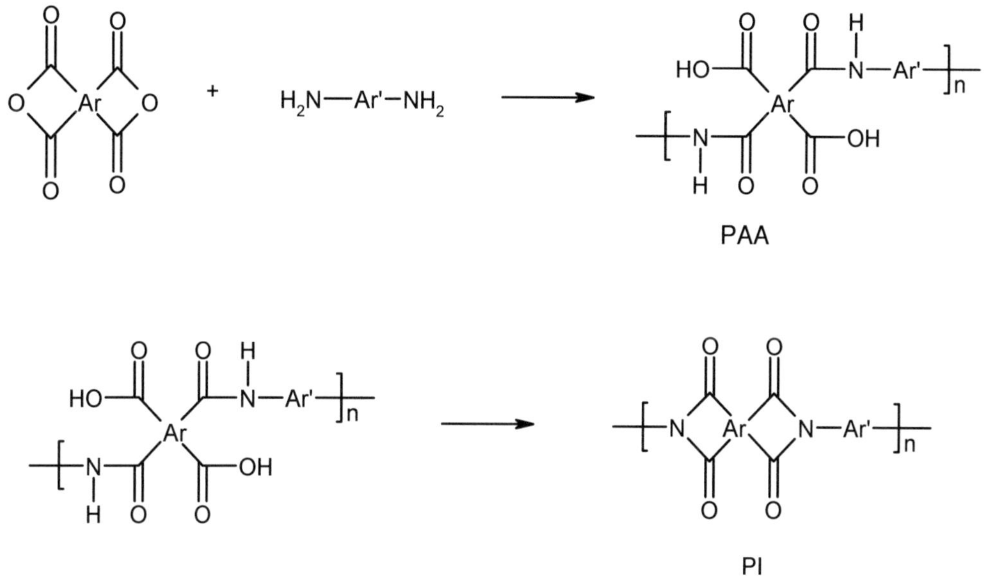

Figure 2. Two-step preparation of polyimides (PI) via a polyamic acid (PAA) precursor

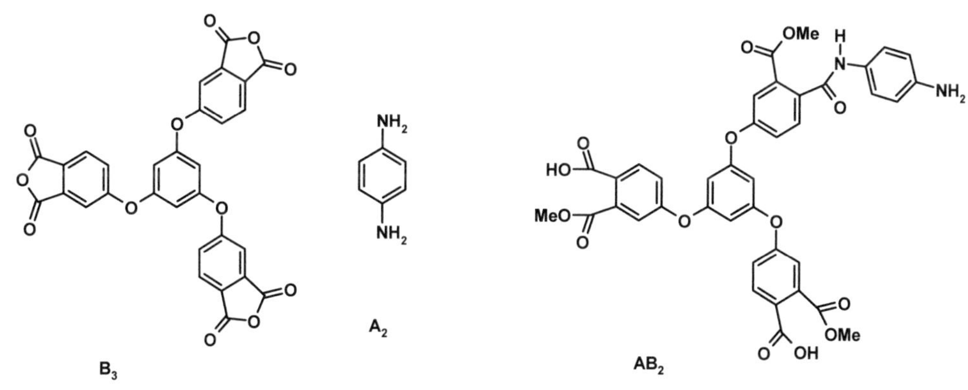

Figure 3. Chemical structure of the monomers A_2, B_3 and AB_2 used in [6]

Okamoto [7] studied the HBPI synthesis at which the different dianhydrides and tris(4-aminophenyl)amine were used (Figure 4). By using a dianhydride and triamine in molar ratio 1:1 the amine end-capped HBPI and in the ratio 2:1 the anhydride end-capped HBPI were formed. The degree of branching of the amine end-capped HBPI was in the range 0.64-0.72 in the dependence on the kind of dianhydride and this value approached 1 for the anhydride terminated HBPI (independently on a dianhydride). HBPI prepared from 4,4'-(hexafluoroisopropylidene)diphthalic anhydride (Figure 4 a) and tris(4-aminophenyl)amine in their molar ratio 1:1, showed a weight average molar mass 37000 $g.mol^{-1}$ and polydispersity (non-uniformity) 5.8 and the product based on 3,3',4,4'-diphenylsulfonetetracarboxylic dianhydride (Figure 4 b)) and tris(4-aminophenyl)amine in their molar ratio 2:1 150000 $g.mol^{-1}$ and a very high polydispersity 18.

The difficulties with the formation of a three-dimensional product were not observed if 2,4,6-triaminopyrimidine (Figure 5) was employed as a monomer for the preparation of HBPI probably due to different reactivities of amino groups in the position 2 and positions 4 and 6 [8]. But, as a consequence of it, a less regular HBPI strucure is expectable.

The application tests of hyperbranched polymers have namely focused on their employment as non-linear optical polymers, polymer electrolytes, biomaterials, supramolecular components (nanomaterials), photolithographic materials, coatings, modifiers and additives and polymeric membranes so far [3].

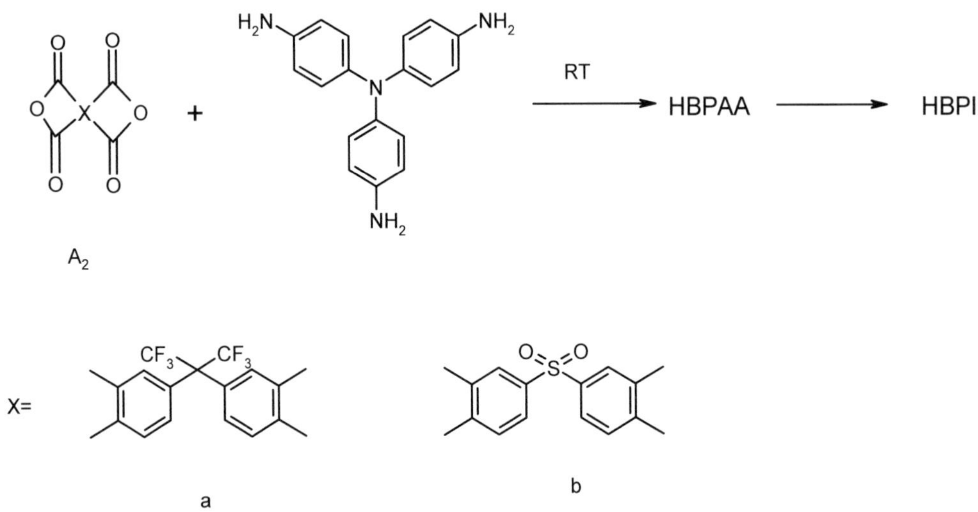

Figure 4. HBPI synthesis from tris(4-aminophenyl)amine and different dianhydrides [7]

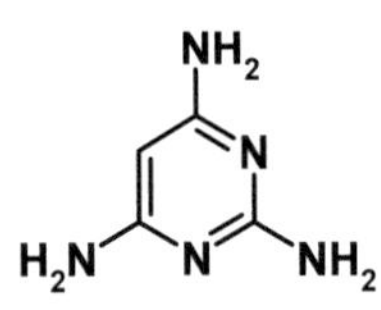

Figure 5. Chemical strucure of 2,4,6-triaminopyrimidine

Membrane separation processes belong to a relatively new type of mass-separation techniques. They have been tested or even applied in various technological processes, e.g., of chemical and food industries. Membrane processes could be used, e.g., instead of a cryogenic distillation of air for the production oxygen and nitrogen [9]. Their importance increases also in environmental technologies reducing or eliminating emissions, wastes and pollutants in air and water. Inorganic membranes have high thermal and chemical stabilities, which make them attractive for separations at high temperatures and in aggressive environments. However, inorganic membranes still have technical limitations and suffer from problems such as brittleness and lack of surface integrity [10]. On the contrary, polymeric membranes are becoming increasingly important for the separation of gas and liquid mixtures because of their low cost and the ease of their production. But two key technical challenges exist in this field. The first of these challenges is to achieve higher ability to separate mixtures with at least equivalent productivity. The second challange is to mantain these properties in the presence of complex and aggresive feeds (polymeric membranes are traditionally less stable against chemicals and temperature in comparison with inorganic membranes) [10].

Mass transport through a nonporous polymeric membrane is characterized by permeability, diffusion and solubility coefficients which depend particulary on the nature of both the membrane and the penetrant. A level of the separation depends also on additional factors (e.g. temperature). Generally, the permeability (flux) of membranes with excellent separation properties (selectivity) is usually not very high. For practice it is very important to have membranes with high permeability and with sufficient selectivity, too. The permeability in a polymer depends on the solubility and the difffusivity of the permeating species in that polymer. When there exists a large difference in the solubilities of two species mostly rubbers are taken for separation one of other (for example organic vapours recovery from air). When the difference in solubility is low mostly glassy polymers are used so that the separation is more based on the difference in diffusivity which is also governed by the free volume of polymer [9].

From the point of view of a controllable free volume, hyperbranched polymers are a very attractive candidate to the membranes with convenient transport parameters. It is reported that according to the results of computer simulation there are many accesible cavities of atomic size (or slightly larger) in the rigid hyperbranched polymers [11]. As mentioned above the polymers with overall stability are needed for producing some membranes used in separation technologies. Polyimides are available for this purpose.

Non-porous, flat polyimide membranes show high separation factors (selectivities) in separation of gas mixtures but low permeability both of gases and organic vapours. Very high selectivities organic vapours/gases were reached when non-porous flat membranes based on polyimides crosslinked with polyesters and polyethers were employed [12]. Linear aromatic polyimides were sucessfully used for separation of gases (hydrogen, helium etc.). Composite membranes, assymetric membranes and hollow fibre membranes based on polyimides were employed for separation of organic vapours from air [9].

Recently, hyperbranched polyimides were auspiciously tested as polymeric membranes for gas separation [11]. Suzuki [13] monitored gas transport properties of the membranes made of 1,3,5-tris(4-aminophenoxy)benzene(Figure 6) and 4,4'-(hexafluoroisopropylidene)diphthalic anhydride and compared these parameters with those for LPI with a similar chemical composition. The permeability, diffusivity and solubility of the gases were higher in the membranes based on HBPI. It seems that the favourable effect of both the chain character (rigidity, interactions, arrangements, higher free volume) and selective gas sorption (cavities, end-capping groups) contributes to this increase. The O_2/N_2 selectivity reached up to 6.2.

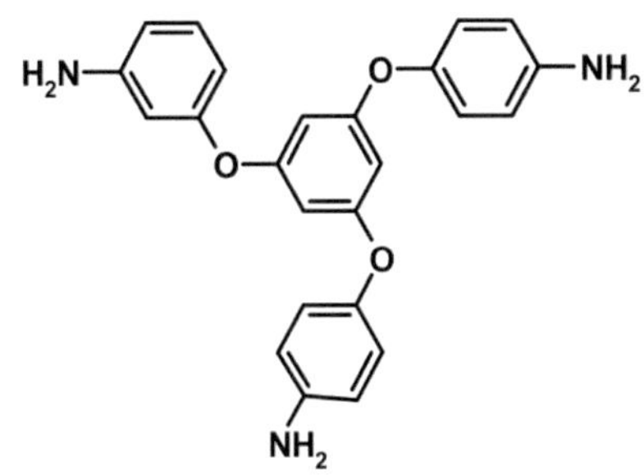

Figure 6. Chemical structure of 1,3,5-tris(4-aminophenoxy)benzene [13]

An influence of the HBPI end-groups on their gas transport properties was studied. The amine-endcapped HBPI showed the higher CO_2, N_2 and O_2 permeability coefficients than anhydride-endcapped HBPI. The amino groups of HBPI can probable create the stronger interactions with some gases, especially with CO_2 [14]. The introduction of bulky fragments (e.g. via 3,5-di(trifluoromethyl)aniline as the HBPI endgroups) also increased the gas permeabilities but their selectivities were decreased. The membranes prepared from hyperbranched polyimides based on commercially available 4,4′,4′′-triaminotriphenylmethane were prepared and some of their transport properties tested [15]. The permeability coefficients of hydrogen, carbon dioxide, oxygen, nitrogen and methane in the membrane from HBPI were 2-3.5 times higher than those in the membrane from LPI at comparable selectivities [16]. Fang [11] prepared membranes from HBPI based on tris(4-aminophenyl)amine and a few dianhydrides followed by their crosslinking. Crosslinking was realized by a coupling agent enabling us to link up macromolecules with chemical bonds by using their end-groups. The crosslinking also influenced the transport properties of membranes. The use of coupling agent ethylene glycol diglycidylether resulted in the increase of CO_2 permeability in comparison of that of N_2. On the contrary, by using a more rigid agent terephthalaldehyde the permeabilities of both gases were comparable. The selectivity CO_2/N_2 reached the values up to 32. Crosslinking brings both the reinforcement of resulted polymer structure and the partial loss of compactness of the polymer globular structure.

Nevertheless, studies concerning transport properties of the membranes made of those materials have showed that principal requirements of the membrane technologies on increasing permeability at sufficiently high selectivity are still topical. Generally, it seems that polymeric, non-porous, flat membranes reached their limit.

A promising route to membranes of improved transport characteristics consists in incorporation of inorganic additives with suitable structure (e.g., zeolites, carbon molecular sieves, microporous silica) into polymer matrix [17]. Some of polymer-inorganic composite materials showed much higher permeabilities but similar or even improved selectivity compared to pure polymer membranes. For example, the polysulfone membranes containing mesoporous silica MCM-41 showed higher gas permeabilities in comparison with that prepared from the net polymer only without a significant decrease in their selectivities [18].

The most commonly used preparation techniques for the fabrication of filled/composite materials are [17]:

1. Solution blending - a polymer is first dissolved in a solvent to form a solution, and then inorganic particles are added into the solution and dispersed by stirring.
2. In situ polymerization - particles are mixed well with organic monomers, and then the monomers are polymerized.
3. Sol-gel - organic monomers, oligomers or polymers and inorganic nanoparticle precursors are mixed together in the solution. The inorganic precursors then hydrolyze and condense into well-dispersed nanoparticles in the polymer matrix.

Polyimides exhibit outstanding dielectric and mechanical properties at elevated temperatures. Nevertheless, relatively high values of water sorption (up to 3-4 wt%) and coefficients of thermal expansion ($5x10^{-5}$ K^{-1}) impede (micro)electronic applications, e.g., forming stress-free films on silicon substrates. From this point of view silica (SiO_2), that exhibits very low values of water sorption and coefficients of thermal expansion ($5x10^{-7}$ K^{-1}), would be more suited for (micro)electronic applications but dielectric properties and planarizability are inferior to PI [17].Combined materials exhibiting favourable properties of both polyimides and silica are therefore in great demand.

From this point of view, the preparation and characterization of materials combining these components appear as highly topical. Therefore, the new mixed matrix materials prepared from HBPI based on the commercially available MTA and mesoporous silica MCM-41 or nanoparticles of silica, whose inorganic and organic phases are linked upby covalent bonds in selected cases, were made and their properties were studied in this work.

2. Experimental

4,4′-Oxydiphthalic anhydride (ODPA) (Figure 7 b) was heated to 170 °C for 5 h in a vacuum before use. 4,4′-Methylenedianiline (MDA) (both Aldrich, Czech Republic) (Figure 7 c) and 4,4′,4′′-triaminotriphenylmethane (MTA) (Dayang Chemicals, China) (Figure 7 a) were used as received. Mesoporous silica MCM-41 (M-SiO_2) having a pore size 2.5-3 nm and nanosilica (N-SiO_2) having a particle size 10-20 nm and surface area 160 m^2 g^{-1} (both Aldrich) were heated to 120 °C for 3 h in an oven before use. 3-Glycidoxypropyltrimethoxysilane (GPTMS) (Aldrich) was used as received.

1- Methyl-2-pyrrolidone (NMP; Merck, Czech Republic) was distilled under vacuum over phosphorus pentoxide, and stored in an inert atmosphere. Gases in the gas cylinders (Siad, Czech Republic) were used as received (nitrogen 99.99 vol%, oxygen 99.5 vol%, methane 99.0 vol%, carbon dioxide 99.0 vol%, hydrogen 99.90 vol%).

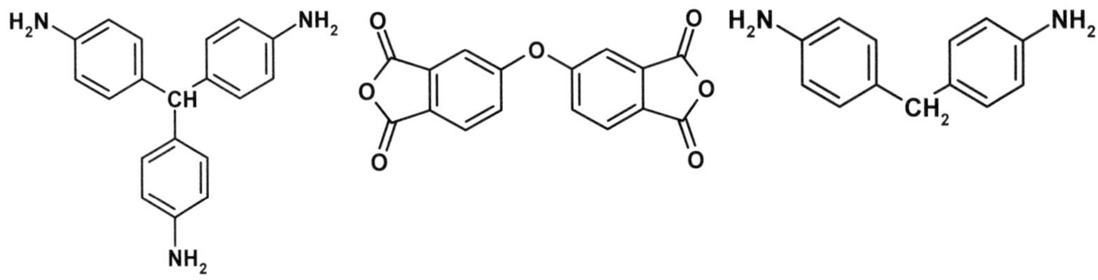

Figure 7. Chemical structure of a) 4,4′,4′′-triaminotriphenylmethane, b) 4,4′-oxydiphthalic anhydride and c) 4,4′-methylenedianiline

HBPAA and LPAA were prepared in a two necked flask equipped with a magnetic stirrer and a nitrogen inlet/outlet. At room temperature ODPA solution in NMP was added dropwise to a solution of MTA or MDA in NMP. This reaction mixture was then stirred at room temperature for 24 h. The general procedure for the preparation of amine end-capped HBPAA based on ODPA and MTA [HBPAA(ODPA-MTA)11] is followed: At room temperature, a solution of 4.973 g (0.016 mol) ODPA in 65 ml of NMP was added dropwise to a solution of 4.562 g (0.016 mol) of MTA in 86 ml of NMP for approximately 1 h. This reaction mixture was then stirred at room temperature for 24 h.

A mixture of HBPAA with M-SiO_2 or N-SiO_2 was prepared by adding the calculated amount of silica particles to the 4 wt% solution of polyimide precursor in NMP under stirring with a magnetically driven stirrer. The preparation of mixture (suspension) under stirring proceeded for 12 h. Some of M-SiO_2 and N-SiO_2 were reacted with GPTMS before use.

For the preparation of membranes based on HBPI or LPI or HBPI and silicaa solution of either HBPAA or LPAA or a suspension of silica in HBPAA solution were spread onto a glass substrate (treated with chlorotrimethylsilane). The resulting thin layer was kept at 60 °C for 12 h, 100 °C for 1h, 150 °C for 1 h, 200 °C for 2 h and 250 °C for 1 h. The thickness of the self-standing films obtained was about 50 - 100 μm.

^{1}H NMR spectra were taken on Bruker Avance DRX 500 at 500 MHz in d_6-dimethylsulfoxide. IR spectra were recorded on a Nicolet 740 spectrometer using a reflective mode. Thermogravimetric measurements (TGA) were performed in air using a TG-750 Stanton-Redcroft. The glass transition temperatures were found by using a dynamic mechanical analysis (an instrument DMA DX04T (RMI, Bohdanec, Czech Republic) operating at 1 Hz and in the temperature range 25-400 °C with a temperature gradient 3 °C min^{-1}). Scanning electron microscopy was made by using a microscope Jeol JSM-5500 LV. The material cross section areas were covered by a deposited platinum layer prior to measurement. The following procedure was used to test the chemical resistance of the materials in the NMP, as well as in methanol and toluene: each film was dried at 100 °C for 3 h, weighed, and immersed in an appropriate solvent. After 35 days the weight change was determined. The kinematic viscosities of the HBPAA and LPAA solutions were measured using a capillary viscometer at 20 °C, and their intrinsic viscosities measured in NMP at 25 °C. The permeation measurements were conducted using the self-developed manometric integral apparatus, the detailed description of which is given in [16]. The values of permeation coefficients vary within the accuracy of measurement up to 20% if experiments are repeated (especially in the case of methane due to its extremely low permeation).

3. Results and discussion

In this work, the commercially available monomers - bifunctional 4,4′-oxydiphthalic anhydride (ODPA) and trifunctional 4,4′,4′′-triaminotriphenylmethane (MTA) - were used for the hyperbranched polyimide (HBPI) preparation. Besed on its composition the final hyperbranched polyimide (HBPI) is designated as HBPI(ODPA-MTA) or as the case may be with an additional specification of the monomer molar ratio [e.g. HBPI(ODPA-MTA)11 if ODPA and MTA were used in the molar ratio 1:1]. The use of the commercially available monomers is reasonable from the point of view of a shift to the larger (industrial) scale.

Due to the MTA structure the identical reactivity ofits three amino groups is expectable which is one of the basic requirements of the regulary branched polymeric structure[8]. In the case of ODPA it was shown that the presence of the ether bridge (-O-) decreases the glass transition temperature of the final PI in comparison with those cases where different dianhydrides were employed [15]. The lower rigidity of polymer chains can favourably influence the preparation of self-standing films. Their successful formation is important for the evaluation of the thermo-mechanical and transport properties of these materials. Knowledge of these properties is substantial for making conclusions concerning potential applications. For the better evaluation of the specific HBPI(ODPA-MTA) properties the linear polyimide (LPI) based on ODPA and bifunctional amine 4,4′-methylenediamine (MDA) [LPI(ODPA-MDA)11] in their molar ratio 1:1 was also prepared and characterized. 1-Methyl-2-pyrrolidone (NMP) was used as a solvent. It dissolves both monomers and polyimide precursors (PAA). Due to its hygroscopic character it was distilled in vacuum in the presence of a phosphorus pentoxide.

The MTA quality (purity) was determined by using ^{1}H NMR (Figure 8). It is obvious a very good relation between the integral intensity of signals belonging to the aromatic protons (6.7-6.4 ppm) and those to the protons of amino groups (4.87-4.82 ppm). Unfortunately, the integration of the signal of methine (-CH-) group is not possible due to its coincidence with another signal.

The two-step preparation via an intermediate (HBPAA or LPAA) was chosen for the preparation both HBPI (Figure 9) and LPI [5]. Taking into consideration the gel (crosslinked structure) formation at definite reaction extent during the reaction of a bifunctional and a trifunctional monomer [15], our attention was also devoted to the choice of optimal reaction conditions of the HBPI(ODPA-MTA) preparation. The maximal monomer concentrations in NMP and the way of their combination were especially found. The dianhydride solution was added drop-to-drop to the diamine solution. This monomer order also decreases a probability of the dianhydride hydrolysis during an intermediate preparation.

Two monomer molar ratios were used for the HBPAA preparation. The ODPA:TMA ratio 1:1 gave an amine end-capped HBPAA and that 2:1 an anhydride end-capped HBPAA (Figure 10).The maximally available HBPAA concentrations – to avoid a gel formation – varied to some extent in the dependence on their character (Table 1) [15]. The higher maximal concentration for theamine end-capped HBPAA (0,2 mol.l^{-1}) in comparison with that terminated with anhydride groups (0,075 mol.l^{-1}) is also reported in [7]. The combination of the monomer solutions at their molar concentration ratio 1:1 was also used for the preparation of LPAA.

PAA specification	PAA content[a] (wt%)	N[b] ($m^2.s^{-1}$)	[η][c] ($ml.g^{-1}$)
HBPAA(ODPA-MTA)11	4	$3.17x10^{-5}$	54
HBPAA(ODPA-MTA)21	1	$2.23x10^{-6}$	36
LPAA(ODPA-MDA)11	4[d]	$2.83x10^{-5}$	28

[a]maximal concentration of HBPAA in the final solution without an immediate gel formation
[b]kinematic viscosity of the solution at 20 °C
[c]limiting viscosity number (intrinsic viscosity) of HBPAA (in 0.075 mol.l^{-1} LiCl in NMP at 25 °C)
[d]content of solids was chosen

Table 1. Polyamic acids (PAA) and their characterization

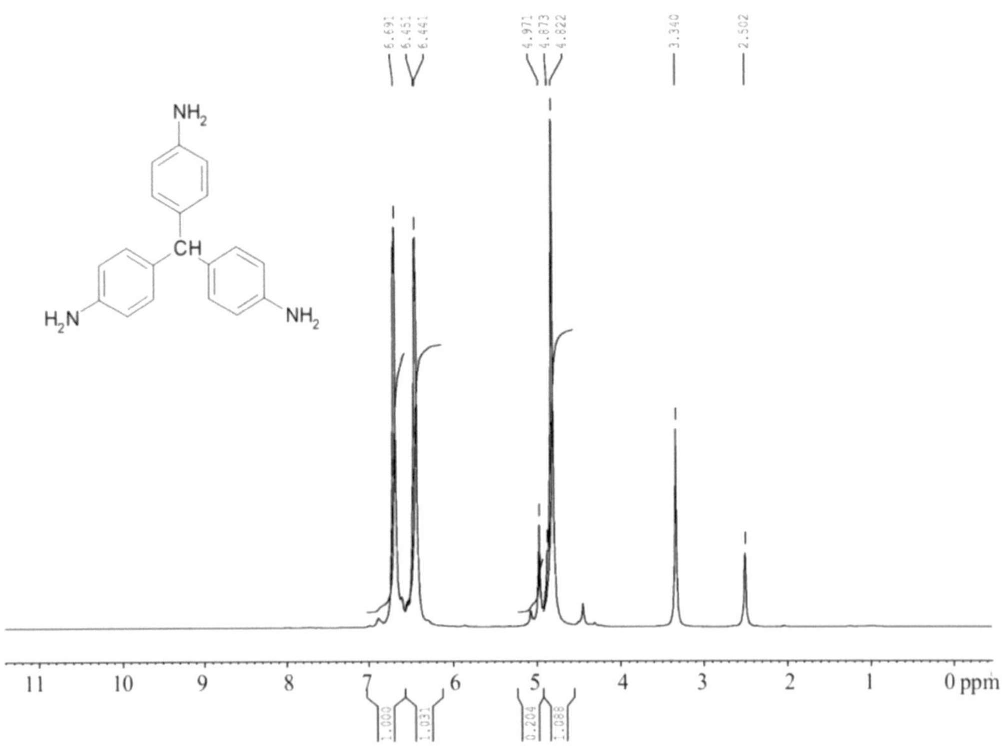

Figure 8. ^{1}H NMR spectrum of MTA

In the case of LPAA(ODPA-MDA)11 the maximal solution concentration is not a limiting factor of their preparation and it was chosen 4 wt% in agreement with that for HBPAA(ODPA-MTA)11. Actually, HBPAA(ODPA-MTA)11 was only employed in further work. The final product HBPI(ODPA-MTA)11 provides self-standing films with a good mechanical stability in contrast to HBPI(ODPA-MTA)21. In addition to it, amino end-groups served as reactive sites for linking up the polyimide matrix with silica particles by using the coupling agent 3- glycidyloxypropyltrimethoxysilane.

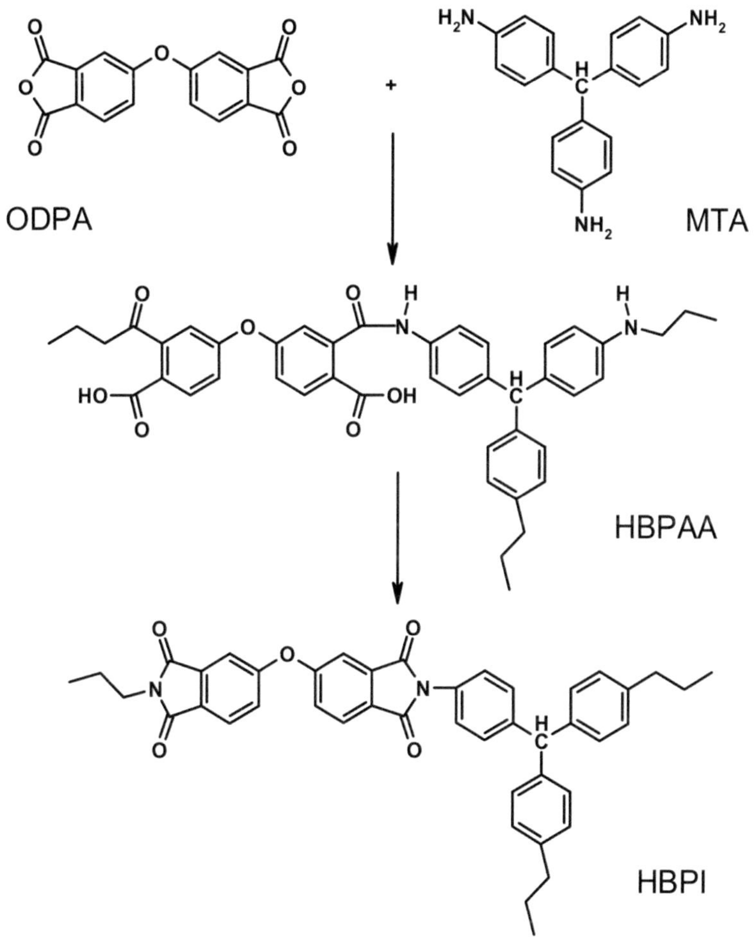

Figure 9. Two-step preparation of HBPI(ODPA-MTA)

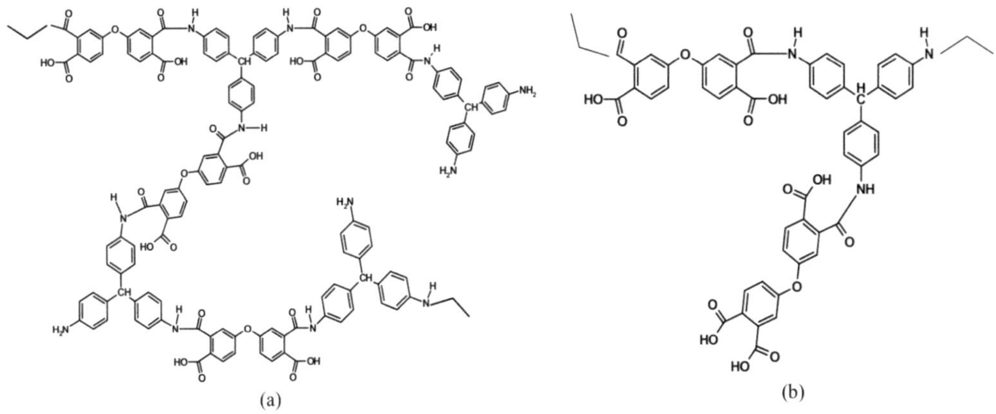

Figure 10. Chemical strucure of a) amine and b) anhydride terminated HBPAA(ODPA-MTA)

The kinematic viscosities of the PAA solutions of a defined (given) concentration and limiting viscosity numbers (intrinsic viscosities) of PAA are also given in Table 1. It is obvious that kinematic viscosities are proportional to the solution concentrations (cf. 4 wt%) solution of HBPAA(ODPA-MTA)11 and 1 wt% solution of HBPAA(ODPA-MTA)21). But it is not a significant difference between the solution of hyperbranched [HBPAA(ODPA-MTA)11] and linear [LPAA(ODPA-MDA)11] PAA having the same theoretical solution concentration (4 wt%). Nevertheless, it was found that the kinematic viscosities of these solutions changes to a certain degree with their storage (i.e. between its preparation and analysis). For example, the kinematic viscosity of HBPAA(ODPA-MTA)11 was changed about of 10% during 50 days. Therefore, the solutions were processed (transformed to the corresponding polyimides (PI)) immediately after their preparation. It is expected that both HBPAA formation and degradation (hydrolysis) reactions occur (take place) simultaneously [5]. Limiting viscosity number [η] serves as an indicator of a viscosity average molecular mass M_v. It can be calculated for the given system a polymer – solvent at a defined temperature by using a Mark –Houwink equation $[\eta] = K.M_v^a$, where a, K are parameters. These parameters are tabulated for common polymers, but for PI exceptionally only.In addittion to it, majority of PI is not soluble and therefore polyimide precursors can only be characterized by viscometry. Limiting viscosity number is determined from the linear dependence of the reduced viscosity on solution concentration as an extrapolated value at the zero concentration. To obtain the linear dependence a polyelectrolytic effect must be avoided. For this purpose, lithium chloride in the concentration of 0.075 $mol.l^{-1}$ was added into NMP, which was used as the solvent for viscometry measurements [15]. From Table 1 it is obvious, that [η] of HBPAA(ODPA-MTA)11 (i.e. amine end-capped HBPAA) is about 50% higher than that for HBPAA(ODPA-MTA)21 (i.e. anhydride end-capped HBPAA).

This difference in [η] should also be given by the different degree of branching HBPAA(ODPA-MTA)11 and HBPAA(ODPA-MTA)21. Degree of branching is determined from the fraction of dendritic, linear and terminal units included in the given polymer structure (see above). It is necessary to find a group of atoms occuring in all these structures and monitore it by using a suitable instrumental technique. In the case HBPAA(ODPA-MTA)11 andHBPAA(ODPA-MTA)21 the position of methine group (-CH-) coming from MTA was monitored in ^{1}H NMR spectra of given HBPAA. The values of shifts for the dendritic, linear and terminal units were found by helping of the ^{1}H NMR analysis of model reactions of MTA and phthalic anhydride in the molar ratio 1:1, 1:2 and 1:3 and p-toluidine (4-aminotoluene) with phthalic anhydride in the molar ratio 1:1. The model reactions were carried out in deuterated dimethylsulfoxide to avoid the isolation and re-dissolution of these products for NMR analyses.

In the ^{1}H NMR spectrum of MTA (Figure 8) there are the signals corresponding to the amino group protons (4.82 ppm) and to the methine group proton (4.97 ppm) very close each other and their integration is accompanied with problems. In the spectra of model compounds (i.e. MTA with phthalic anhydride) the signals of methine protons shift to 5 - 6 ppm and they are separated from those of unreacted amino groups at 6.5 - 6.7 ppm markedly (as seen in a model reaction of p-toluidine with phthalic anhydride also).From the ratio among signals in the 5 - 6 ppm range the conclusion was made that the signal with the highest shift belongs to the linear units and with the lowest shift to the terminal units. ^{1}H NMR spectra of HBPAA(ODPA-MTA)11 andHBPAA(ODPA-MTA)21 are collected in Figure 11. The broad signal at about 13 ppm belongs to the protons of carboxylic groups, the signal at 10.4 ppm to protons of amide groups, at 8 – 7 ppm to aromatic protons. The degree of branching was determined as ca. 0.7 for HBPAA(ODPA-MTA)11 and 0.9 for HBPAA(ODPA-MTA)21.

The transformation of HBPAA or LPAA into corresponding final products was conducted by using of a thermal imidization in solid phase. This technique is commonly employed for the preparation of insoluble PI. The products are often the layers having a thickness in the range of 1-100 μm. The thickness of the films in this study was about 50 μm. The layer (with a thickness about 1 mm) of the HBPAA or LPAA solution was cast on a glass substrate. A majority of the solvent escaped during heating it in the oven in (circulating) air atmosphere at 60 °C. Then, the temperature was gradually increased, finally to 230 – 250 °C, to reach a full imidization and avoid a formation of holes and bubbles in the film (Figure 12). As already mentioned, the mechanical stability of HBPI(ODPA-MTA)21, prepared from 1 wt% solution of (HBPAA(ODPA-MTA)21 (see Table 1), was very poor and it was not used in further studies.

The level of transformation of HBPAA(ODPA-MTA)11 and LPAA(ODPA-MDA)11 into corresponding PI was analyzed by IR spectroscopy (Figure 13). The absorption bands at about 1780 and 1720 cm^{-1} (symmetric and asymmetric stretching of the ring carbonyl groups), together with the band at 1380 cm^{-1} (stretching of the ring C-N bond), are distinct in the spectrum of PI and characterize the formation of imide structures. The absence of the band at 1680 cm^{-1} [amide group of polyamic acid - see the IR spectrum of HBPAA(ODPA-MTA)11] supports our notion that thermal treatment leads to almost complete imidization. The spectrum of the HBPI(ODPA-MTA)11 whose amino end-groups were reacted with phthalic anhydride THBPI(ODPA-MTA) was also taken. This material was prepared to judge the influence of the amino end-groups on its behaviour. The band of terminal amino groups at ca 1620 cm-1 is overlapped in the spectra of both corresponding HBPAA and HBPI. Nevertheless, it is distinct in the IR spectra of model reactions of TMA with phthalic anhydride (see above). Its intensity is proportional to the molar ratio of these components (Figure 14).

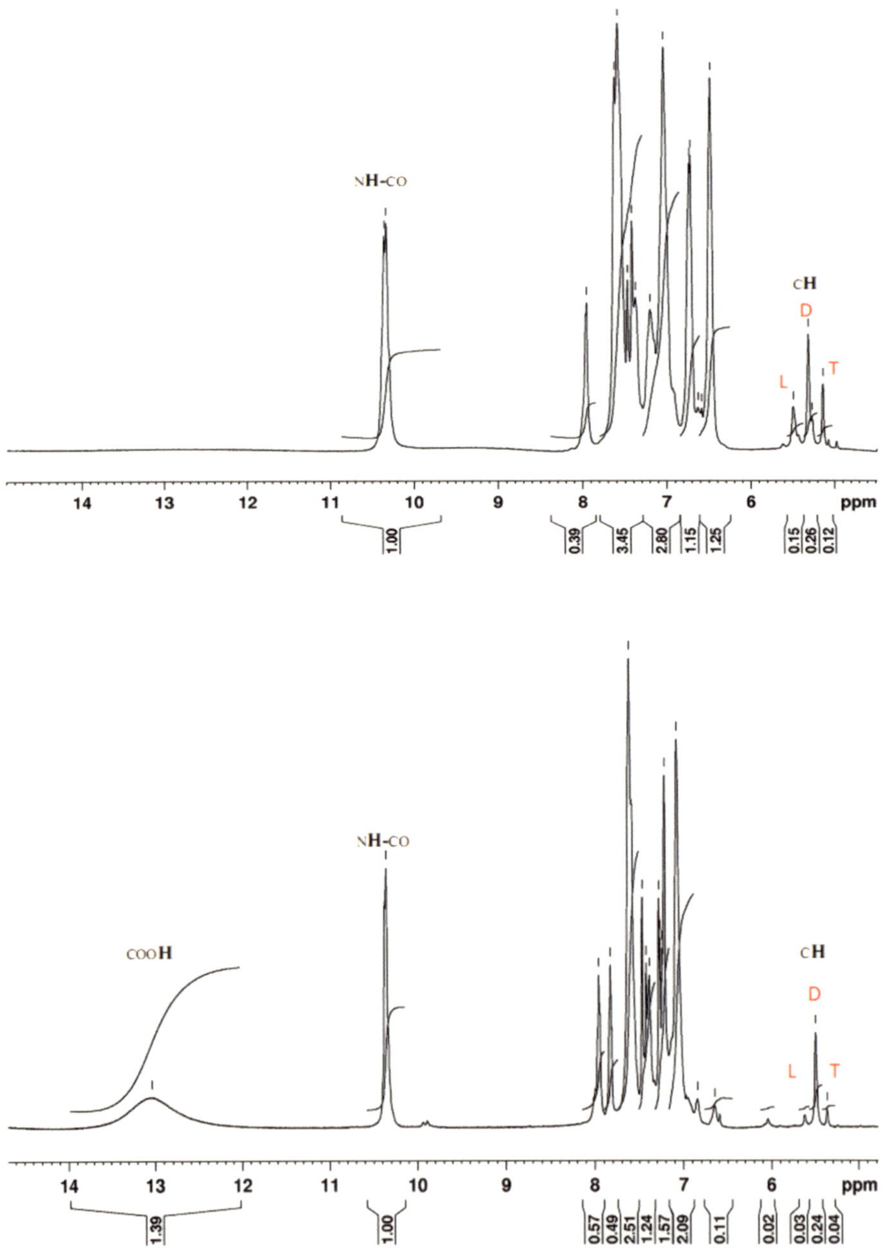

Figure 11. 1H NMR spectra of (top)HBPAA(ODPA-MTA)11 and(bottom) HBPAA(ODPA-MTA)21

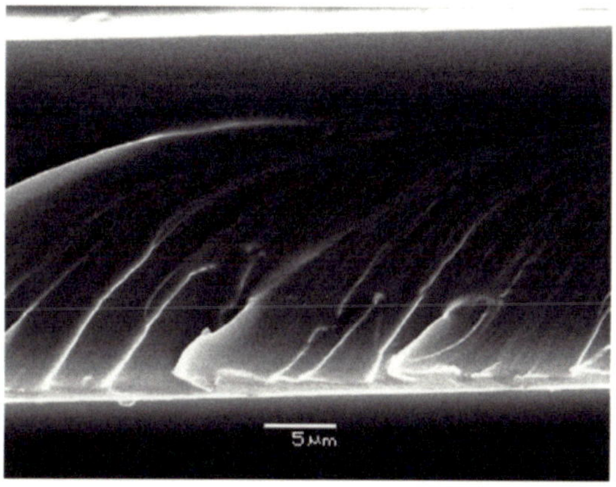

Figure 12. SEM photograph of HBPI(ODPA-MTA)11

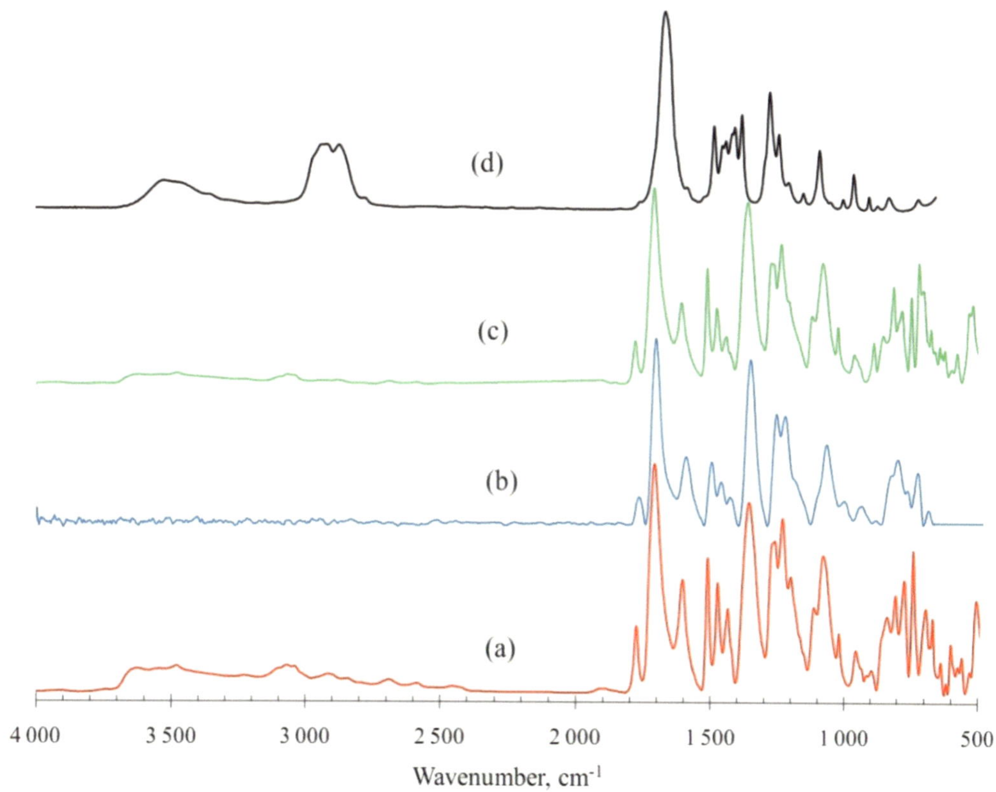

Figure 13. IR spectra of a) LPI(ODPA-MDA)11, b) HBPI(ODPA-MTA)11, c) THBPI(ODPA-MTA)11, d) HBPAA(ODPA-MTA)11

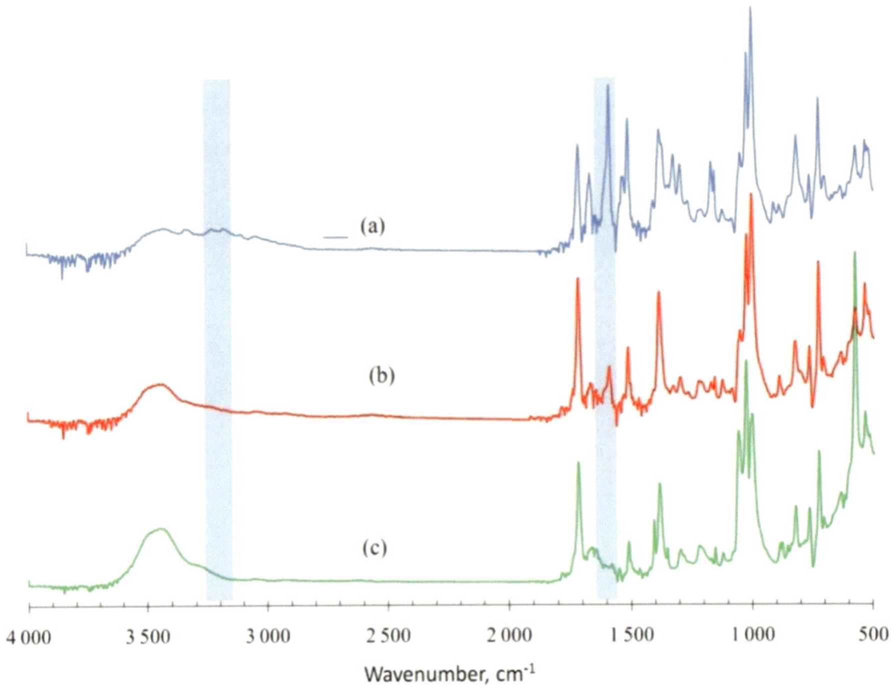

Figure 14. IR spectra of model compounds prepared from MTA and phthalic anhydride in molar ratio (a) 1:1, (b) 1:2 and (3) 1:3

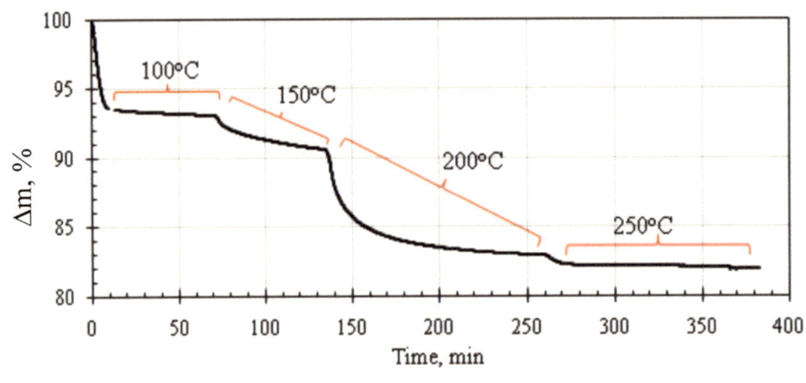

Figure 15. TGA record simulating the HBPAA(ODPA-MTA)11 imidization

The choice of the final imidization temperature is limited on the one side by practically full imidization and on the other side by the request of the stability of the product under these conditions. To support the second request, HBPAA(ODPA-MTA)11 was processed by using a thermogravimetric analysis simulating the imidization proces. It is obvious from Figure 15 that the product is stable (a weight loss is practically unchanged) at 250 °C.

Materials HBPI(ODPA-MTA)11 and LPI(ODPA-MDA)11 were analyzed by using a dynamic-mechanical analysis (DMA). The records (tan δ) vs temperature of both materials, which are very different, are shown in Figure 16 a,b.

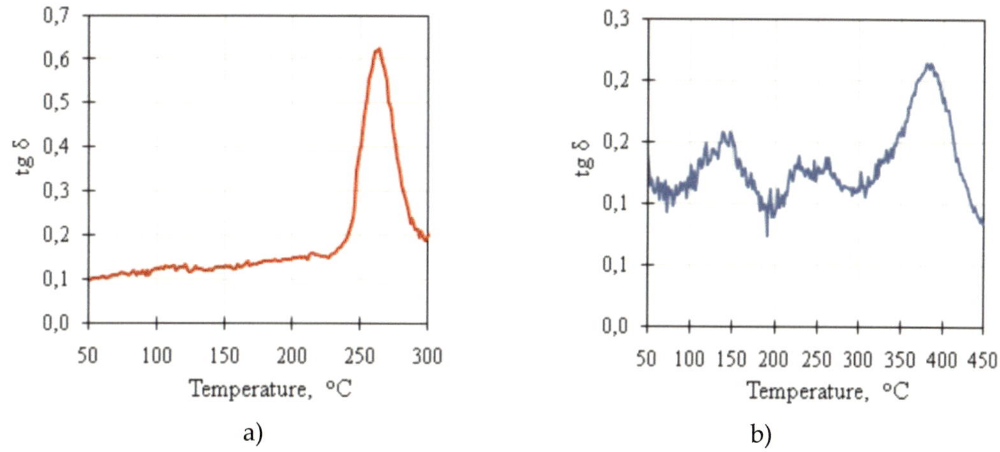

Figure 16. DMA records of a) LPI(ODPA-MDA)11 and b) HBPI(ODPA-MTA)11

DMA record of LPI(ODPA-MDA)11 provides one rather broad band with a maximum at 263 °C corresponding to its glass transition temperature. On the other hand, HBPI(ODPA-MTA)11 record is very complicated. It is created by a few partly overlaped bands in the temperature region from 50 to 400 °C. The band at above 250 °C is bringing by a linear portion and thatabove 370 °C by a hyperbranched portion of the product. The accuracy of these DMA measurements is about ± 5 °C.

To evaluate activation energies of thermooxidative decomposion the materials HBPI(ODPA-MTA)11, THBPI(ODPA-MTA)11 and LPI(ODPA-MDA)11 were analyzed by a thermogravimetric analysis using heating rates 2,5; 5; 10 and 15 °C.min^{-1}. The obtained dependences are collected in Figure 17 a, b.

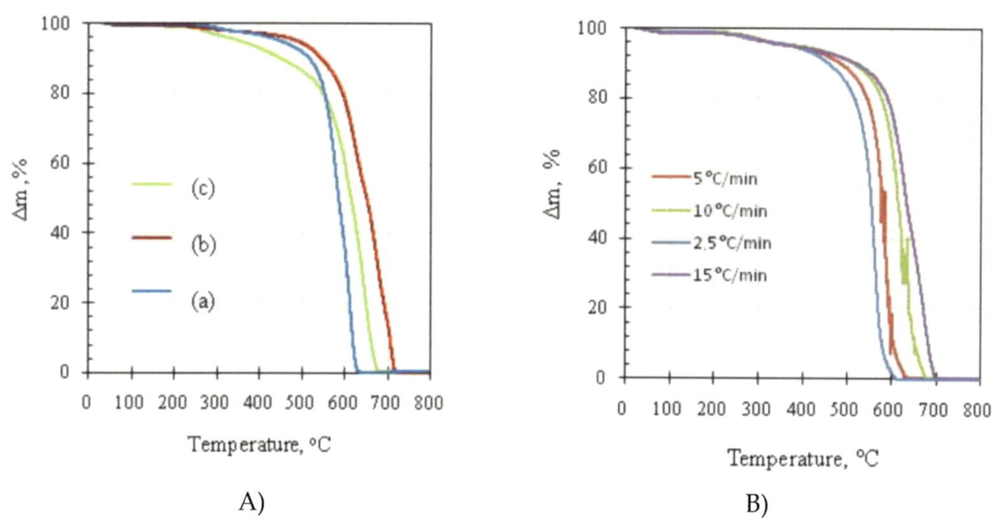

Figure 17. TGA records in air atmosphere of A) HBPI(ODPA-MTA)11 (a), THBPI(ODPA-MTA)11 (c), LPI(ODPA-MDA)11(b) at heating rate 10 °C.min^{-1}; B) HBPI(ODPA-MTA)11 at heating rates 2,5; 5; 10 and 15 °C.min^{-1}

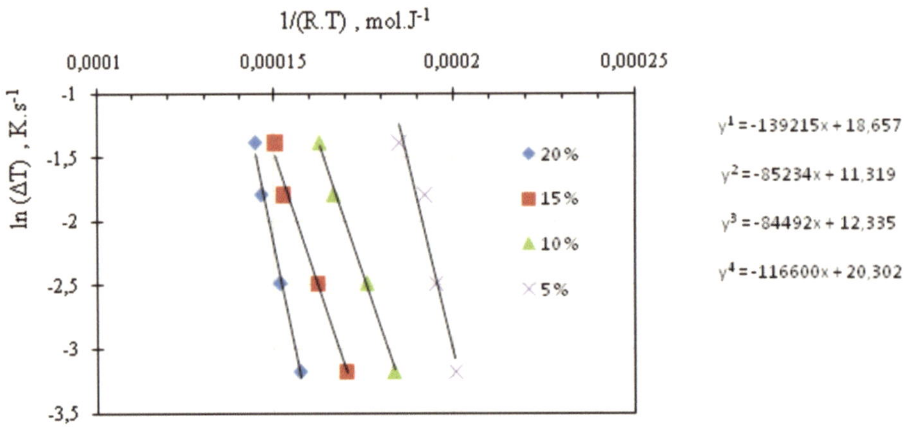

Figure 18. Graphical evaluation of the activation energies of different levels (from 5 to 20 wt%) of the thermooxidative decomposition of THBPI(ODPA-MTA)11 (regression equations are also given in this figure).

Material	E_a (kJ mol^{-1})
HBPI(ODPA-MTA)11	127
THBPI(ODPA-MTA)11	106
LPI(ODPA-	184

Table 2. Activation energies of thermooxidative decomposition of HBPI(ODPA-MTA)11, THBPI(ODPA-MTA)11 and LPI(ODPA-MDA)11

The values of temperature corresonding to the weight loss from 5 to 20 wt% were used for the evaluation of activation energy of thermooxidative decomposition E_a by using an Arrhenius equation

$$\ln(k) = \ln(A) - E_a \cdot \frac{1}{R \cdot T},$$

where k is a heating rate, R is a gas constant, T is an absolute temperature.

The activation energies for the weight losses from 5 to 20 wt% were determined from the dependence of ln k on reciprocal value T (1/T). Itis shown for THBPI(ODPA-MTA)11 in Figure 18. The average values of activation energies in the region 5 – 20 % are collected in Table 2.

The E_a of HBPI(ODPA-MTA)11 is about 30 % lower than that of LPI(ODPA-MDA)11 and the blockage of amino end-groups decreases the E_a of THBI(ODPA-MTA)11 in comparison with HBPI(ODPA-MTA)11 about 20 kJ mol^{-1}. The order of these values supports the relation of the temperatures corresponding to ,e.g., 10 wt% loss in air at a heating rate 10 °C.min^{-1} (544 °C for LPI(ODPA-MDA)11, 513 °C for HBPI(ODPA-MTA)11 and 448 °C for THBPI(ODPA-MTA)11 (also see Figure 17 a).

It is clear that the blockage of amino end-groups of HBPI(ODPA-MTA)11 worsens its thermooxidative stability. It is probably a consequence of a non-existence of interactionsof these groups [19].

The deeper information concerning a thermooxidative decomposition was obtained by using thermogravimetric analysis combined with an IR analysis of degradation products. It is obvious from the IR spectra taken at 670 °C (Figure 19) that the main degradation product is carbon dioxide (band at 2350 cm^{-1}). The carbon monooxide (ca 2150 cm^{-1}), water (3500 cm^{-1}) and aromatic fragments (1500 cm^{-1}) were also identified in the spectra.

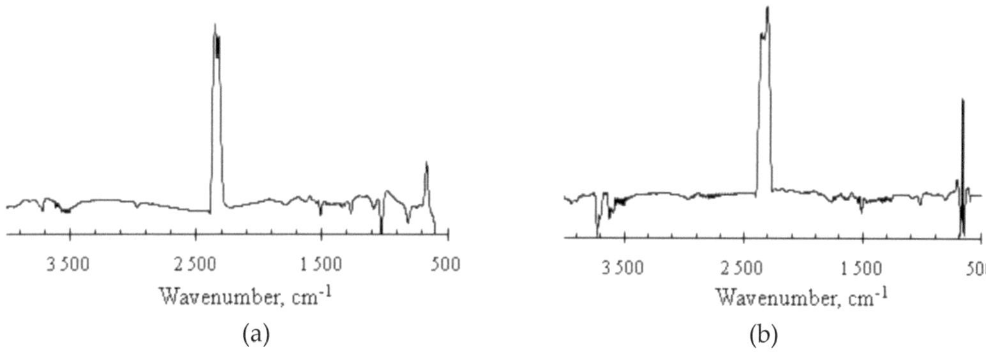

Figure 19. IR spectra of the degradation products of thermooxidation of a) HBPI(ODPA-MTA)11, b) LPI(ODPA-MDA)11 at 670 °C

Both the HBPI and LPI were not soluble in NMP, toluene and methanol. It means that the hyperbranched architecture of polymer chains does not contribute to the solubility. The permeability coefficients of hydrogen, oxygen, nitrogen and methane in the membrane prepared from HBPI were ca 2 and of carbon dioxide 3.7 times higher than those in the membrane from LPI at comparable selectivities (e.g., selectivity oxygen/nitrogen about 6) [16].

As an additive to the hyperbranched polyimide matrix the silica was chosen. It has a lot of favourable properties and it is available in a lot forms. These forms were chosen with a view of improving permeability/selectivity relation of the membranes made of these materials. Compact (non-porous) additives generally decrease a gas permeability due to their behaviour as a barrier for a gas molecule trajectory [17]. Therefore, the mesoporous silica (M-SiO_2) and non-porous nanosilica (N-SiO_2) were chosen for this work. M-SiO_2 with a particle size about 1 μm shows a pore diameter about 2.5 nm, i.e. the diameter larger than that of gases whose permeability was tested in this work (0.29 nm for H_2, 0.33nm for CO_2, 0.35 nm for O_2, 0.36 nm for N_2 and 0.38 nm for CH_4 [18]). In the case of N-SiO_2 with a particle size about 15 nm, a large interfacial area, having different properties than both polyimide matrix and silica, can be expected. By using a titrimetric determination (weakly acidic Si-OH groups were treated with NaOH and an unreacted portion of NaOH was titrated with HCl) the silanol group concentrations (Si-OH) 1.3 and 0.3 mmol.g^{-1} were found for M-SiO_2 and N-SiO_2, respectively. The higher Si-OH concentration in the case of M-SiO_2 corresponds with its larger surface in comparison with N-SiO_2.

In the selected cases (samples), the organic and inorganic phases were bound by the covalent bonds by using a coupling agent 3- glycidyloxypropyltrimethoxysilane (GPTMS). Its epoxy groups react with amino end-groups of HBPI(ODPA-MTA)11 (Figure 20) and its alkoxy groups (after their hydrolysis to hydroxyl groups) with the surface hydroxyl groups of silica (it is schematically shown in Figure 21). The consequence of covalently bound phases could be also a better distribution of the silica particles in a polymer matrix [17,20].

The IR spectra of the non-functionalized and GPTMS functionalized M-SiO_2 are shown in Figure 22. The bands at about 1100 cm^{-1} correspond with Si-O-Si atom arrangement. The completeness of functionalization is supported by the disappearing of the broad band at 3400 cm^{-1} corresponding to silsanol groups (Si-OH).

The calculated amounts of unmodified or modified silicas were mixed with HBPAA(ODPA-MTA)11 to obtain the mixtures with 5 – 20 wt% content of additive. The layers of these mixtures (with a thickness about 1 mm) on a substrate were gradually heated, finally at 250 °C for 2 h. The IR spectra of HBPI(ODPA-MTA)11 containing theoretically 16 wt% of M-SiO_2 or N-SiO_2 are shown in Figure 23. In addition to the peaks characterizing polyimide moieties (see above) the broader bands with a maximum about 1100 cm^{-1} are present. It confirms that silicas were built in the polyimide matrix.

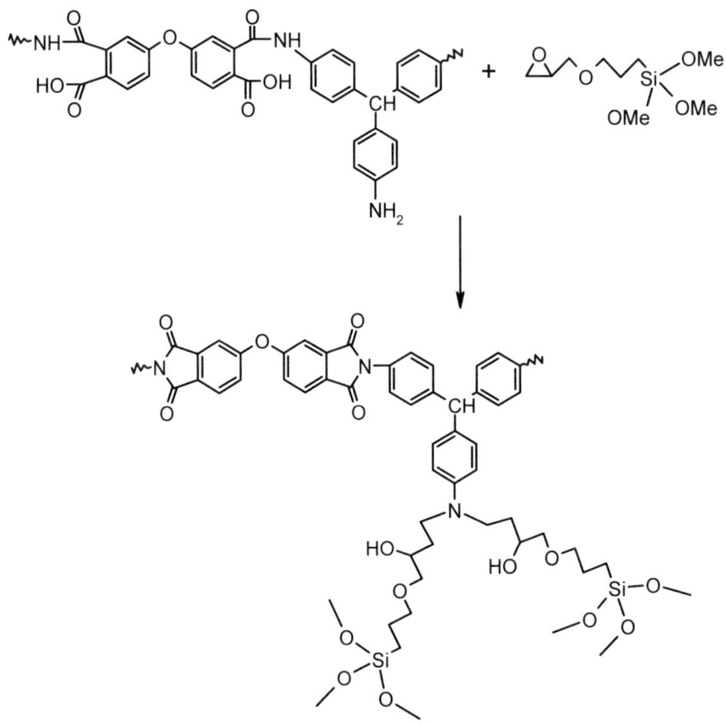

Figure 20. Schematic reaction of the coupling agent 3- glycidyloxypropyltrimethoxysilane with amine end-capped HBPAA(ODPA-MTA)11

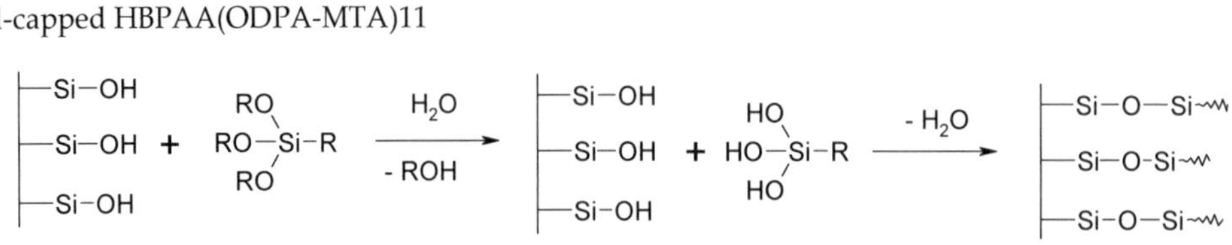

Figure 21. Reaction scheme of the surface hydroxyl groups of silica with alkoxysilanes (OR is an alkoxygroup, R is, e.g., glycidyloxypropyl or aminopropyl group)

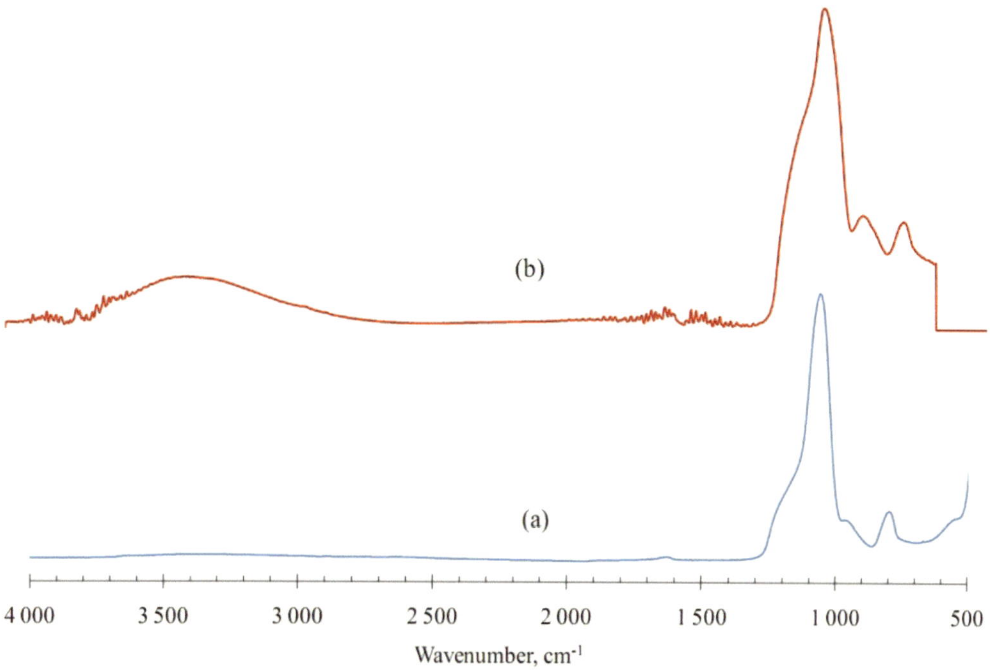

Figure 22. IR spectra of (a) functionalized M-SiO_2-and (b) non-functionalized M-SiO_2 with GPTMS

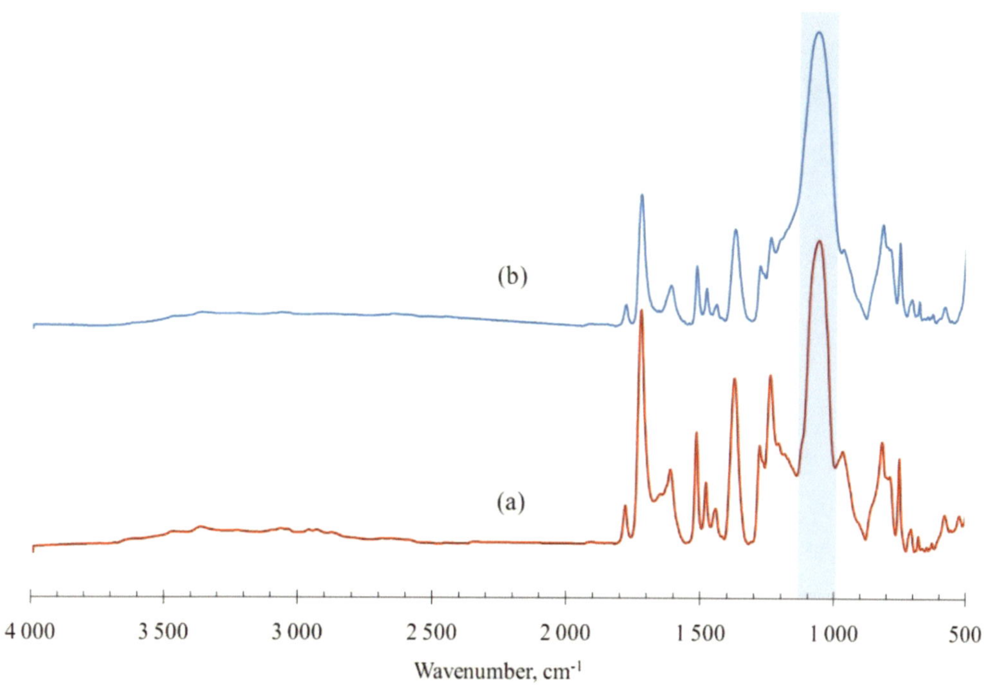

Figure 23. IR spectra of HBPI(ODPA-MTA)11 containing 16 wt% of (a) M-SiO_2, (b) N-SiO_2

The typical photographs of cross-section areas of the mixed matrix materials containing 10 wt% of N-SiO_2 and 16 wt% of M-SiO_2 are shown in Figure 24. It is obvious that the silica particles distribution in the polymer matrix is not regular. Nevertheless, it seems that the porous character of M-SiO_2 (i.e., also its lower specific weight) influenced slightly favourably a silica dispersation. The use of the coupling agent has not brought a more significant improvement. So, these products do not show the morphology of nanomaterials.

The glass transition temperature (Figure 25) and temperatures corresponding to 10 wt% weight loss (Figure 26) change very slightly with both M-SiO_2 and N-SiO_2 content in comparison of these values for pure HBPI(ODPA-MTA)11 (see Figures 16 and 17). It means that the rigidity of polymer chains and the interactions among them are not practically influenced by the silica presence. It is also supported with very similar values of parameter d obtained by using wide-angle X-ray diffraction measurements.What concerns transport properties of the flat membranes based on HBPI(ODPA-MTA)11 containing N-SiO2 our attention was mainly focused on oxygen/nitrogen separation which is connected with a lot problems [9]. It was shown that at the theoretical N-SiO2 content 10 wt% the permeability coefficients of oxygen and nitrogen decreases semiquantitatively in harmony with a relation

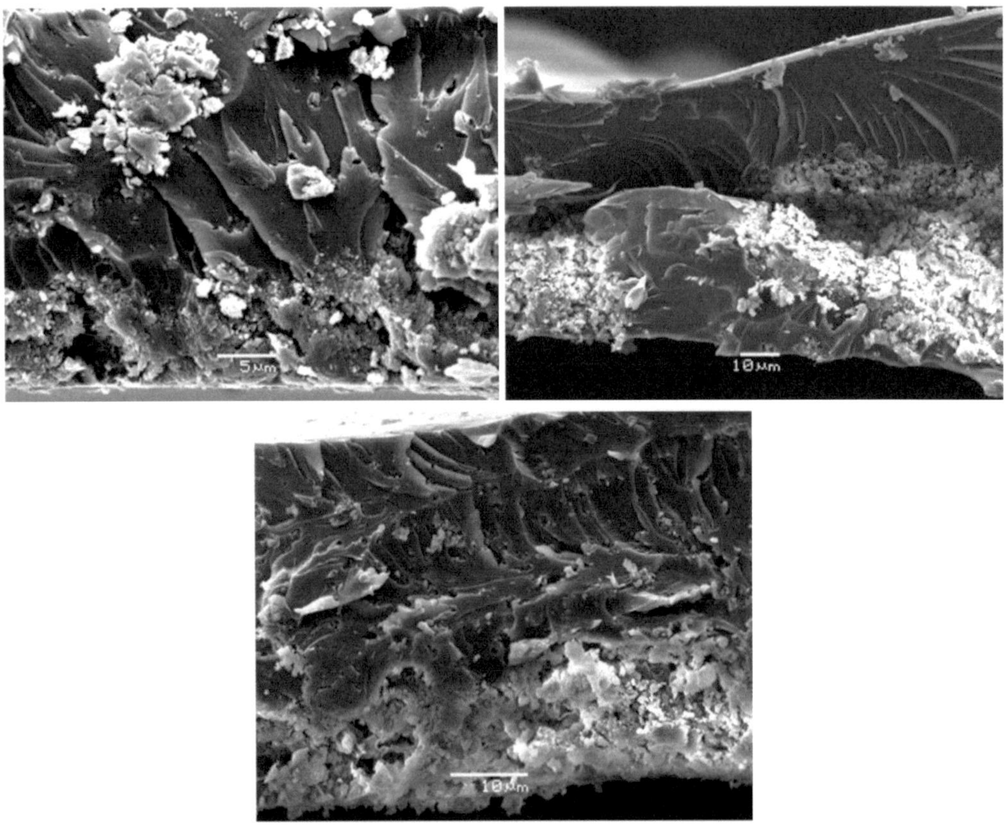

Figure 24. SEM photographs of mixed matrix materials based on HBPI(ODPA-MTA)11 containing (top left) 10 wt% of N-SiO_2, (top right) 10 wt% of M-SiO_2 and (bottom) 16 wt% of M-SiO_2

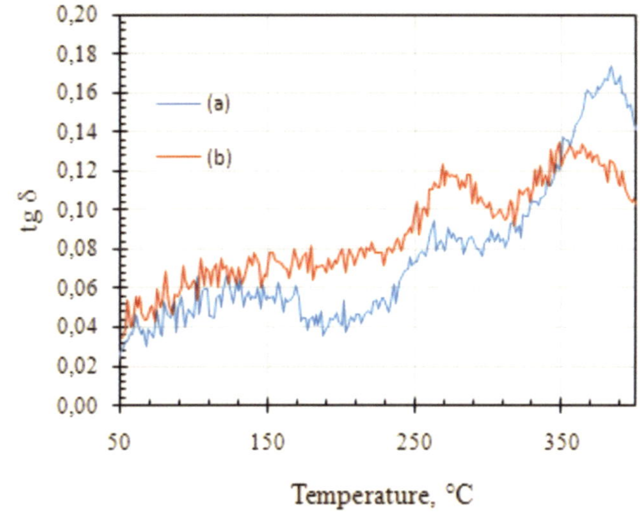

Figure 25. DMA records of mixed matrix materials containing (a) 10 wt% of M-SiO_2; (b) 10 wt% of N-SiO_2

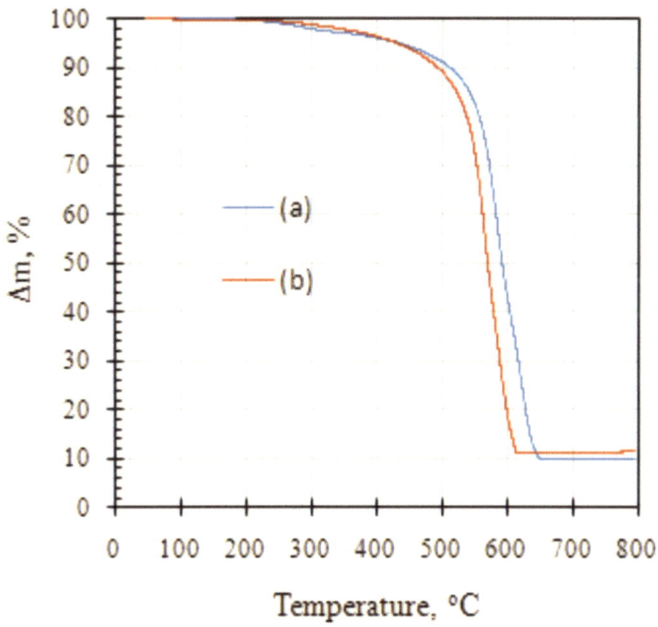

Figure 26. TGA:records of mixed matrix materials based on HBPI(ODPA-MTA)11 containing (a) 10 wt% of M-SiO_2; (b) 10 wt% of N-SiO_2

$P_C = P_{PI}[(1-\phi_{SiO2})/(1+0.5\phi_{SiO2}))]$, where P_C is the permeability coefficient of a gas in the matrix mixed membrane, P_{PI} is the permeability coefficient of gas in the pure polyimide matrix and ϕ_{SiO2} is the silica volume fraction [21]. Due to the fact that the decrease was deeper for nitrogen, the oxygen/nitrogen selectivity reaches the values higher than 9. Koros [22] explains this difference in oxygen and nitrogen permeabilities by the various levels in their jump frequency reduction. The significant increase of both permeability coefficients (about 80 %) and diffusion coefficients of these gases without change in oxygen/nitrogen selectivity was reached if the organic and inorganic (10 wt%) phases were bound by using 3-glycidyloxypropyltrimethoxysilane. In a such case the upper-bound for the couple oxygen-.nitrogen is approached [23]. The importance of the coupling agent (GPTMS) is explained in such case by its contribution to the better distribution of silica particles in the polymer matrix (i.e. a larger interphase area with specific gas separation properties [24].

The HBPI(ODPA-MTA) are also intended as starting materials for the preparation of membranes for theseparation of racemic mixtures.

4. Conclusion

As the main findings of this work are considred: It has been shown in detail, that

1. the thermooxidative stability of HBPI(ODPA-MTA)11 is lower in comparison with LPI(ODPA-MDA)11
2. the flat membrane prepared from mixed matrix materials based on HBPI(ODPA-MTA)11 and 10 wt% of N-SiO_2 with covalently bound phases approaches the upper bound for the couple oxygen-nitrogen.

Author details

Evgenia Minko, Petr Sysel* and Martin Spergl
Department of Polymers, Institute of Chemical Technology, Prague, Czech Republic

Petra Slapakova
Central Laboratories, Laboratory of Thermal Analysis, Institute of Chemical Technology, Prague, Czech Republic

Acknowledgement

This work was supported by the grant GA CR P106/12/0569 and MSM 6046137302. The financial support from specific university research (MSMT No. 21/2010-013 and 21/2010-011) is also acknowledged.

5. References

[1] Odian G (2004) Principles of Polymerization. New York: John Wiley & Sons, Inc.

---

* Corresponding Author

[2] Tomalia D A (2005) Birth of New Macromolecular Architecture: Dendrimers as Quantized Building Blocks for Nanoscale Synthetic Polymer Chemistry. Progress in polymer science 30:294-324.

[3] Gao C, Yan D (2004) Hyperbranched Polymers: from Synthesis to Applications. Progress inpolymerscience. 29:183-275.

[4] Jikei M, Kakimoto M (2004) Dendritic Aromatic Polyamides and Polyimides. Journal of polymerscience: part A: polymerchemistry42: 1293-1309.

[5] Hergenrother P M (2003) The Use, Design, Synthesis, and Properties of High Performance/High Temperature Polymers: an Overview. High performance polymers 1: 3-45 .

[6] Hao J, Jikei M, Kakimoto M (2003) Hyperbranched Polyimides Prepared by Ideal A_2 + B_3 Polymerization, Non-ideal A_2 + B_3 Polymerization and AB_2 Self-polymerization. Macromolecular symposia 199: 233-242.

[7] Fang J, Kita H, Okamoto K (2000) Hyperbranched Polyimides for Gas Separation Applications. 1. Synthesis and Characterization. Macromolecules 33:4639-4646.

[8] Bershtein V A, Egorova L M, Yakushev P N, Sysel P, Hobzova R, Kotek J, Pissis P, Kripotou S, Maroulas P (2006) Hyperbranched Polyimides Crosslinked with Ethylene Glycol Diglycidyl Ether: Glass Transition Dynamics and Permeability. Polymer 47: 6765-6772.

[9] Bernardo P, Drioli E, Golemme G (2009) Membrane Gas Separation: A Review/State of the Art. Industrial and engineering chemistry research 48: 4638-4663.

[10] Chung T S, Juany L Y, Li Y, Kulprathipanja S (2007) Mixed Matrix Membranes Comprising Organic Polymers with Disperse Inorganic Fillers for Gas Separation. Progress in polymer science32: 483-507

[11] Fang J, Hidetoshi K, Okamoto K: (2001) Gas Separation Properties ofHyperbranched Polyimide Membranes. Journal of membrane science 182: 245-256.

[12] Sindelar V, Sysel P, Hynek V, Friess K, Sipek M, Castaneda N (2001) Transport of Gases and Organic Vapours through Membranes Made of Poly(amide-imide)s Crosslinked with Poly(ethylene adipate). Collection of Czech Chemical Communication 66: 533-540.

[13] Suzuki T, Yamada Y, Tsujita Y (2004) Gas Transport Properties of 6FDA-TAPOB Hyperbranched Polyimide Membranes. Polymer 45: 7167-7171.

[14] Suzuki T, Yamada Y (2007) Effect of End Group Modification on Gas Transport Properties of 6FDA-TAPOB Hyperbranched Polyimide-Silica Hybrid Membranes High performance polymers 19: 553-564.

[15] Sysel P, Minko E, Cechova R (2009) Preparation and Characterization of Hyperbranched Polyimides Based on 4,4′,4′′-Triaminotriphenylmethane. E-Polymers no. 081.

[16] Friess K, Sysel P, Minko E, Hauf M, Vopicka O, Hynek V, Pilnacek K and Sipek M (2010) Comparison of Transport Properties of Hyperbranched and Linear Polyimides. Desalination and water treatment 14: 165-169.

[17] Zou H, Wu S, Shen J (2008): Polymer/Silica Nanocomposites: Preparation, Characterization, Properties, and Applications. Chemical reviews 108: 3893-3957.

[18] Reid B D, Ruiz-Trevino F A, Musselman I H, Balkus Jr K J, Ferrari P (2001) Gas Permeability Properties of Polysulfone Membranes Containing the Mesoporous Molecular Sieve MCM-41. Chemistry of materials 13: 2366-2373.

[19] Voit B I, Lederer A (2009) Hyperbranched and Highly Branched Polymer Architecture-Synthetic Strategies and Major Characterization Aspects. Chemical reviews 109:5924-5973.

[20] Sysel P, Minko E, Hauf M, Friess K, Hynek V, Vopicka O, Pilnacek K, Sipek M (2011) Mixed Matrix Membranes Based on Hyperbranched Polyimide and Mesoporous Silica for Gas Separation. Desalination and water treatment 34: 211-215.

[21] Cong H, Radosz M, Towler B F, Shen Y (2007) Polymer-Inorganic Nanocomposite Membranes for Gas Separation. Separation and purification technology 55:281-291.

[22] Moaddeb M, Koros W J (1997) Gas Transport Properties of Thin Polymeric Membranes in the Presence of Silicon Dioxide Particles. Journal of membrane science 125: 143-163.

[23] Robeson L M (2008) The Upper Bound Revisited. Journal of membrane science 320: 390-400.

[24] Chung T S, Jiang L Y, Li Y, Kulprathipanja S (2007) Mixed Matrix Membranes Comprising Organic Polymers with Dispersed Inorganic Fillers for Gas Separation. Progress in polymer science 32:483-507.

Section 2

Chemical and Physical Properties

Chapter 4

Chemical and Physical Properties of Polyimides: Biomedical and Engineering Applications

Anton Georgiev, Dean Dimov, Erinche Spassova, Jacob Assa,
Peter Dineff and Gencho Danev

Additional information is available at the end of the chapter

http://dx.doi.org/10.5772/53918

1. Introduction

The development of chemistry and solid matter physics lead to improved technologies in producing polymers and nanosized films such as vacuum deposition and solid state reactions. Over the past 20 years because of intensive research in chemistry and physics of organic materials, place the development of the new area in materials chemistry - organic electronics and photonics (polytronics). Polymer materials are becoming more widely used to improve the quality of human life: from social to high-tech - household appliances, textiles, insulation materials in industry and construction, medical implants, materials for optoelectronics, the formation of nanosized films and more over. Important classes of polymeric materials responded to certain technical requirements are polyimides (PI). They contain in molecules functional group $CO-NR_2$ called imide. The presence of n-π conjugation between non-pair electron of nitrogen atom and π electrons of the carbonyl group makes them resistant to chemical agents and moisture. Mainly the type of hydrocarbon residues (arenes, aliphatic) and the presence of other functional groups (Cl, F, NO2, OCH3, etc) determine their physical properties and their application in practice. In this chapter will be discussed preparation of PI, the mechanisms of reactions, possible competitive reactions, physical properties, their application as biomedical materials (implants, functionalized nucleic acids) and their application as optical materials for the producing of nanocomposite layers matrix including metallic or dielectric clusters as "guest". One of the achievements in obtaining PI layer is the application of microwave (MW) irradiation combined with low temperature treatment compared to the application of pure thermal imidization. This approach is particularly effective imidization for inclusion of PI layers in multilayer lithography or optical systems in which the formed nanocomposite films are not stable in thermal imidization condition.

1.1. Classification of polyimides

Depending on the polymer chain, the type of hydrocarbon residues and the presence of other functional groups, polyimides can be classified as follows:

1.1.1. Linear

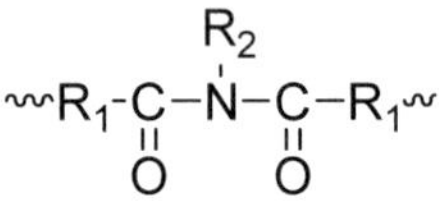

1.1.2. Cyclic

a. Aromatic (main-chain)

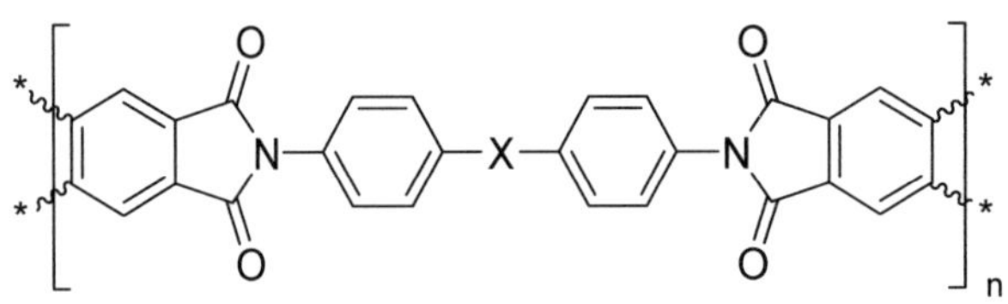

X = -O-, >C=O, $>CH_2$, -N=N-, $-SO_2-$ и др.

b. Aliphatic – aromatic (main-chain)

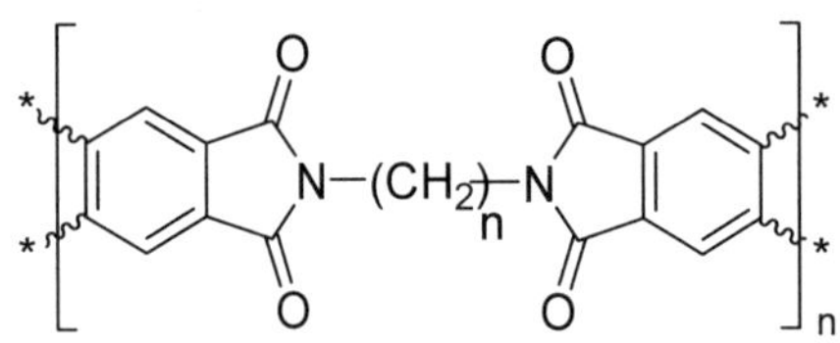

1.1.3. Side-chain

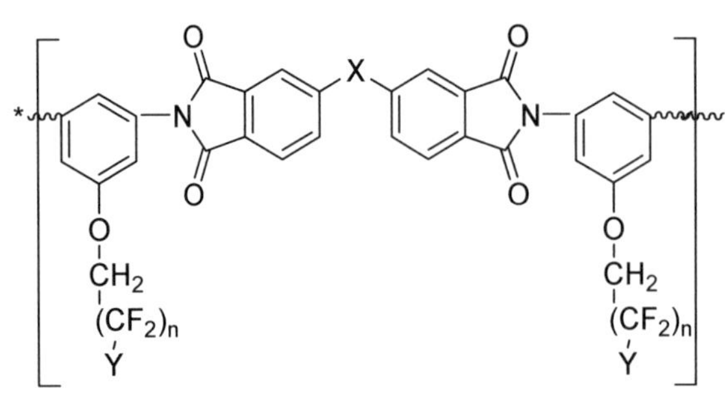

n = 1,2,3,4,5; Y= F, H

1.1.4. Functionalizing

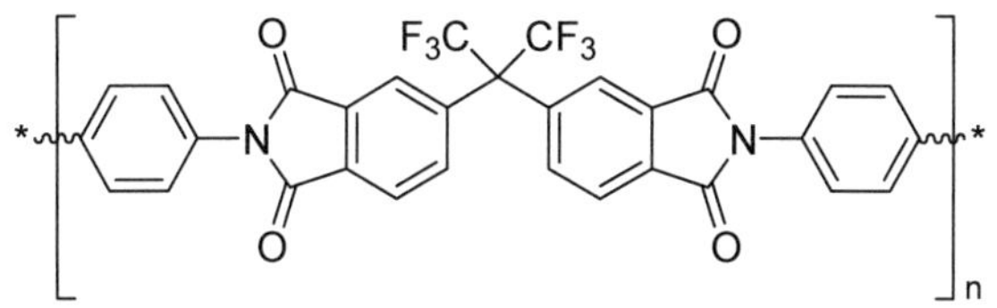

Many polyimide investigations have mainly been concentrated on aromatic polyimides, and little information is available about aliphatic polyimides that are also potential candidates for engineering and biomedical applications.

2. Chemical properties of polymides

2.1. Preparation and structure of polyimides

Aromatic polyimides generally prepared by a two-step procedure from aromatic diamines and aromatic tetracarboxylic dianhydrides. The chemistry of polyimides is a specific area with large variety of monomers available and several methodologies for synthesis [1,2,3]. The subtle variations in the structure dianhydride and diamine components have tremendous effect on the properties of the final polyimide. The most widely practiced procedure in polyimide synthesis is two-step process via poly(amic acid). The reaction between dianhydride (*pyromellitic dianhydride* PMDA) and diamine (*4,4'-oxydianiline* ODA) is required ambident conditions in dipolar aprotic solvents, such as N-methylpyrrolidone (NMP) or N,N-dimethylacetamide (DMAc). The next polycyclodehydration reaction of poly(amic acid) depending on ratio of precursors lead to final polyimide with different molar mass (Scheme 1) [4,5].

Reactions between cyclic anhydrides and primary diamines run as $S_{N2}Ac$ mechanism (bimolecular nucleophillic acyl substitution). The reaction running in two steps, the first is attaching of nucleophillic reagent to electrophillic carbonyl C-atom. The intermediate poly(amic acid) is formed by the nucleophilick attack of the amino group on the carbonyl carbon of the anhydride group. Thus reaction is irreversible, because the amino group is strong nucleophillic agent, consequence is not good leaving group than the hydroxyl group from the carboxylic acid [1,2,5,6]. In additions, anhydride cycle have not good resonance stability and charge delocalization, because oxygen atoms have equal electronegativity and electron structure. The second step is nucleophillic ring closure due to dehydration and imide ring formation. One of the disadvantages of this method is unavoidable presence of solvents and need for their removal [7]. During the polycyclodehidration side reaction is run that defy stereoregular control [1]. In figure 1 shows the structural formulas of the side products of the reaction for obtaining PI.

These side compounds caused defects in the synthesized PI film. They are unwanted impurity in production of nanostructured films with chromophores as a "guest" in the matrix, since they lead to low quality of dielectric and optical properties of layers. At temperatures above 200 °C by intramolecular rearrangement isoimide convert to imide (Scheme 2) [9,10].

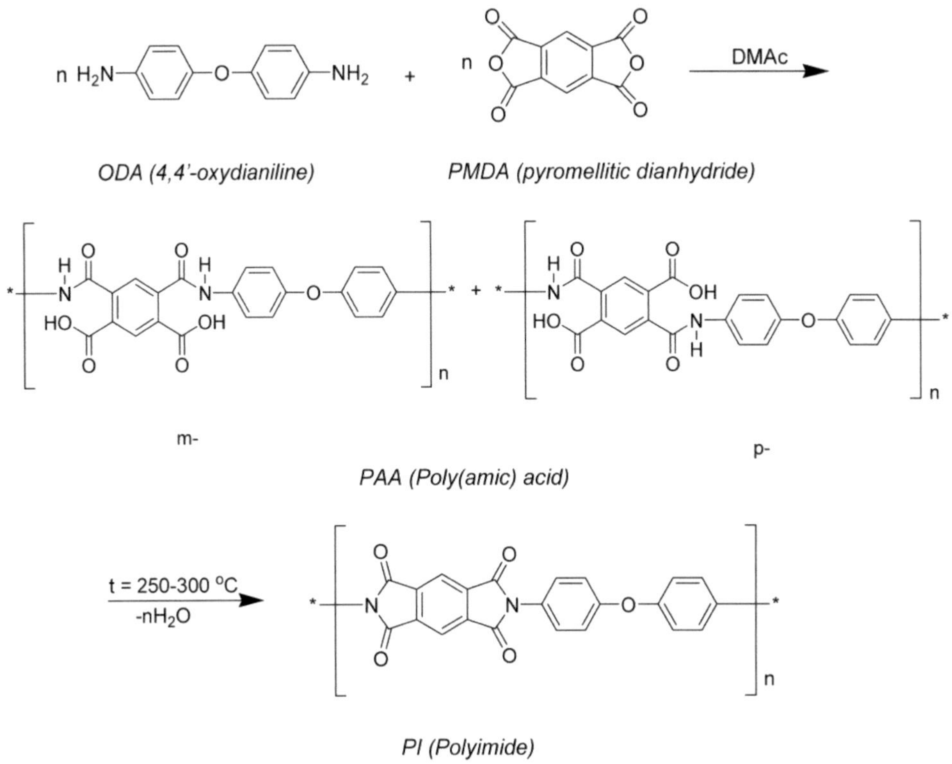

Scheme 1. Scheme 1. Reaction between ODA and PMDA to PAA with following cyclodehidration to PI.

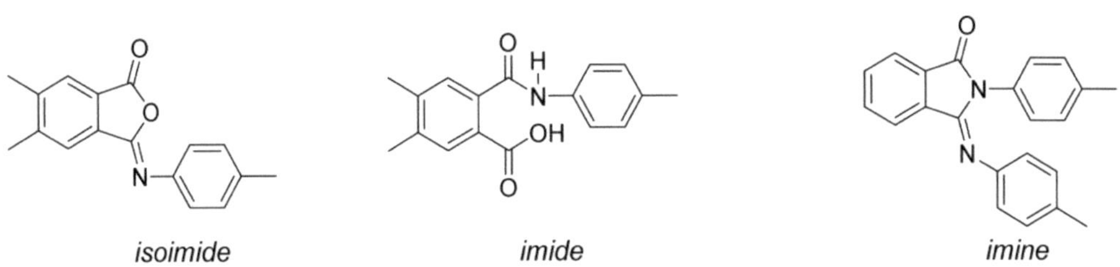

Figure 1. Structural formulas of side compounds of the reaction for preparation of PI.

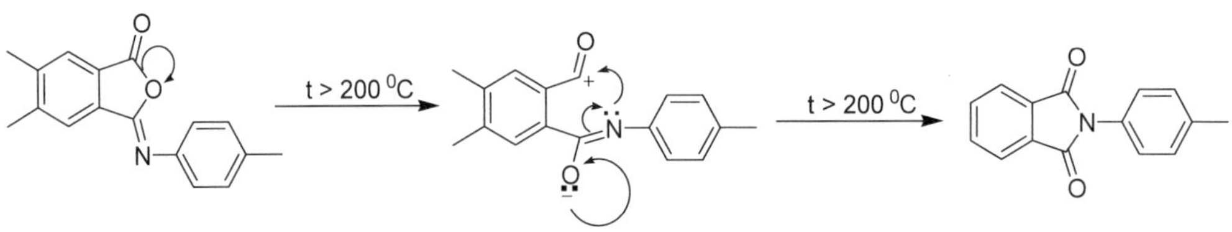

Scheme 2. Intramolecular rearrangement of isoimide to polyimide.

Another way for synthesis of polyimides from Nylon-Salt-Type Monomers has been reported from Imai [4]. The method based on the melt polycondensation of diamine and pyromellitic acid half diester (pyromellitic acid diethyl ester) (Scheme 3).

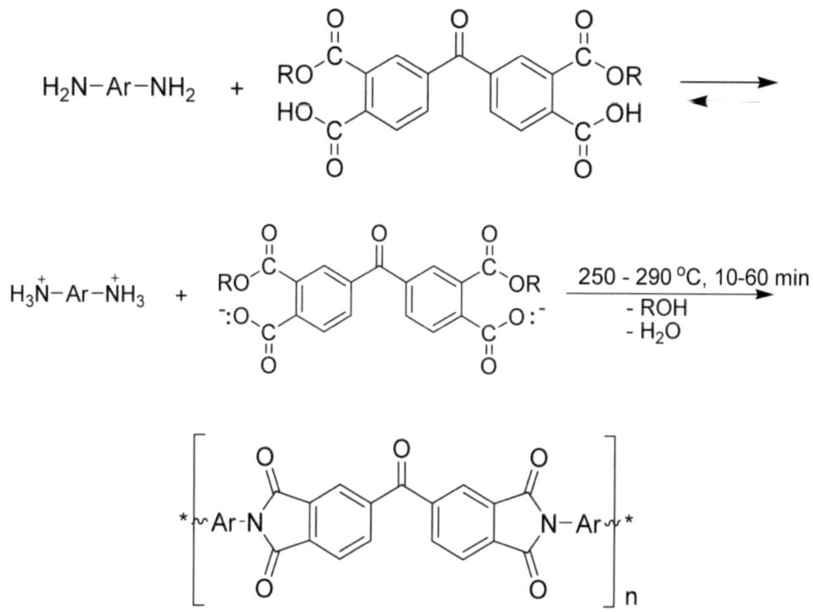

Scheme 3. Synthesis of PI from pyromellitic acid half diester and diamine.

These salt monomers have been prepared as white crystalline solids by dissolving an equimolar amount of each individual diamine and tetracarboxylic acid half diester in hot ethanol (or methanol), and subsequently cooling the resultant solution. The author has been found that imidization (polycondensation) of salt run to polyimide for 10 min at 250 °C. The high-pressure polycondensation of the salt monomer has been applied. The pressure affect on the temperature and reaction time that directly afforded high molecular-weight polyimide. This method is useful for the synthesis of the polyimides having well defined structures, compared with the other synthetic methods [4].

2.2. Vapour deposition and solid state reaction

The development of the chemistry of polymers and their application in nanotechnology many researchers have been seen alternative forms for obtaining nanosized films. Our studies have focused primarily on vapour deposition of precursors and solid state imidization reaction [7,10,11,12]. Vapour deposition processes of organic layers play an important role in polytronics. They allow the construction of systems without solvents, based on principle of bottom-up and have significant role in the formation of intermediate and protective layers. Important advantages of vapour processes are follows: (i) vacuum deposition is basically a cleaning process from impurities and resulting deposited layers have a much higher purity. This reduces problems with local anisotropy properties, polymorphism, etc.; (ii) vapour process allows the run of additional activated or modified processes as resulting deposited films are changed and improved properties: higher physical density, polymerization, high chemical purity and others. This is related take plasma processes, electronic flow, microwave irradiation or photon interaction. All these additional processes generate enough number of active particles as free radicals and ions in the gas phase. The disadvantage of these so-called vapour assistance deposition processes is the need for very precise control [7,13].

In the vacuum deposition of the precursors PMDA and ODA at temperature of 120-145 ^{0}C reaction of polycondensation to PAA with opening of the anhydride ring of PMDA takes place (Ac-S$_N$2 -reaction). These processes are to great extent accelerated and controlled in the thermal treatment of the condensed solid phase which represents PAA, with regard to their transformation to PI by means of reaction of polycyclodehydration in solid state to linear PI [7,10,11]. The FTIR spectra of individual films of PMDA, ODA and PAA are shown in figure 2.

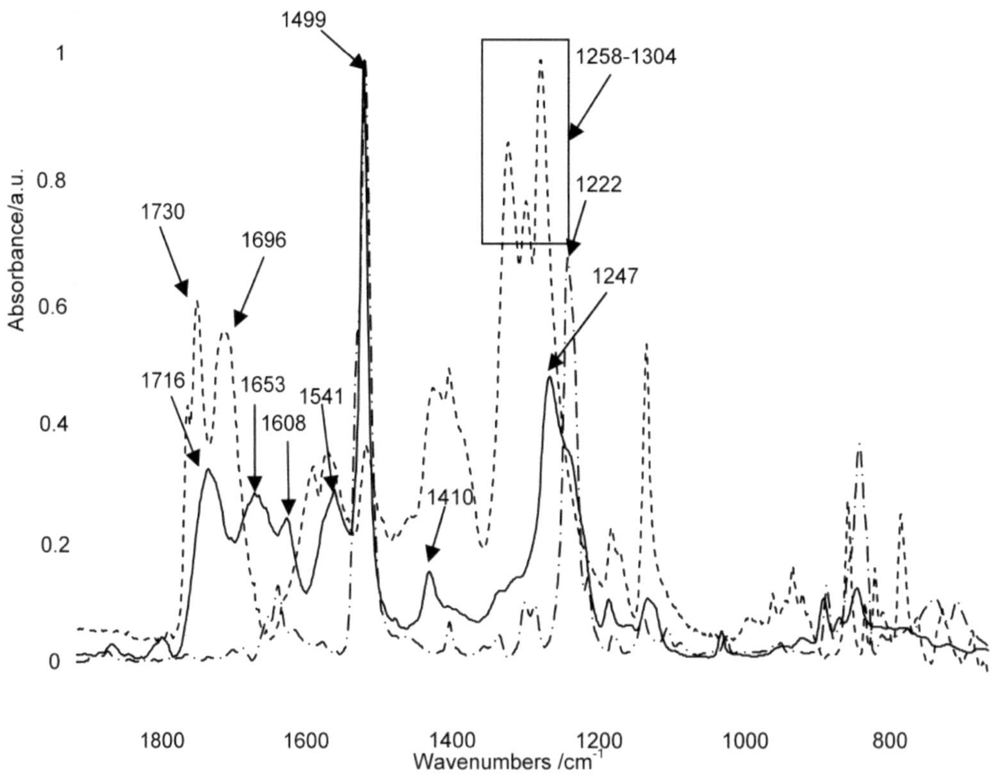

Figure 2. FTIR spectra of the vacuum deposited films of: - - - - - - - PMDA; – · – · – · – ODA; ———— simultaneously deposited both precursors in a mole ratio of PMDA : ODA= 1:1(PAA).

The band ν 1716 cm^{-1} for the >C=O group in the PAA spectrum related to acid. The presence of hydroxyl group of the acid (C-OH) is corresponded at δ 1247 cm^{-1}. The amide bond is identified by the bands at ν 1653 cm^{-1} (>C=O amide I band) and δ 1608 cm^{-1} (N-H amide II band).

Compared to the classical methods for producing films from PAA, in which the acid is preliminarily obtained in a solution and after that deposited as a thin film usually by spin coating, in the vacuum deposition method this process is performed in only one step. More and more researcher accept that the method of vacuum deposition provides for a greater degree of purity in the thin film production, opportunity for controlling and computerizing the processes of heating, imidazion and layer formation designed for obtaining standard PI or nanocomposite products of reproducible composition, thickness, structure and properties [12,14]. In figure 3 are compared the spectra of PAA and solid imidization film of PAA to PI obtained after thermal treatment.

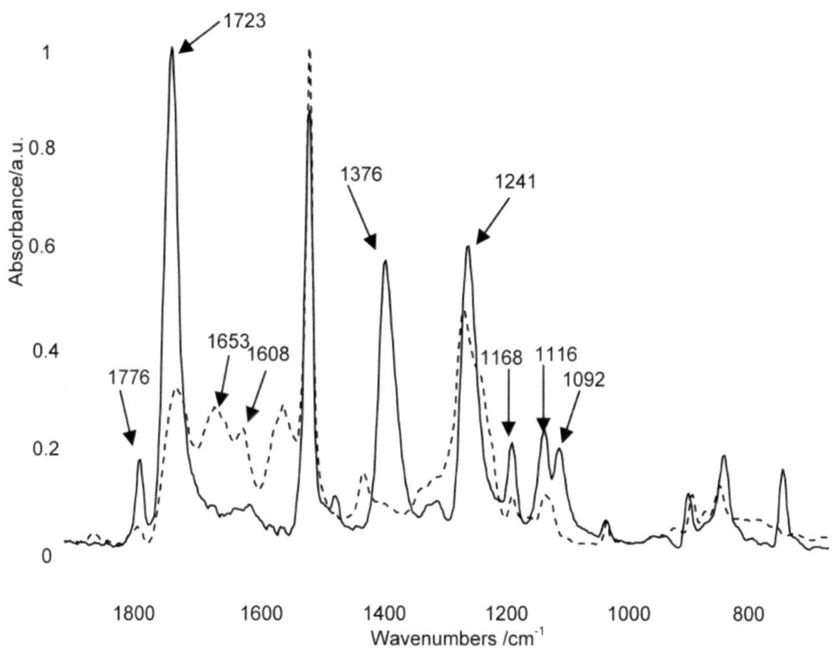

Figure 3. FTIR spectra of vacuum deposited films with thickness 200 nm and PMDA:ODA(PAA)=1:1
- - - - - - - thermally untreated; -------------thermally treated 1 h at 300 ºC.

The bands at ν_s 1776 and ν_{as} 1723 cm^{-1} characterize the >C=O groups of the imide ring. The imidization confirmed by the ν_{C-N} 1376 cm^{-1} imide III band where a minimum is observed in the spectrum of the untreated layer in this frequency. The exits of new bands in the area deformation vibrations for C-O and C-N bonds respectively at 1241, 1168, 1116 and 1092 cm^{-1} is observed [7,9,10].

Similarly, the vacuum deposition gives the opportunity to syntheses of azo-polyimide, type "main-chain". In this case, as starting precursors were used PMDA and DAAB (*4,4 '-diaminoazobenzene*), which are then vacuum deposition in a thin layer subjected to thermal and microwave treatment for imidization reaction to polyimide (Scheme 4) [7,15]. As is seen the azo-chromophore is covalently bonded in the polymer matrix. So a preparation of colored polymer is realized without to incorporate in the matrix an additional chromophore.

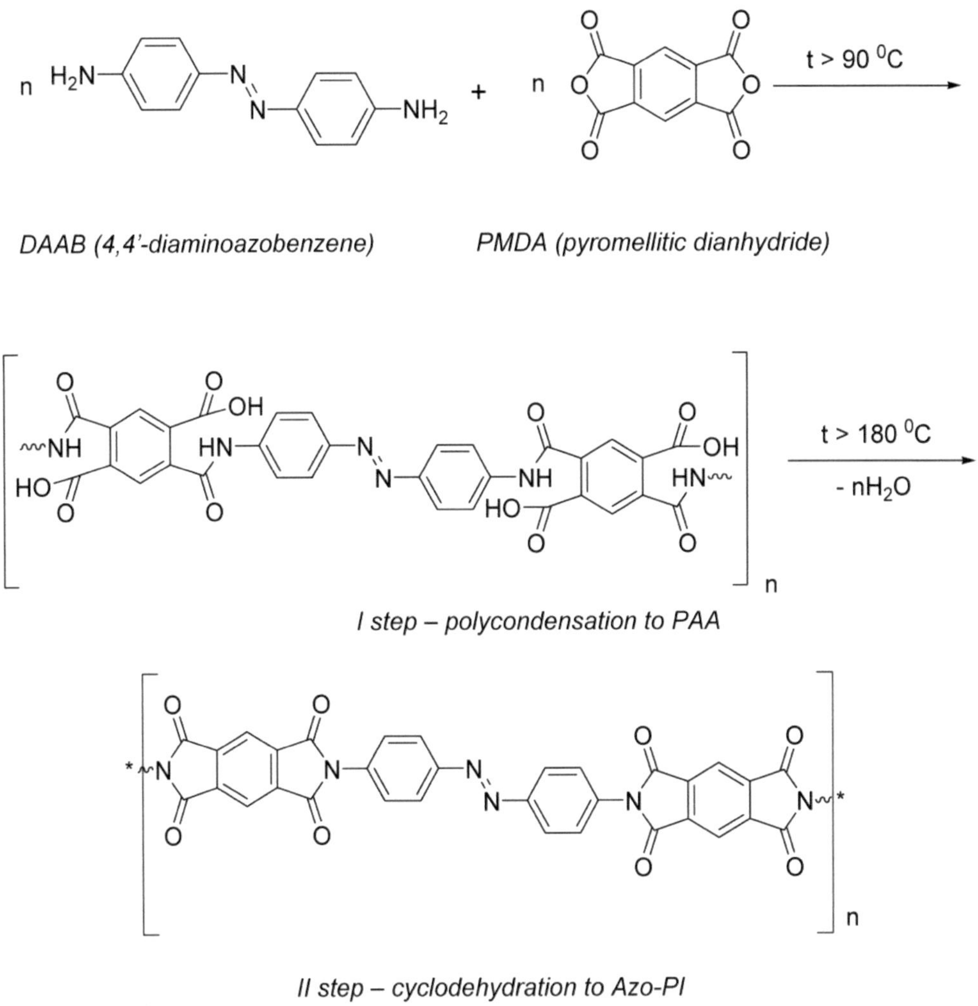

Scheme 4. Reaction between PMDA and DAAB to Azo-PI

The resulting films were characterized by FTIR and UV-VIS spectroscopies before and after imidization. It was studied kinetics imidization at different temperatures. The results show that optimal conditions for imidization are temperature 300 °C for 1 h. It is known that azobenzene derivatives possess optical properties associated with optical anisotropy due to photoisomerisation photoorientation of azochromophore, perpendicular to the direction of the polarized beam. These properties of azobenzene derivatives are important for their application in nonlinear optics and nanotechnology, optical modulators, optical recording media and other devices [16,17,18,19].

2.3. Microwave synthesis

One of the achievements of polymer chemistry is development of microwave synthesis. It is new and promising methods in organic synthesis, that time of the reaction manifold shorten. The literature describes that reaction occurs within hours or days, by microwave synthesis reaction time is greatly reduced - from 5 to 30 min [19,20,21]. Microwaves are distributed evenly in the reaction mixture, which makes the temperature field uniform (homogeneous) and radio frequency radiation provides the energy required for the reaction, and molecules get more energy (Ea) and the reaction speed increases manifold [20,21,22,23].

Our investigation showed that after combined treatment - MW and thermal, the imidization reaction take place for 5 to 15 min that is confirmed by FTIR spectroscopy. The quality of the film is identical with the one of PI obtained only after a thermal treatment for 1h at 300 ^{0}C. The method carried out allowed for PI production to be optimized by the involving of MW treatment of vacuum deposited films [10, 24].

It was found that the prerequisite for obtaining films of PI without mechanical defects is the establishment impact energy. It was investigated and established optimum conditions of MW and thermal treatment. Our suggestion that the type of energetic treatment power and time can be reduced to the substitution of different parameters: time, power and intensity of impact is not confirmed, i.e. it was not prepared the PI films with the same imidization degree in terms variation of treatment. Experimentally proved that 5 hours additional heat treatment at 170 ^{o}C is not offset one hour treatment at 250 °C. By applying the DTA (Differential Thermal Analysis) and FTIR spectroscopy was found to be necessary to achieve clearly defined temperatures for better imidization to PI. These results are also valid in MW treatment. Only the combination of both MW and heat, leading up to the high imidization. It was found that higher imidization degree achieved for 15 min. after application of MW and thermal treatment at 250 °C compared with only thermal treatment [24,25,10]. Therefore, the mechanism of the reaction according Scheme 1 and 6 requirements not only affect energy and same temperature achievement in the layer, but also requires strict established parameters: the type of treatment, power and time of action to achieve a high imidization degree.

2.4. Kinetic of the imidization

The kinetics of imidization is measured by the degree of transformation of PAA into PI by given output parameters - temperature and time of reaction. Appropriate method for determination of imidization degree is FTIR spectroscopy. The band at 1376 cm^{-1} is called imide III (C-N-C vibrations) band and using for qualitative and quantitative estimations of imidization degree in the present of internal standard at 1500 cm^{-1} (C-C$_{Ar}$ vibrations) [26,27]. The imidization degree of PAA to PI was determined using eq 1.

$$Degree\ of\ imidization = \frac{(peak\ area\ at\ 1376cm^{-1})time/(peak\ area\ at\ 1500cm^{-1})time}{(peak\ area\ at\ 1379cm^{-1})300\ ^{\circ}C \Big/ (peak\ area\ at\ 1500cm^{-1})300\ ^{\circ}C} \tag{1}$$

Yang et al. was used Seo model to correlate of imidization with time [28]. In Seo's approach the rate constant was proposed as $k(t) = b \times sech(-at)$. Inserting into the first-order rate equation, the relationship between the degree of imidization and curing time is obtained as:

$$-\ln(1-x) = -\frac{2b}{a}\tan^{-1}e^{-at} + const. \quad (2)$$

The constant in eq. (2) was obtained by fitting with the experimental data to the original Seo model. However the constant in eq. (2) was found to be 0.785 by fitting the experimental data at the initial condition ($t = 0, x = 0$). Hence, the kinetic equation was expressed as eq (3)

$$-\ln(1-x) = -\frac{2b}{a}\tan^{-1}e^{-at} - 0.785. \quad (3)$$

The parameters a and b were expressed in the form of Arrhenius expressions in this study, which are shown as eqs (4) and (5)

$$a = A_a \exp(-E_a/RT) \quad (4)$$

$$b = A_b \exp(-E_b/RT) \quad (5)$$

The constants A_a, A_b and E_a were calculated from Arrhenius plots of the parameters a and b. The assumption that the parameter b in the original Seo model is independent of temperature conflict obviously with the fact that the rate constant of first-order reaction $k(t)$ is a function of temperature since $k(t)$ approaches b at the initial time, $t \rightarrow 0$. The corresponding activation energy obtained from parameter a should be the energy barrier for the transition state and parameter b represents the rate constant for the imidization reaction.

3. Physical properties of polyimides

3.1. Intramolecular interaction

The formation of charge transfer complex (CTC) has been studied from many researches [29,30]. They reported that CTC formated between dianhydride and diamine groups in polyimides being an important reason for high glass transition temperature (Tg) of polyimides [29,30]. The increased interchain attractive forces due to such interactions were proposed as effectively increasing the chain rigidity and hence the Tg. It was also proposed that the presence of any bridging group in the dianhydride had a strong influence on the glass transition as it changed its electron affinity and hence promoted the possibility of CTC formation. Figure 4 shows the idealized form of such an interaction between the dianhydride and diamine groups.

In terms of electronic structure of PI is a prerequisite for strong intermolecular interactions and charge transfer. The nitrogen atom is electrondonor to carbonyl group which is electronacceptor, this leads to electron move and charge interactions. On the other hand they are interactions between aromatic rings with their π electronic sextet, which lead to parallel and planar orientation of individual chains to each other. Many of the properties of the PI's dependent on intramolecular interactions due to the use of vapour processes and solid state reactions to obtain essential to create packing of the polymer chain and effective CTC.

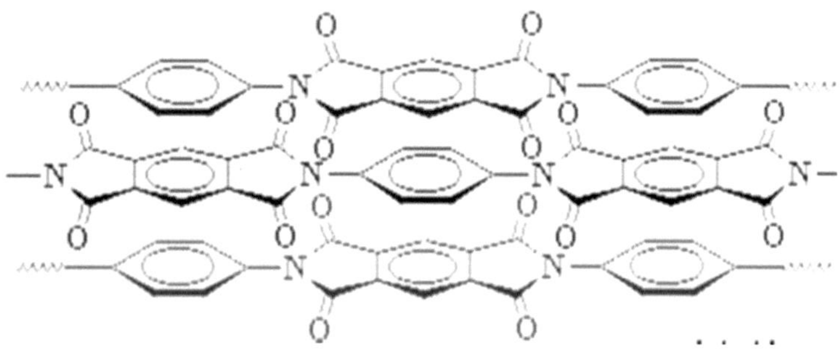

Figure 4. Idealized charge transfer complex formation between dianhydrides and diamines

3.2. Thermal stability

Thermo-Gravimetric Analysis (TGA) is an appropriate method for estimation of thermal stability of polymers. In figure 5 TGA curves presented our experiments after thermal treatment of ODA and PMDA mixture [11].

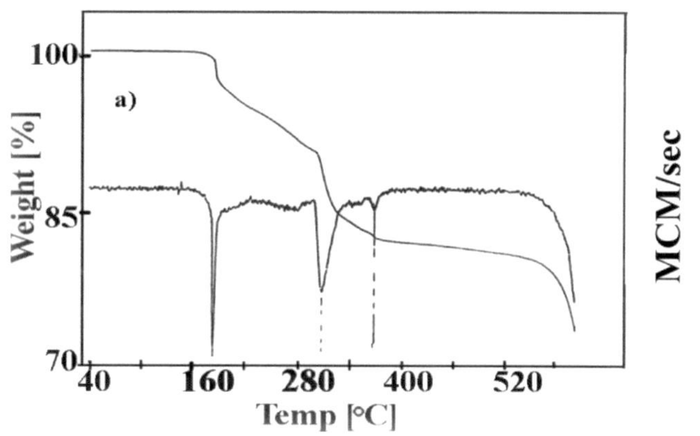

Figure 5. TGA curves of ODA and PMDA powder mixture (1:1 ratio)

Two drastically changes of the TGA curves are observed at about 180°C and about 280°C, which indicate relatively weight loss of 3% and 9%. These values correspond to one or three mols of water. It is well known that the polycondenzation reaction of ODA and PMDA (Sheme 1), takes place by separate of water. Therefore is quite logical to accept that the weight loss is a consequence of the process starting at lower (180°C) temperature till the moment the PAA is obtained. With increasing the temperature the interaction process continues and at higher (280°C) temperature the dehydratation leads to imidization and PI formation. It is also clearly seen (Fig. 5), that the rest of the sample remains stable up to 520°C. As the rest most probably is consisting of PI, it can be underline that the PI considering the destruction or weight loss is stable at temperatures higher than 500°C.

3.3. Optical transmittance

High light transmittance for a wide spectral range in combination with dielectric and chemical resistance properties is an attractive feature for potential applications as encapsulating layers, protective films and intermediate for micro- and optoelectronics. In figure 6 presents a series of spectra of PI obtained under different conditions. It is seen that the conditions for the PI films do not significantly affect the values of transmittance in the range 80 to 92 % in a relatively wide spectral range λ 420-900 nm.

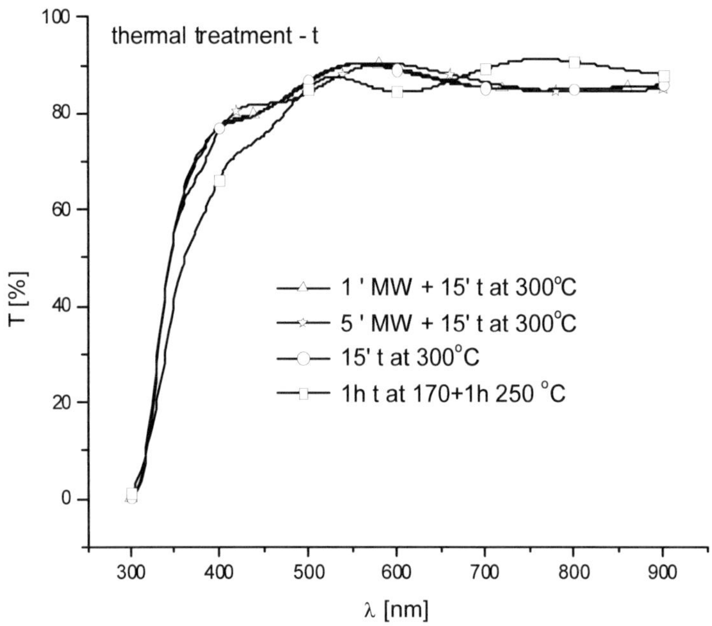

Figure 6. Spectral dependence of transmittance (T) for PI (PMDA-ODA) films 0,22 μm prepared under different conditions.

3.4. Electrical properties

The possibility of treatment condensation of the precursors in Ar atmosphere during vapour deposition can be interpreted as incorporation of pores, i.e. change the density or porosity of the condensed films of ODA and PMDA respectively, and the influence of pores on the permittivity PI layers. There was also depending on the reduction of dielectric constant with increasing imidization degree which is higher for higher temperatures of treatment in view of running imidization reaction [31,32,33]. Tables 1 to 3 are summarized data for conductivity, dielectric constant and capacity of the PI films with different composition, thickness, condition of the substrate or energy treatment and influence the conditions for obtainig PI or composite layers with embedded nanosized carbon particles.

Impressive ability to drastically change the conductivity by incorporation of carbon particles. The measured differences in conductivity depending on the concentration of "guests" - carbon 0,55% and 1,4% vol, respectively filling factor can be interpreted as a condition of embedded clusters. Therefore, there is the possibility of one side by vacuum deposition of the precursors and conditions of thermal treatment to influence the parameters examined. On the other, through the incorporation one type particles created opportunities to change individual parameters in a wide range in identical composition, for example different conduction with or without embedded carbon [30,31].

Thickness d [μm]	Thermal treatment [°C]	Permittivity [ε]
0.1	1 h 170° C + 1 h 250° C	3.0 - 3.2
0.1	1 h 170° C + 1 h 350° C	2.8
0.22	1 h 170° C + 1 h 250° C	3.1 – 3.3
0.22	1 h 170° C + 1 h 350° C	2.9

Table 1. Thermal dependence of polyimide permittivity.

Vacuum deposited layer 500 nm	Capacity [pF]	Permittivity [ε]
Planetary rotating substrates	9	3.2
Linearly moving substrates	11	3.4

Table 2. Polyimide capacity (C) end permittivity (ε) depending on the dynamic state of the substrates during the vapour depositions of precursors.

Type of layer formation	Conductivity (σ), [$ohm^{-1}m^{-1}$]
Without Ar (residual pressure 10^{-4}Pa)	1.9×10^{-8}
With Ar (residual pressure 10^{-2}Pa)	0.7×10^{-14}
With 0.55 vol.% Carbon	2.0×10^{-5}
With 1.4 vol.% Carbon	4.0×10^{-1}

Table 3. Polyimide conductivity (σ) depending on the type of layer formation

4. Biomedical and engineering application

4.1. Biomedical

Polymer materials have been established as excellent materials in chemistry, automotives and electronics to interconnect different components, to electrically insulate conductors and to survive harsh corrosive environments. Polyimides are the most common material class for substrate and insulation materials in combination with metals for interconnection wires and electrode sites. Therefore, it is quite natural that the medical device industry has focused its attention to polyimides for medical devices in general and especially for encapsulation and insulation of active implants. Neural implants are technical systems that are mainly used to stimulate parts and structures of the nervous system with the aid of implanted electrical circuitry or record the electrical activity of nerve cells. Their application in clinical practice has given rise to the fields known as ''neuromodulation'' and ''neuroprosthetics'' or neural prostheses. Neuromodulation, namely the stimulation of central nervous system structures to modulate nerve excitability and the release of neurotransmitters, alleviates the effects of many neurological diseases. Deep brain stimulation helps patients suffering from Parkinson's disease to suppress tremor and movement disorders. It is also a treatment option for severe psychiatric diseases like depression and obsessive-compulsive disorder. Neural prostheses aim to restore lost functions of the body, either sensory, motor or vegetative. All neural implants have to fulfill general requirements to become approved as a medical device - they must not harm the body and should stay stable and functional over a certain life-time which is in most cases in the range of decades. Generally polymides are used as an insulation or passivation layer, polyimides provide protection for underlying circuitry and metals from effects such as moisture absorption, corrosion, ion transport, and physical damage. Furthermore, it acts as an effective absorber for alpha particles that can be emitted by ceramics, and as a mechanical stress buffer. Key properties are thermoxidative stability, high mechanical strength, high modulus, excellent insulating properties, and superior chemical resistance. Devices made of polyimide have elicited only mild foreign body reactions in several applications in the peripheral and central nervous system showing good surface and structural biocompatibility. They have proven to be biostable and functional for months in chronic *in vitro* and *in vivo* studies. Most often used polyimide is BPDA-PPD as biomaterial and commercially available under the trademark of DuPont's PI2611 or UBE's U-Varnish-S (figure 7) [34]

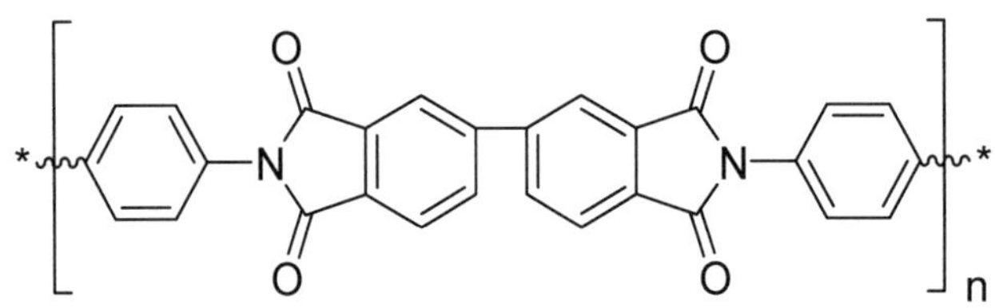

Figure 7. Structure of BPDA-PPD polyimide.

Advanced microtechnologies offer new opportunities for the development of these active implants. Biocompatible materials such as titanium and polyimide are potential candidates in encapsulating implant devices. A. Main et al. have been tested laser joined titanium-polyimide samples for bond strength, andwas observed that the laser bonding parameters such as laser power and feed speed affect the bond strength of all material systems [35]. The applicability of laser joining for a specific material combination and the selection of the appropriate approach depends on the optical properties of the materials. The need to join dissimilar materials occurs as the encapsulation includes functional elements such as electrodes used for neural stimulation or modules with defined micropores for the effusion of drugs. Currently, the implants such as pacemakers and cochlea implants are being used to treat cardiac and hearing disorders, respectively. Examples of other implants include neural and muscular stimulators for the treatment of seizures caused by epilepsy, and implantable drug delivery systems to locally effuse chemicals and drugs such as pain medication, hormones and other pharmaceutical compounds. Research is underway to develop subretinal implant devices that will replace damaged photoreceptor cells. These damaged cells fail to send electrical signals to the rest of the eye and eventual the brain, that results in blindness [36]. The study of Humayun et al. has showed successful results to restore vision by permanently implanting a retinal prosthesis in the blind eye. These active implants utilize microelectronic devices that are developed on biologically toxic materials such as silicon. Jong-Mo Seo et al. have been development biocompatibilities of polyimide microelectrode array for retinal stimulation [37]. They development a platinum electrode- embedded polyimide film and tested in feline eye. The authors used flexible polyimide selected as the substrate material of gold microelectrode array to minimize the damage during ophthalmic surgery and to get better contact to retina. To evaluate the feasibility of polyimide microelectrode array as a retinal prosthesis, in vitro and in vivo biocompatibility have been shown that retinal pigment epithelial on gold electrode-embedded polyimide film in a monolayer after 10 days of culture, and showed good affinity to it. Electroretinography revealed no difference between the transplanted eye and the healthy eye. Gold electrode-embedded polyimide film showed good biocompatibility in vitro and in vivo test and was suitable as a candidate biomaterial for the fabrication of retinal stimulator in visual prosthesis system.

Development of nanotechnology and chemistry of polyimide have been applied in biology and medicine as nano-biosensor device, biochips and thin film DNA immobilize. Biochips, particularly those based on DNA, are powerful devices that integrate the specificity and selectivity of biological molecules with electronic control and parallel processing of information. Examples of current applications of DNA chips include genomic analysis to screen and identify single nucleotide polymorphisms or to sequence gene fragments, pathogen identification, and gene expression profiling. Other possible features of a DNA chip surface include the presence of microelectrodes based on PI matrix thin films, that can be used as sensing devises and to generate electric fields that promote the migration of oligonucleotides, hybridization and covalent binding [39]. Forster et.al. have been purposed the Nanochip™ with a layered polyimide structure that supports the controlled electrophoretic transport of oligonucleotides [40]. They shown show that the biased electrodes preserve the integrity of the DNA by performing an electronic reverse-dot blot hybridization assay after electrophoretic transport of the target oligonucleotides [40].This nanodevice in future can serve as the basis for a low cost specific biomolecular detection tool in clinical diagnostics.

4.2. Engineering

The attractiveness and applicability of the PI's are determined not only from single their characteristics like high chemical and thermal stability, high optical transmission, high electrical resistance, but most important is their saving at combination of the cited properties [41,42]. The combination of the high optical transparency by simultaneously high chemical and thermal stability make the PI very attractive material for applying as capsulating or protective layers, films for laser microstructuring in micro- or optoelectronic, etc. [11,42]. Additionally, it could be noted the influence of the fast unlimited choice of the "guest" to enlarge the number of composites for many different purposes – from conductive or insulating transparency or colored films until insulating substrates, optical, chemical and thermal stable coatings [15,43]. Recently, much research interest focuses on polyimide films containing azobenzene fragments. These films are interest because of their potential application as photoactive materials for optical recording, biosensors, reversible optical switch, liquid crystal, photosensitive elements, optical information storage, holographic effect etc. An intensive research effort is being undertaken to use holographic techniques for optical information storage and optical information processing. Holographic storage is technologically very promising because information storage capacity, that can be reached with this technique, is much higher compared with other techniques (the storage capacity of CD is 0.7 GB, of DVD is from 4.7 to 17 GB, of PAP DVD is about 40 GB, of HOLO CD could be 1 TB) [19,44]. Development of holographic technology depends on the properties of recording materials. For such practical applications materials besides of specific requirements viz the principal the presence of photochromic moieties, the thermal stability of orientation alignment, the high optical non-linearity, high damage threshold, chemical resistance, mechanical endurance, they should exhibit the high optical quality ability and feasibility of device fabrication which are determined in wide range by their solubility. Significant efforts have been made to improve solubility of polyimides by designing their structure [45].

5. Conclusions

In this chapter, we have discussed methods for obtaining of polyimides, chemical properties and physical parameters that are related with obtaining nanosized films by vapour deposition. It was discussed possibilities for the solid state synthesis of polyimides in thin film and applied microwave synthesis. The studies show that can be obtained homogeneous films without defects on the surface and volume of layers with controlled density, thickness and dielectric properties. The developments of polymer chemistry produce polyimide films with covalent bonded chromophore to the chain. One of these achievements is our development a method for solid state synthesis of azo-polyimide.

Polyimide layers are suitable matrix for incorporation of metal, salts, chromophores as nanoscale particles to obtain of nanocomposite materials. It was discussed the possibility of use polyimides in materials chemistry and nanomaterials, one of these applications is the use for making biomedical implants for neurology, ophthalmology, biosensor device and chips which are a powerful tool in clinical diagnostics. Another important trend is use in electronics and optoelectronics such as dielectric substrates and intermediate barrier layers, creating nanocomposite films with various nanosized particles such as dyes, metal, dielectric and other clusters.

Author details

Anton Georgiev *
University of Chemical Technology and Metallurgy, Department of Organic Chemistry, Sofia, Bulgaria

* Corresponding Author

Dean Dimov, Erinche Spassova, Jacob Assa and Gencho Danev
Institute of Optical Materials and Technologies "Acad. Jordan Malinovski",
Department of "Nanostructured Materials and Technology", Sofia, Bulgaria

Peter Dineff
Technical University, Sofia, Bulgaria

Acknowledgement

The financial support of the National Fund of Ministry of Education and Science, Bulgaria – contract № DO-02/254 – 18.12.2008 is gratefully acknowledged.

6. References

[1] Strunskus Y, Grunze M (1994) Polyimides—fundamentals and applications. In: Crosh M, Mittal K, editors. New York: Marcel Dekker. pp. 187–205.
[2] Marc J.M. Abadie Alexander L. Rusanov (2007) Practical Guide to Polyimides. Smithers Rapra Technology Limited. pp 45-77.
[3] Bessonov, M.I., Koton, M.M., Kudryavtsev, V.V. and Laius, L.A (1987) Polyimides: Thermally Stable Polymers. Plenum, New York, 2-nd edition. pp. 56-187.
[4] Yoshio Imai (1999) Rapid Synthesis of Polyimides from Nylon-Salt-Type Monomers. In: Progress in Polyimide Chemistry I. H.R. Kricheldorf editor, Springer. pp. 3-20.
[5] Rohit H. Vora, P. Santhana Gopala Krishnan, S. Veeramani and Suat Hong Goh (2003) Poly(amic acid)s and their ionic salt solutions: Synthesis, characterization and stability study. In: Polyimides and Other High Temperature Polymers, Vol. 2. K.L. Mittal editor. VSP. pp. 14-35.
[6] Harris, F.W. (1990) Polyimides. Wilson D., Stenzenberger, H.D., Hergenrother, P.M., Chapman and Hall editors. New York. pp. 23-96.
[7] Anton Georgiev, Erinche Spassova, Jacob Assa and Gencho Danev (2010). Preparation of Polyimide Thin Films by Vapour Deposition and Solid State Reactions. Polymer Thin Films. Abbass A Hashim editor. InTech. pp. 71-92.
[8] D. Sek, A. Wanic (2000) High-temperature polycondensation of six membered dianhydrides with o-substituted aromatic diamines 1. Model compounds investigations. Polymer. 41: 2367-2378.
[9] D. Yu. Likhachev, S.N. Chavlin, Yu. A. Zubov, R.N. Nurmukhamedov and I. Ye. Kardash (1991) Effect of chemical structural defects on morphology of polyimide films. Polymer Science.33(9):1885-1894.
[10] Anton Georgiev, Ilyana Karamancheva, Dejan Dimov, Ivailo Zhivkov, Erinche Spassova (2008) FTIR study of the structures of vapor deposited PMDA–ODA film in presence of copper phthalocyanine. Journal of Molecular Structure. 888: 214-233.
[11] E. Spassova (2003) Vacuum deposited polyimide thin films. Vacuum. 70: 551-561.

[12] KiRyong Ha and John L. West (2002) Studies on the photodegradation of polarized UV-exposed PMDA–ODA polyimide films. Journal of Applied Polymer Science. 86: 3072-3077.

[13] K.S. Sree Harsha (2006) Principles of Physical Vapor Deposition of Thin Films Elsevier, San Jose State University, CA, USA, first edition. pp. 11, 367, 961.

[14] Mitchell Anthamatten, Stephan A. Letts, Katherine Day, Robert C. Cook, Anthony P. Gies, Tracy P. Hamilton and William K. Nonidez (2004) Solid-state amidization and imidization reactions in vapor-deposited poly(amic acid). Journal of Polymer Science: Part A: Polymer Chemistry. 42: 5999-6010.

[15] A. Georgiev, I. Karamancheva , D. Dimov, E. Spassova, J. Assa, G. Danev (2008) Polyimide coatings containing azo-chromophores as structural units. Journal of Physics: Conference Science. 113: 012032.

[16] Masashi Takahashi, Takashi Okuhara, Tomohiro Yokohari, Koichi Kobayashi (2006) Effect of packing on orientation and cis–trans isomerization of azobenzene chromophore in Langmuir–Blodgett film. Journal of Colloid and Interface Science. 296: 212-219.

[17] Cristina Cojocariu and Paul Rochon (2004) Light-induced motions in azobenzene-containing polymers. Pure Applied Chemistry. 76:1479-1497.

[18] Kiyoaki Usami, Kenji Sakamoto, Norio Tamura, Akihiko Sugimura (2009) Improvement in photo-alignment efficiency of azobenzene-containing polyimide films. *Thin Solid Films*. 518: 729-734.

[19] Ewa Schab-Balcerzak, Lukasz Grobelny, Anna Sobolewska and Andrzej Miniewicz (2006) Cycloaliphatic–aromatic polyimides based on diamines with azobenzene unit. European Polymer Journal. 42: 2859–2871.

[20] Brittany L Hayes (2002) *Microwave Synthesis: Chemistry at the Speed of Light.* CEM Publishing. pp. 11, 77, 95.

[21] Michael D., Mingos P. (2005) Theoretical aspects of microwave dielectric heating. Microwave Assisted Organic Synthesis. Tierney J., P., & Lidstroem P. editors. Blackwell Publishing Ltd. pp. 63-179.

[22] C. Oliver Kappe, Doris Dallinger and S. Shaun Murphree (2009) Practical Microwave Synthesis for Organic Chemists. WILEY-VCH Verlag GmbH & Co. KGaA, Weinheim. pp. 11, 87, 161, 203.

[23] Yumin Liu, Y. Xiao, X. Sun, D. A. Scola (1999) Microwave irradiation of nadic-end-capped polyimide resin (RP-46) and glass–graphite–RP-46 composites: cure and process studies. Journal of Applied Polymer Science. 73(12): 2391-2411.

[24] D. Dimov, A. Georgiev, E. Spassova, I. Karamancheva, Y. Shopov and G. Danev (2007) Microwave assisted processes for producing thin layer materials in the field of nanotechnologies. Journal of Optoelectronics and Advanced Materials. 9(2): 494-499

[25] Richard Hoogenboom and Ulrich S. Schubert (2007) Microwave-Assisted Polymer Synthesis: Recent Developments in a Rapidly Expanding Field of Research. Macromolecular Rapid Communication. 28: 368–386.

[26] M.B. Saeed, Mao-Sheng Zhan (2006) Effects of monomer structure and imidization degree on mechanical properties and viscoelastic behavior of thermoplastic polyimide films. European Polymer Journal. 42:1844-1854.
[27] M.B. Saeed, Mao-Sheng Zhan (2007) Adhesive strength of partially imidized thermoplastic polyimide films in bonded joints. International Journal of Adhesion & Adhesives. 27: 9-19.
[28] Chang-Chung Yang, Kuo Huang Hsieh and Wen-Chang Chen (2003) "A new interpretation of the kinetic model for the imidization reaction of PMDA-ODA and BPDA-PDA poly(amic acid)s". In: Polyimides and Other High Temperature Polymers, Vol. 2. K.L. Mittal editor. VSP. pp. 37-45.
[29] Fryd, M (1984) Structure –Tg relationships in Polyimides: Synthesis, Characterization and Properties, Vol. 1. Mittal K.L. editor. Plenum New York. pp. 377-384.
[30] St. Clair, T.L. (1990) In Polyimides. Wilson D., Stenzenberger, H.D., Hergenrother, P.M., Chapman and Hall editors. New York. pp. 187-208.
[31] D. Dimov, E. Spassova, J. Assa and G. Danev (2009) Ion beam assisted physical deposition of polyimide. Journal of Optoelectronics and Advanced Materials.11(10): 1436 - 1439.
[32] V. Strijkova, D. Dimov, A. Paskalevaa, I. Zhivkov, E. Spassova, J. Assa, G. Danev (2005) Electrical properties of a thin layer polyimide matrix Journal of Optoelectronics and Advanced Materials. 7(3):1319-1322.
[33] F.-Y. Tsai, Y.-H. Kuo and D. R. Harding (2006) Properties and structure of vapor-deposited polyimide upon electron-beam irradiation. Journal of Applied Physics 99: 064910.
[34] Christina Hassler, Tim Boretius, Thomas Stieglitz (2011) Polymers for Neural Implants. Journal of Polymer Science: Part B: Polymer Physics. 49: 18–33.
[35] A. Mian, G. Newaz, l. Vendra, N. Rahman, D.G. Georgiev, G. Auner, R. Witte, H. Herfurth (2005) Laser bonded microjoints between titanium and polyimide for applications in medical implants. Journal of Materials Science: Materials in Medicine. 16: 229– 237.
[36] Keekeun Lee, Amarjit Singh, Jiping Heb, Stephen Massia, Bruce Kima, Gregory Rauppc (2004) Polyimide based neural implants with stiffness improvement. Sensors and Actuators B. 102: 67–72.
[37] Humayun, M. S.; de Juan Jr, E.; Weiland, J. D.; Dagnelie, G.; Katona, S.; Greenberg, R.; Suzuki, S (1999) Pattern electrical stimulation of the human retina. Vision Research. 39: 2569–2576.
[38] Jong-Mo Seoa, Sung June Kimb, Hum Chunga,b, Eui Tae Kimb, Hyeong Gon Yua, Young Suk Yua (2004) Biocompatibility of polyimide microelectrode array for retinal stimulation. Materials Science and Engineering C. 24: 185–189.
[39] F. Fixe, A. Faber, D. Gongalves, D.M.F. Prazeresl, R. Cabea , V. Chu , G. Ferreira and J.P. Conde (2002) Thin film micro arrays with immobilized DNA for hybridization analysis. Materials Research Society. 723: 125-130.

[40] Anita H. Forster, Michael Krihak, Paul D. Swanson, Trevor C. Young and Donald E. Ackley (2001) A laminated, flex structure for electronic transport and hybridization of DNA. Biosensors & Bioelectronics 16: 187–194.

[41] G.Danev, E. Spassova, J. Assa, J. Ihlemann, D. Schumacher (2000) Excimer laser structuring of bulk polyimide material. Applied Surface Science. 168: 162-165.

[42] G. Danev, E. Spassova, J. Assa (2005) Vacuum deposited Polyimide – A Perfect Matrix for Nanocomposite Materials. Journal of Optoelectronics and Advanced Maters. 7(3): 1379-1390.

[43] M. Hasegawa and K. Horie (2001) Photophysics, Photochemistry, and optical properties of polyimides. Progress in Polymer Science. 26: 295-335.

[44] E. Grabiec, E. Schab-Balcerzak, D. Sek, A. Sobolewska, A. Miniewicz (2004) New polyamides with azo-chromophore groups. Thin Solid Films. 453–454: 367–371.

[45] M. Moniruzzaman, P. Zioupos, G.F. Fernando (2006) Investigation of reversible photo-mechanical properties of azobenzene-based polymer films by nanoindentation. Scripta Materialia. 54: 257–261.

Section 3

Applications

Chapter 5

Controlling the Alignment of Polyimide for Liquid Crystal Devices

Shie-Chang Jeng and Shug-June Hwang

Additional information is available at the end of the chapter

http://dx.doi.org/10.5772/53457

1. Introduction

Liquid crystal (LC) devices have been popular for use in photonics products, such as displays for mobile phones, televisions and computers. As shown in Fig. 1, a typical LC device consists of a thin LC layer sandwiched between a pair of indium tin oxide (ITO) conducting glass substrates with a cell gap of few micrometers. Polyimide (PI) as a polymeric material characterized by its outstanding mechanical, thermal and electrical properties at moderate high temperature has been widely applied in the LC displays (LCDs) industry as the alignment layers to align LC molecules in a certain orientation and conformation with a specific pretilt angle, the angle between the director of the LC molecules and the PI alignment layer. The pretilt angle is very important and required for LC devices to obtain a defect-free alignment and also to improve their electro-optical performance, such as driving voltage, response time, color performance and viewing angle. However, the applications of conventional PIs in LCDs are limited by a small tuning range of pretilt angle (~ few degrees) either by controlling the rubbing depth or the number of rubbings (Paek et al., 1998).

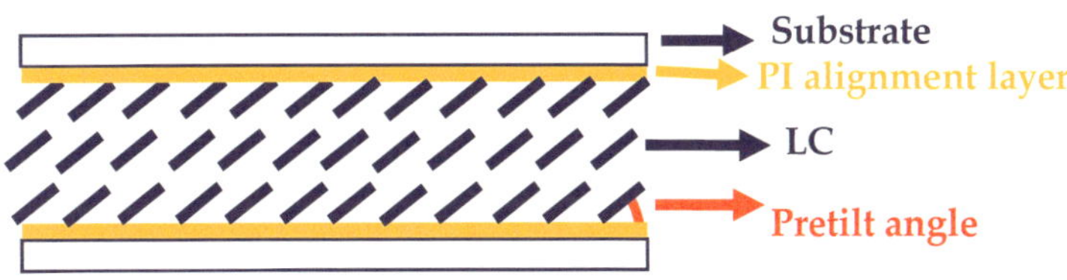

Figure 1. The typical structure of a liquid crystal device.

The pretilt angle of an LCD is either near zero degrees or 90 degrees for using the conventional homogeneous and homeotropic PI materials, respectively. The technique of producing homogeneous and homeotropic PI is mature in the LCDs industry. For example, the introduction of long alkyl groups into the aromatic diamines has been applied for increasing the pretilt angle from near zero degrees to near 90 degrees (Tsuda, 2011). However, the required specific pretilt angles of LCDs depend on their operation modes, for example, near zero degrees for in-plane switching (displays for iPhone and iPad), several degrees for the twisted nematic mode (displays for Laptop computers), more than 5° for the supertwisted nematic mode, 45° ~ 60° for no-bias optically-compensated bend (OCB) mode (LCDs with fast response time and wide viewing angle) and the bi-stable bend-splay (BBS) mode (LCDs requiring memory effect, such as e-books), and near 90 degrees for the vertical alignment mode (LCDs for TVs). Among of them, the OCB LCD has attracted much attention in recent years due to its fast response time and inherent wide view angle property (Acosta et al., 2000), and the memory effect of a bistable LCD is highly suitable for electronic-paper applications due to the low power consumption. The OCB LCD is operated between the bend state and the homeotropic state. A bias voltage should be applied to transfer a conventional OCB mode from the initially splay state to the bend state. Using a high pretilt angle of ~50° is sufficient to provide a stable bend state in an OCB mode at no bias voltage (Yeung & Kwok, 2006a), and it has been successfully demonstrated to reach the no-bias OCB operation. In general, an alignment layer with a high pretilt angle of around 45° ~ 58° is required for a BBS LCD providing the same splay and bend deformation energies (Yu & Kwok, 2004); therefore bistability between bend and splay states are reached.

In order to obtain a demanded pretilt angle for different LCD applications, there are many methods have been developed in the last two decades for controlling the pretilt angle of LC with a wide range, such as: obliquely-evaporated silicon monoxide (Uchida et al., 1980; Janning, 1992), polymer-stabilized alignment (Chen & Chu, 2008), hybrid mixture of two materials (Yeung et al., 2006b; Vaughn et al., 2007; Wu et al., 2008; Ahn et al., 2009), nano-structured surfaces (Komitov, 2008), chemical synthesis (Nishikawa, 2000; Tsuda, 2011), and stacked PI alignment layers (Lee et al., 2009). However, the reliability, the mass production capability and ease of material synthesis for those developed techniques are questionable. For example, the complicate molecular design and synthetic processes are required for a new development of alignment material.

Recently, we have developed a new approach to align LC vertically by adding polyhedral oligomeric silsesquioxane (POSS) nanoparticles in LCDs (Jeng et al., 2007; Hwang et al., 2009). POSS nanoparticles with nano-sized cage structures have been incorporated into polymers for improving their thermal, mechanical and oxidation resistance (Xiao et al., 2003; Yei et al., 2004). A typical structure of the POSS molecule is shown in Fig. 2, with R indicating the functional group (Sigma-Aldrich Corp., St. Louis, MO, USA). In our recent work, we have demonstrated that the pretilt angle of LC molecules can be continuously controlled by using conventional homogeneous PI alignment material doped with different concentrations of POSS nanoparticles (Hwang et al. 2010). The addition of POSS in the homogenous PI decreases the surface energy of the PI alignment layer and generates the controllable pretilt angle θ_p in a range of $0° < \theta p < 90°$ on demand. This method utilizes the conventional PI materials, the manufacture processes and facilities, therefore it can readily be adopted by the current LCD industry.

The pretilt angle of a PI alignment layer can also be controlled by ultra-violet (UV) irradiation (Lu et al., 1996; Yoshida & Koike, 1997; Ichimura, 2000; Chigrinov et al., 2008). UV irradiation on PI alignment films can produce extensive physical and chemical changes, such as photo-isomerization, photo-dimerization and photo-decomposition (Ichimura, 2000; Chigrinov et al., 2008), for PI materials with or without photo-functional groups. Techniques of photo-alignment by polarized and non-polarized UVs had been developed in the 90s and have recently matured for producing large LCDs (Miyachi et al., 2010). UV-modified PI films have also been applied for fabricating LC photonics devices, such as single-cell-gap transflective LCDs and LC Fresnel lens (LCFL) recently (Fan et al., 2004; Jeng et al., 2010; Hwang et al. 2012). The required pretilt angle can be achieved by using UV irradiation for controlling the surface energy of a PI film. Therefore, any LC device requiring patterned alignment areas with different pretilt angles can be obtained by this method.

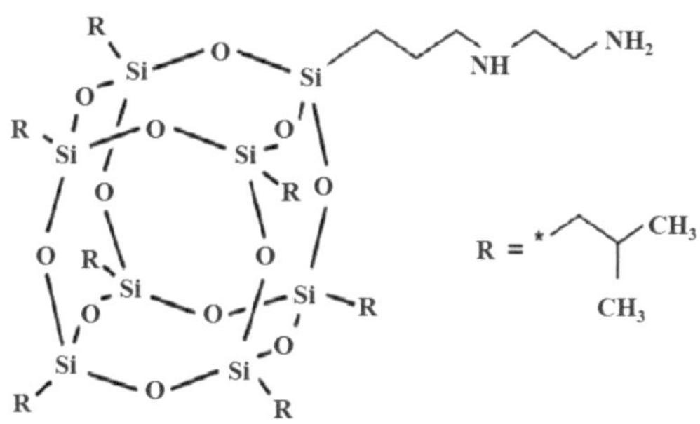

Figure 2. The structure of a PSS-(3-(2-Aminoethyl)amino)propyl-Heptaisobutyl substituted POSS nanoparticle (Sigma-Aldrich Corp., St. Louis, MO, USA).

In this chapter, we reported our developed non-synthetic techniques for controlling the pretilt angles of LC molecules either by doping homogeneous PI with POSS nanoparticles or treating homeotropic PI with UV. The surface energy of POSS-doped PI and UV-treated PI alignment layers were studied to investigate the mechanism of pretilt control. The LC devices, such as no-bias OCB LCDs, LC Fresnel lenses and LC phase gratings, were fabricated and presented in this chapter.

2. Sample preparation

2.1. POSS-doped polyimide

Several commercial homogeneous PI materials from Chimei, Daily Polymer and Nissan Chemical have been tried in the experiments, and some of them work. The POSS nanoparticle, PSS-(3-(2-Aminoethyl)amino)propyl-Heptaisobutyl substituted POSS (SIGMA-ALDRICH) as shown in Fig. 2, was purchased and used in the experiment without further treatment and purification. A powerful ultrasonic processor (S4000, Misonix) for producing the mixture of the POSS/PI was applied for obtaining a good dispersion of 0.2 wt% POSS doped in PI. The mixture was then filtered through a 200 nm syringe filter. The mixture was further diluted with pure PI in order to generate different concentration of POSS in PI. The POSS/PI films were cast on ITO conducting glass substrates by spin coating at 1000 rpm for 5 s and at 4000 rpm for 60 s. They were then prebaked at 100 °C for 10 minutes and post-baked at 180 °C for 4 hours in a vacuum oven to cure the POSS/PI mixture for forming the alignment layers. The manufacture parameters used here may depend on the PI materials. The surface of the PI alignment layer was then mechanically buffed by using a nylon cloth in such a way that the alignment layer was rubbed once in each direction. Prior to film casting, the ITO glass substrates were cleaned with distilled water, 2-propanol and acetone in an ultrasonic bath and dried at 80 °C for 1 hr.

2.2. UV-treated polyimide

Several commercial homeotropic PI materials from Chimei and JSR have been tried in the experiments, and they required different curing parameters of UV dosage. As shown in Fig. 3, the homeotropic PIs were first spin-coated on the ITO substrates in order to fabricate the UV-treated PI alignment layer. Then, they were pre-baked at 80°C for 10 minutes and post-baked at 210°C for 35 minutes in a vacuum oven. Subsequently, an UV light irradiated the homeotropic PI film through a designed photo mask. For those LC devices requiring precision alignment patterns, the substrates were laminated before UV irradiation as shown in Fig. 3. After finishing UV irradiation, the homeotropic PI film became tilted with a specific pretilt angle. The pretilt can be controlled by the UV dosage, exposure time or intensity. The parameters of UV dosage to reach a specific pretilt angle depended on the PI materials and UV light source. Fig. 4 illustrates a typical UV irradiation spectra used in this work (SP-9,Ushio).

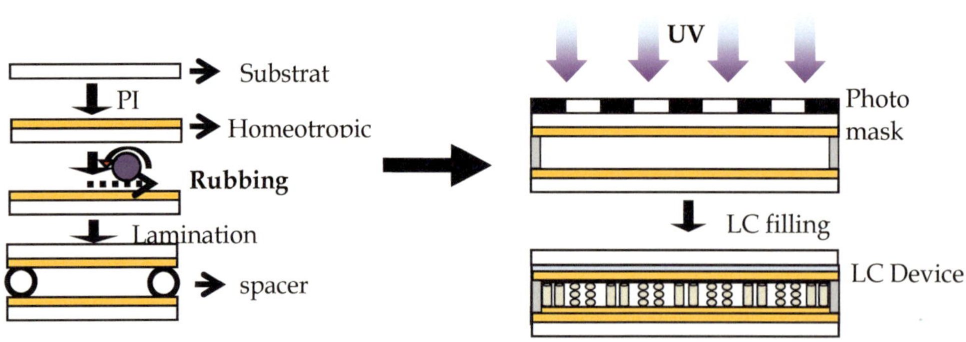

Figure 3. The processes of fabricating an LC device with patterned UV-treated homeotropic PI alignment layers.

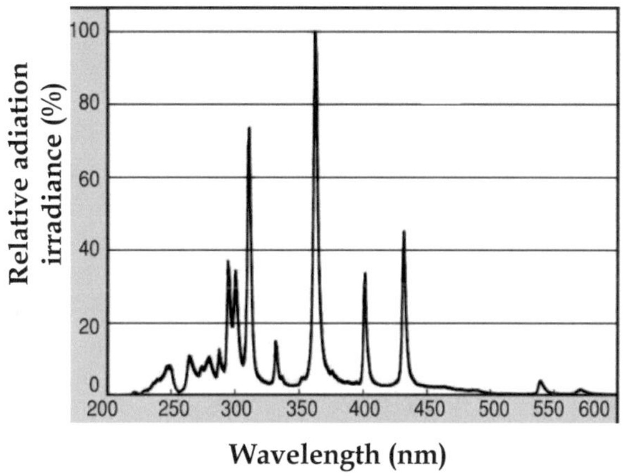

Figure 4. Radiation spectra of an UV light source (SP-9,Ushio).

2.3. Characterization of PI alignment layers

The surface energy of the PI alignment film mainly contributed by the polar part was determined by measuring the contact angle of distilled water on the alignment layers according to the Girifalco-Good-Fowkes-Young model (http://www.firsttenangstroms.com /pdfdocs/SurfaceEnergy.pdf). The contact angle was measured using a contact angle meter (CAM-100, Creating-Nanotech Co.)

To determine the pretilt angle and the polar anchoring energy (PAE) of LC molecules on PI alignment layers, anti-parallel LC test cells were fabricated with a cell gap of ~ 6 μm and were capillary filled with the positive dielectric anisotropic LC molecules (E7, $\Delta\varepsilon$ = 14.1, $\varepsilon_{\perp}$= 5.2, γ=223 mPa.s, K_{33}= 19.5 pN). Several methods have been developed to determine the pretilt angle of LC molecules on PI alignment layers (Baur et al., 1976; Scheffer & Nehring, 1977; Han et al., 1993). Among of them, the crystal rotation method has widely been adopted because of its simple and rapid measurement. However, it is difficult to precisely determine the symmetry point of transmission in the rotation-angle-dependent transmission curve for a LC device with a thin cell gap or a medium pretilt angle. A modified crystal rotation method combined with the common path heterodyne interferometer was used in this work (Hwang & Hsu, 2006; Li et al., 2008). Due to the designed common-path configuration, the phase retardation of the LC cell can be accurately determined in terms of the phase difference of the optical heterodyne signal. The polar anchoring energy (PAE) of the POSS-doped PI and UV-treated PI alignment layers was measured by using the high electric field method (Yu et al., 1999; Nie et al., 2005). Ultraviolet–visible (UV–vis) spectra of PIs were measured with an UV–vis spectrophotometer.

2.4. Fabrication of LC devices

Several LC devices, such as no-bias OCB LC cells, LC Fresnel lenses, and LC phase gratings, requiring medium pretilt angles and patterned pretilt angles were demonstrated.

2.4.1. No-bias OCB LC cells

One traditional low-pretilt OCB LCD using PI without POSS dopant (θ_p~ 2°) and one high-pretilt OCB LCD using PI with 0.05 wt% POSS dopant (θ_p~ 65°) were fabricated for comparison. As shown in Fig. 5, the pretilt angle of an OCB LC cell in the PI alignment layers is in the opposite direction. The cell gap maintained by ball spacers was kept around 5.1 μm in this work. The cells were capillary filled with the positive dielectric anisotropic LC molecules E7.

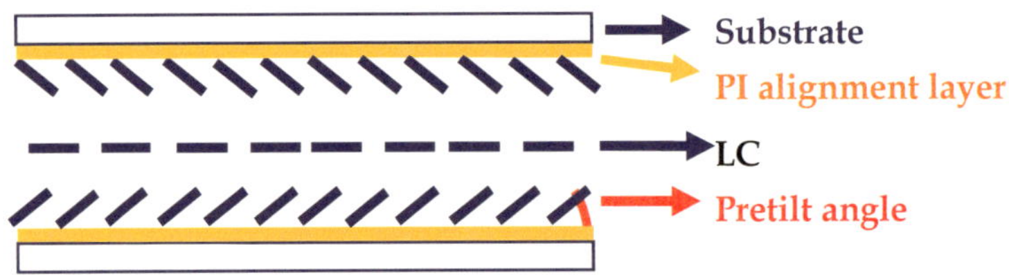

Figure 5. The liquid crystal director configuration in an OCB LC cell.

2.4.2. Fresnel LC Lenses

A binary liquid crystal Fresnel lens (LCFL) can be fabricated by the UV-induced modifications in the pretilt angle of the homeotropic PI films as shown in Fig. 6. An alternating pattern of hybrid-aligned and vertically-aligned LC cells was achieved by irradiating UV on one of the homeotropic PI films through a photo mask with Fresnel zone patterns. Following UV irradiation, the surface of homeotropic PI was modified to become aligned with a specific tilt angle in the even zone areas. The Fresnel zone plate used here is a photo mask with the circular opaque odd zones and transparent even zones. The designed radius r_1 of the innermost zone is 0.4 mm and the radius of the nth zone (r_n) is given by $r_n^2 = nr_1^2$, where n is the zone number. The Fresnel zone plate has 100 zones in approximately a 1 cm aperture, and it has a primary focal length f ~25 cm at λ= 632.8 nm.

As shown in Fig. 6, the polarization-dependent and polarization-independent LCFLs can be fabricated by buffing the UV-treated PI films horizontally and circularly, respectively. Both buffed top (homeotropic PI) and bottom ITO glass substrates (UV-treated homeotropic PI) were then assembled into an LC cell with a cell gap of ~10 μm maintained by spacers. The positive LC material (E7) was then injected into the empty cell.

The image quality and voltage-dependent diffraction efficiency of the LCFLs were measured by using an expanded He-Ne laser light source to approximately 1 cm in diameter corresponding to the active area of the LCFLs as shown in Fig. 7. The first-order diffraction efficiency is defined as the ratio of the first-order diffraction intensity at the primary focal point to the total transmitted intensity through the LCFL. The polarization direction of the incident light with respect to the horizontal buffing direction of the LCFLs was tuned by a linear polarizer and a half wave plate. The focusing properties of the LCFL were recorded by using a CCD camera or a detector, set ~25 cm from the LCFLs.

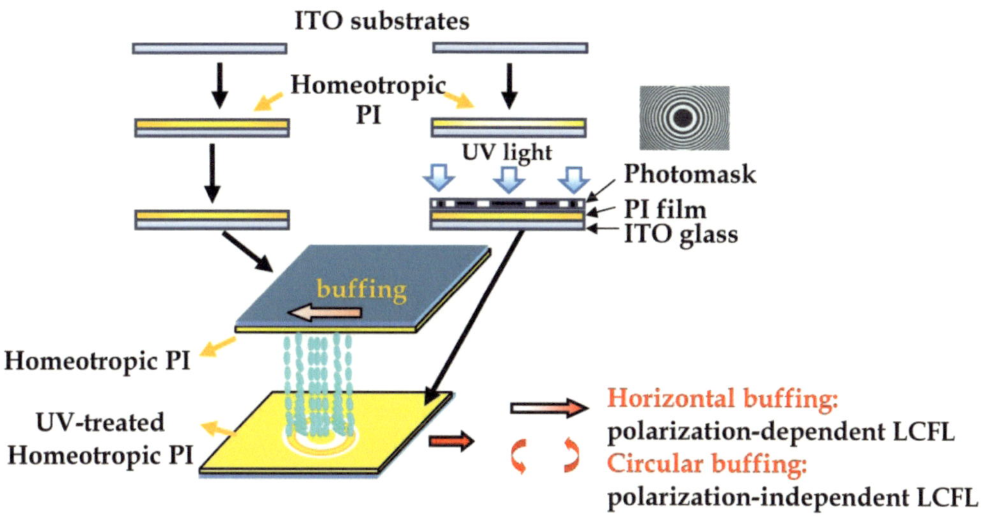

Figure 6. The fabrication process of the LC Fresnel lenses.

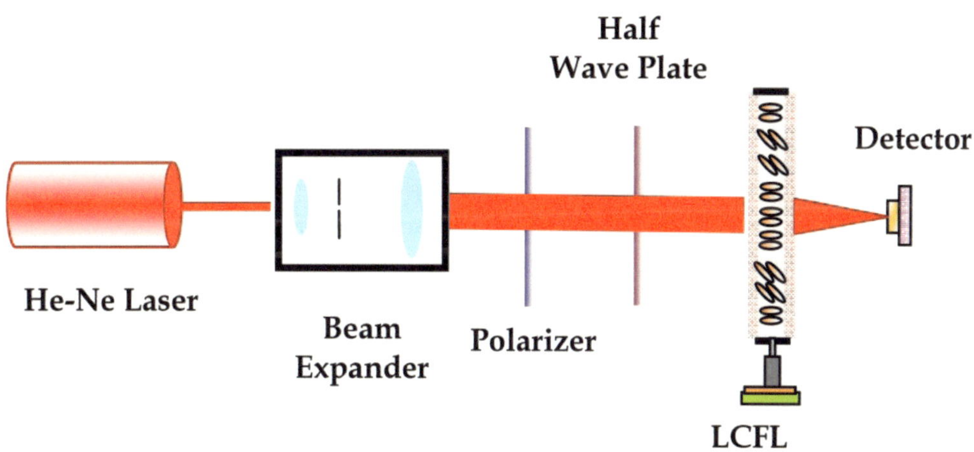

Figure 7. The experimental setup for measuring the focus properties of the LC Fresnel lens.

2.4.3. LC Phase gratings

In order to fabricate an LC phase grating, the laminated empty LC cell was irradiated by UV through a photo mask with the spacing around 200 μm in this work as shown in Fig. 3 and Fig. 8. The empty LC cell shows a strong UV absorption for the wavelength less than 300 nm due to the ITO glasses and PI materials, therefore the wavelength of UV irradiation between 300 nm and 400 nm is used for photo-decomposition of homeotropic PIs. The positive LC material (E7) was then injected into the empty cell for being an LC phase grating.

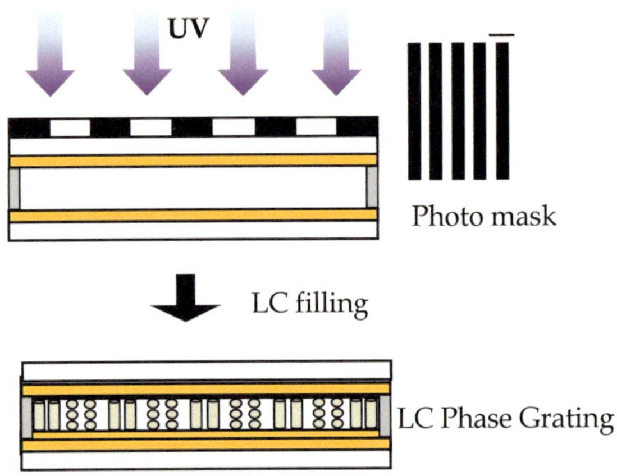

Figure 8. The fabrication process of the liquid crystal phase grating.

3. Results and discussion

3.1. Characteristics of PIs

The results of the surface energy of the POSS-doped PI alignment layer with different POSS wt% doped in PI and the UV-treated PI with different UV irradiation dosage are shown in Fig 9 and Fig. 10, respectively. They both indicate that the pretilt angles of LC molecules depend on the surface energy of the PI alignment layers. As shown in Fig. 9, the addition of POSS nanoparticles in the homogeneous PI mediates and lowers the surface energy of the PI alignment layer. However, the irradiation of UV on homeotropic PI mediates and increases the surface energy of the PI alignment layer as shown in Fig. 10. The pretilt angles of the rubbed PI films are continuously decrease with increasing surface energy as shown in Fig. 9 and Fig. 10 for POSS-doped PI and UV-treated PI alignment layers, respectively. The influence of surface energy of PI alignment layers on pretilt angles had been investigated (Paek et al., 1998; Yu et al., 1999; Ban & Kim, 1999). It showed that an alignment layer with higher surface energy gives the lower pretilt angle due to the increased attractive strength between LC molecules and molecules of the alignment layers.

The PAEs of the homogeneous PI alignment layer doped with different POSS concentration and the homeotropic PI alignment layer irradiated with different UV dosage are shown in Fig. 11 and Fig. 12, respectively. As shown in Fig. 11, the PAE is almost constant around $2.4{\times}10^{-4}$ J/m^2 regardless of the POSS concentration doped in PI. As shown in Fig. 12, The PAE increases from $1.3{\times}10^{-4}$ J/m^2 to $5.5{\times}10^{-4}$ J/m^2 as the dosage of UV irradiation increases to 88 J/cm^2. It indicates the anchoring strength of UV-treated PI films is not degraded by UV irradiation and it may depend on the surface energy of the UV-treated PI. The feature of increased PAE of the UV-treated PI film from homeotropic alignment to homogeneous alignment deserves further study. The novel methods of pretilt control by addition of POSS in homogenous PI and photo-irradiation of homeotropic PI are applicable to fabrication of LCDs requiring a specific pretilt angle with moderate PAE.

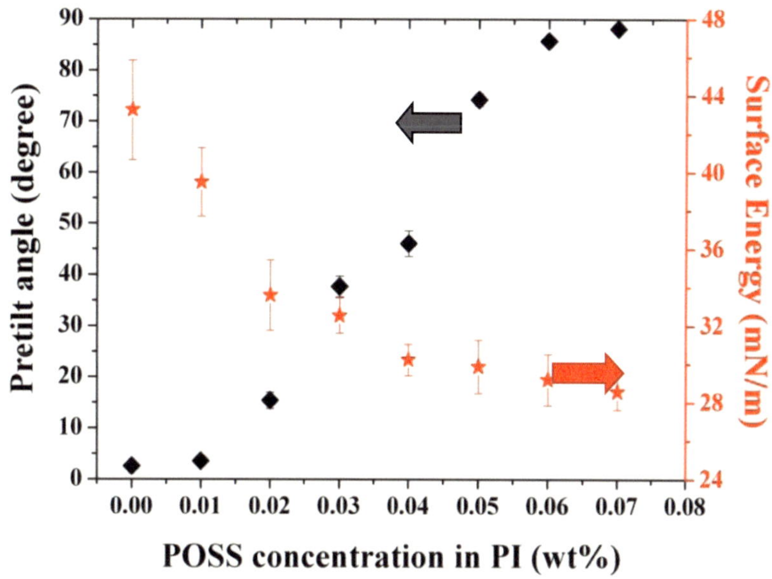

Figure 9. Pretilt angle and surface energy of POSS/PI alignment layers as a function of POSS concentration in PI.

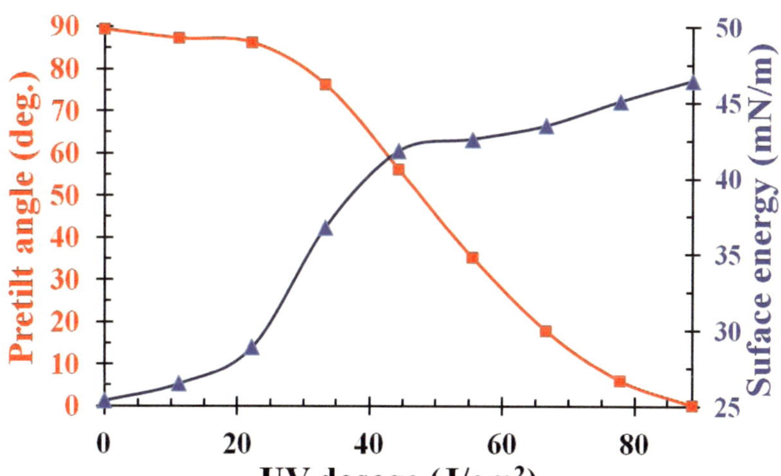

Figure 10. Pretilt angle and surface energy of the UV-treated homeotropic PI alignment layer as a function of UV dosage (Hwang et al. 2012).

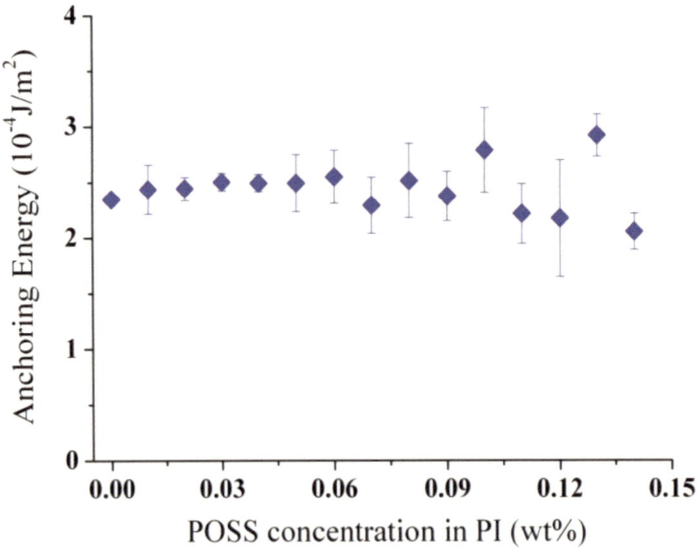

Figure 11. Polar anchoring energy of POSS/PI alignment layers as a function of POSS concentration doped in homogeneous PIs (Hwang et al. 2010).

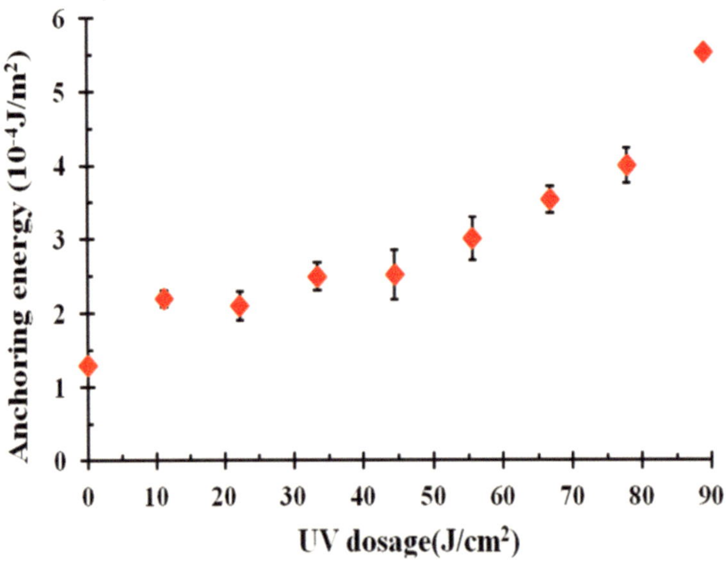

Figure 12. Polar anchoring energy of the homeotropic PI alignment layer as a function of UV dosage (Hwang et al. 2012).

The UV–vis spectra of the POSS doped PIs and UV-treated PIs are shown in Fig. 13 and Fig. 14, respectively. They both show good transparency in the visible range. There is no significant difference in UV-Vis spectra for 0.07 wt % POSS doped in PI as shown in Fig. 13. However, there is a significant increase in absorption and a small blue shift of absorption peak for the UV-irradiated PI film as shown in Fig. 14. It may indicate that the compositions of PIs are changed with UV irradiation. Further investigations, such as FT-IR spectra, may reveal the detailed information.

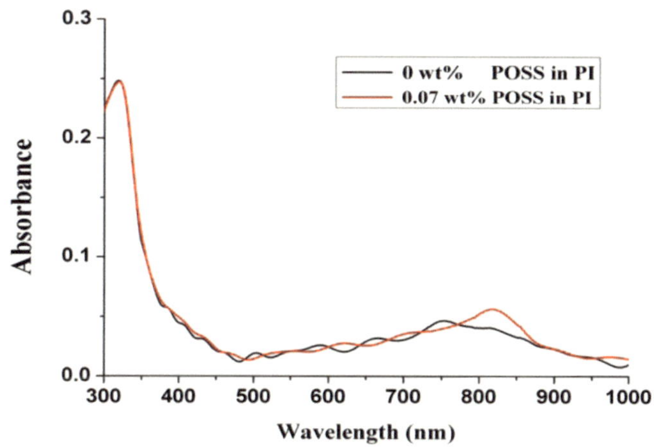

Figure 13. UV-vis spectra of the homogenous PI films with and without POSS dopant.

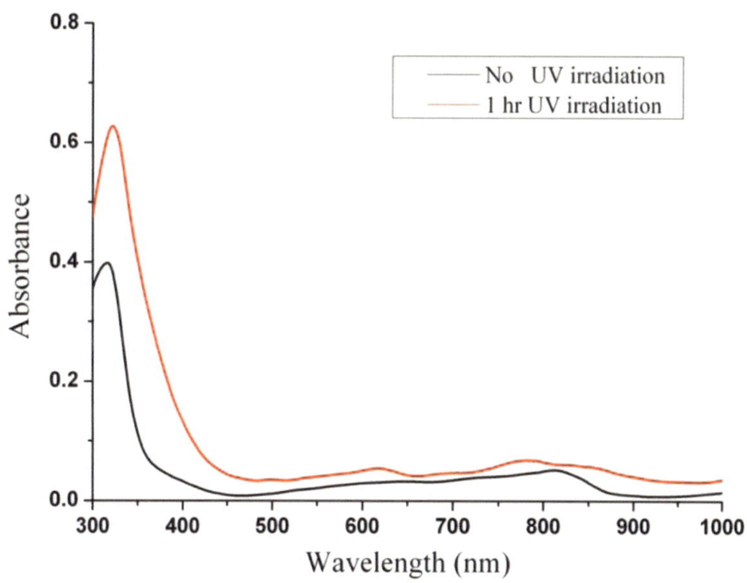

Figure 14. UV-vis spectra of the PI films with and without UV irradiation.

3.2. No-bias OCB results

The transient behaviors of the conventional low-pretilt OCB LC cell and the high-pretilt OCB LC cell under different applied voltages and finally relaxing the bias voltage observed by a polarized optical microscopy (POM) are shown in Fig. 15(a) and Fig. 15(b), respectively. The cell becomes totally dark when applying 10 V for reaching the homeotropic state. The transition from the bend state to the homeotropic state and the subsequent return from the homeotropic state to the initial bend state occur immediately for the high-pretilt OCB LC cell as shown in Fig. 15(b). No intermediate transition modes occur for the high-pretilt OCB LC cell when compared with the low-pretilt OCB LC cell. As a result, our demonstrated high-pretilt POSS-doped PI alignment layer stabilizes the bend state deformation of an OCB LC cell even at zero bias voltage.

The voltage dependent transmission properties of the conventional low-pretilt and the high-pretilt OCB LCDs were measured as shown in Fig. 16, by applying forward voltage (0V to 10 V) and backward voltage (10V to 0 V) where there was no bias voltage applied on the OCB LCDs and the data were recorded for one second at the first second after the voltage was applied. Due to the energy gap between the splay and bend state of a traditional low-pretilt OCB LCD, some transient time (warm-up time) is required between those two states, and that is the reason that the forward voltage dependent transmission curve does not match with the backward voltage one as shown in Fig. 16. The high-pretilt OCB LC cell could overcome the energy barrier and show a better electro-optical property than the low-pretilt one.

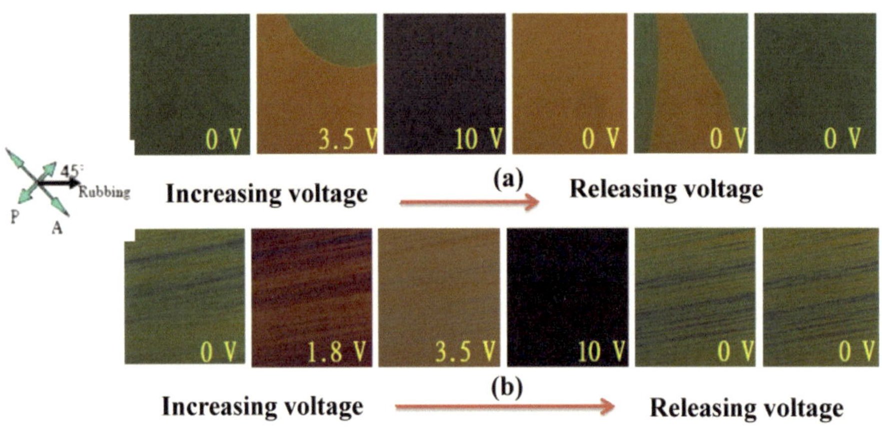

Figure 15. Photographs observed by a polarized optical microscopy. (a) The low-pretilt OCB LC cell and (b) the high-pretilt OCB LC cell under different applied voltages.

The switching properties of the no-bias high-pretilt OCB LC cell with 0.05 wt% POSS doped in PI is shown in Fig. 17 with a driving voltage from 0V to 5V. It can be seen that the switching time of 0.8 ms from bend state to homeotropic state and the relaxation time of 1.5 ms from homeotropic state to bend-state are shown.

3.3. LC Fresnel Lens

The results of the voltage-dependent first-order diffraction efficiency of a polarization-dependent LCFL and a polarization-independent LCFL with linear polarization of the incident light at θ=0°, 45° and 90°, with respect to the buffing direction of PI alignment layer are shown in Fig. 18 and Fig. 19, respectively. The diffraction efficiency of a polarization-dependent LCFL progressively increases to the maximum diffraction efficiency ~ 35% at V

=1.1 V with the linear polarization of the incident light at θ=0° as shown in Fig. 18. The characteristics of the polarization-independent LCFL are shown in Fig. 19, and the maximum diffraction of ~22 % is obtained at ~ 1.2 V regardless of polarization of the incident light. In order to characterize the imaging qualities of the LCFL as function of applied voltage, an expanded He-Ne laser was used as the object, and the focusing images shown in Fig. 20 were recorded with the CCD camera located at focus point, 25 cm behind the LCFL.

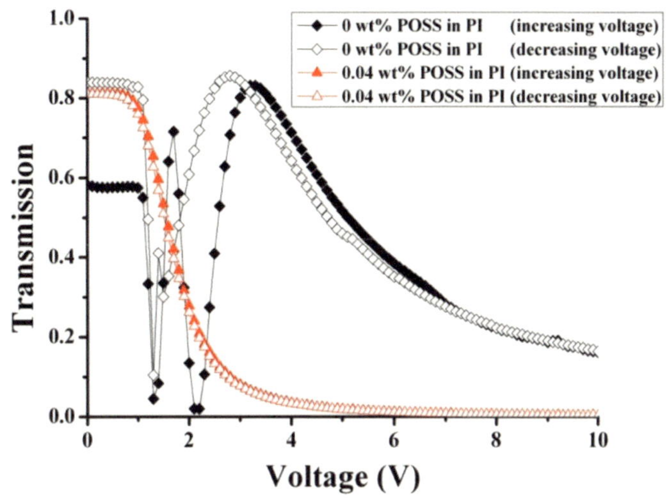

Figure 16. Voltage dependent transmission curves of the traditional low-pretilt and the high-pretilt OCB LC cells by applying forward (0 V to 10 V) and backward (10 V to 0V) voltages.

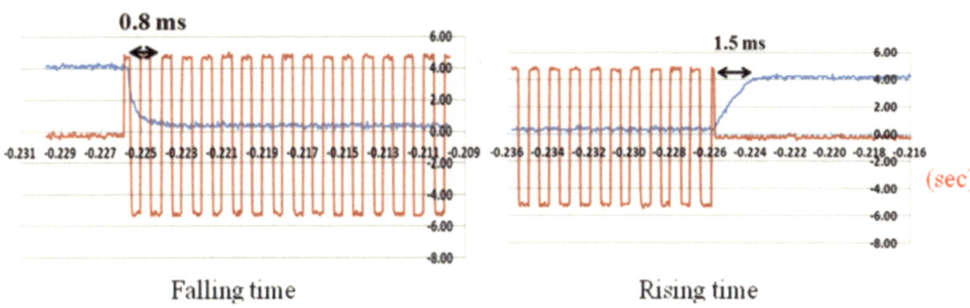

Figure 17. Switching time of the no-bias OCB LC cell.

3.4. LC Phase grating

The tunable diffraction grating has been found in many applications for photonics, such as optical information processing and telecommunication applications. The images of the LC phase grating observed by a POM are shown in Fig. 21. As shown in Fig. 21(a), the dark regions correspond to vertically-aligned LC on PI without UV treatment, and the bright regions correspond to planar-aligned LC on PI with UV treatment. The first-order diffraction efficiency of the LC phase grating as a function of the applied voltage is shown in

Fig. 22. The first-order diffraction efficiency of a phase grating is determined by the relative phase difference between the UV-irradiated and the non UV-irradiated regions of PI alignment layers. If the direction of polarization is parallel to the grating, the phase grating has maximum diffraction efficiency. The diffraction efficiency decreases gradually to zero as the voltage increases, because all of the LC directors are reoriented almost perpendicular to the substrates.

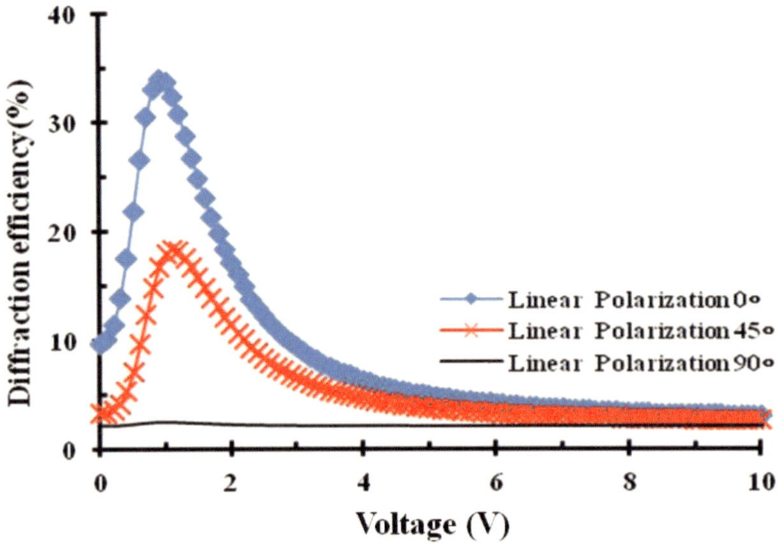

Figure 18. The voltage-dependent diffraction efficiency of a polarization-dependent LC Fresnel lens (Hwang et al. 2012).

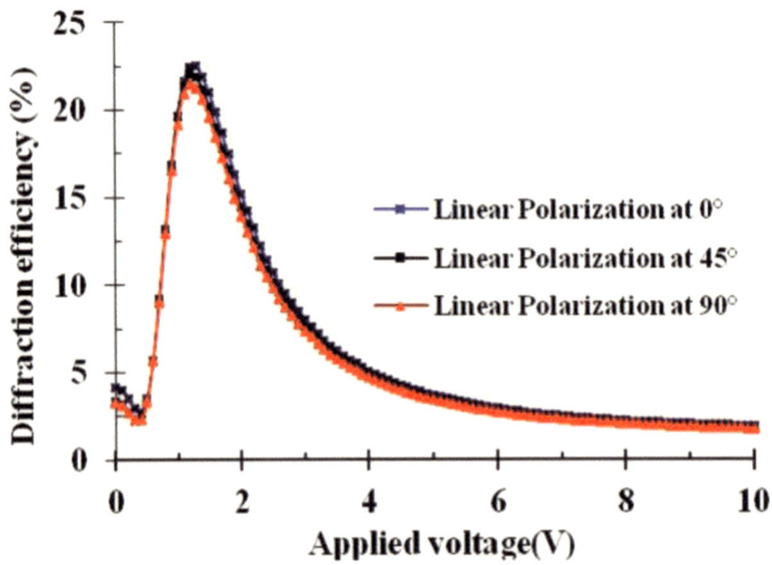

Figure 19. The voltage-dependent diffraction efficiency of the polarization-independent LC Fresnel lens (Hwang et al. 2012).

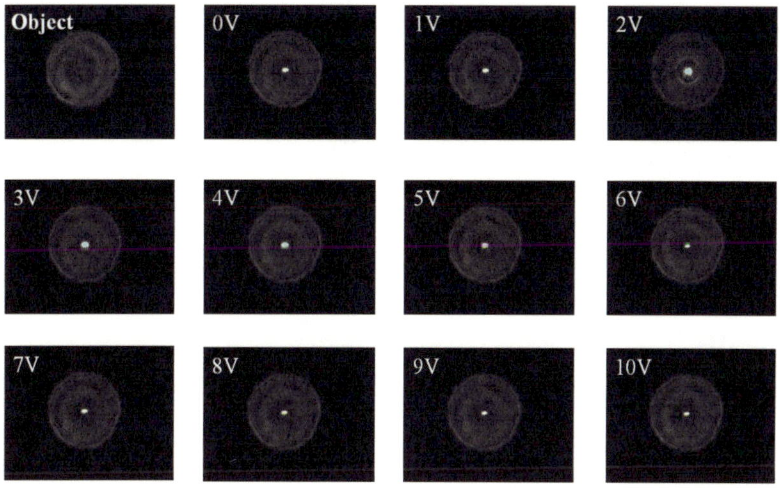

Figure 20. Focusing images of the LC Fresnel lens with different driving voltage recorded by a CCD camera at focal point.

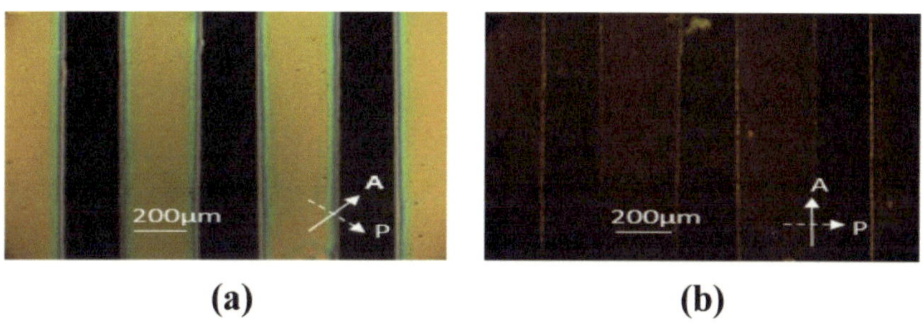

Figure 21. The images of the LC phase grating observed by a POM.

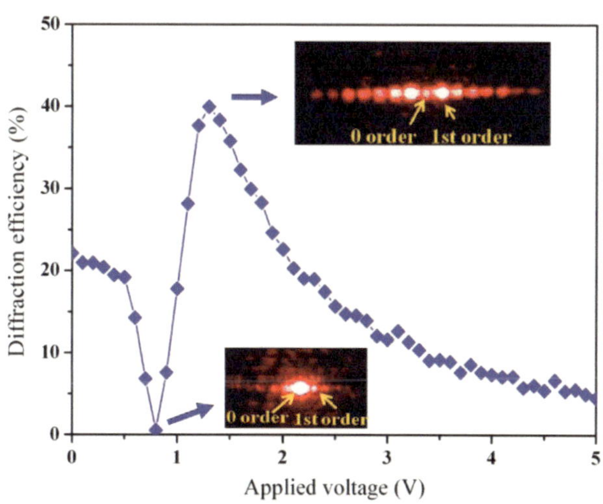

Figure 22. The first-order diffraction efficiency as a function of the applied voltage. Two insets of the diffraction patterns are shown at V=0.8V and V=1.3 V, respectively.

4. Conclusion

We have developed two approaches for controlling the pretilt angle of the PI alignment layers by using the conventional PI materials. The proposed methods are compatible with current methods familiar in the LCD industry. The LC devices, such as no-bias OCB LC cells, LCFLs and LC phase gratings, are demonstrated in this chapter by using the proposed techniques. A future study should examine the long-term thermal stability of these modified PI films.

Author details

Shie-Chang Jeng
Institute of Imaging and Biomedical Photonics, National Chiao Tung University, Taiwan

Shug-June Hwang
Dep. of Electro-Optical Engineering, National United University, Taiwan

Acknowledgement

The authors would like to thank the National Science Council of Taiwan for financially supporting this research under contracts: NSC 98-2112-M-009-020-MY2, NSC 98-2221-E-239-003-MY2, NSC 99RC04, and NSC 100-2112-M-009 -014 -MY3. The authors are grateful to Dr. Huai-Pin Hsueh from Chimei for supporting PI materials and helpful discussions and to I-Ming Hsieh, Tai-An Chen, Han-Shiang Liu and Mu-Zhe Chen for their help with the experiments.

5. References

Acosta, EJ, Towler, MJ & Walton, HG (2000). The role of surface tilt in the operation of pi-cell liquid crystal devices," *Liquid Crystals*, Vol. 27, No. 7, (Aug), pp. 977-984.

Ahn, D, Jeong, YC, Lee, S, Lee, J, Heo, Y & Park, JK (2009). Control of liquid crystal pretilt angles by using organic/inorganic hybrid interpenetrating networks. *Optics Express*, Vol. 17, No. 19, (Sep), pp. 16603-16612.

Ban, BS & Kim, YB (1999). Surface Free Energy and Pretilt Angle on Rubbed Polyimide Surfaces. *Journal of Applied Polymer Science*, Vol. 74, No. 2, (Oct), pp. 267-271.

Baur, G, Wittwer, V & Berreman, DW (1976). Determination of the tilt angles at surfaces of substrates in liquid crystal cells. *Physics Letters A*, Vol. 56, No. 2, (Mar), pp. 142-144.

Chen, TJ & Chu, KL (2008). Pretilt angle control for single-cell-gap transflective liquid crystal cells, *Applied physics letters*, Vol. 92, No. 9, (Mar), pp. 091102.

Chigrinov, VG, Kozenkov, VM & Kwok, HS (July 22, 2008). *Photoalignment of Liquid Crystalline Materials: Physics and Applications*, John Wiley & Sons Ltd, ISBN: 0470065397, West Sussex.

Fan, YY, Chiang, HC, Ho, TY, Chen, YM, Hung, YC, Lin, IJ, Sheu, et al. (2004). A Single-Cell-Gap Transflective LCD. *SID Symposium Digest of Technical Papers*, Vol. 35, No. 1, (May), pp. 647-649.

Han, KY, Vetter, P & Uchida, T (1993). Determination of molecular inclination in rubbed polymer for liquid crystal alignment by measuring retardation. *Japanese Journal of Applied Physics,* Vol. 32, No. 9A, (Sep), pp. L1242–L1244.

Hwang, SJ & Hsu, MH (2006). Heterodyne method for determining the surface tilt angle of nematic liquid-crystal displays. *Journal of the Society for Information Display,* Vol. 14, No. 11, (Nov), pp. 1039-1043.

Hwang, SJ, Jeng, SC, Yang, CY, Kuo, CW & Liao, CC (2009). Characteristics of nanoparticle-doped homeotropic liquid crystal devices. *Journal of Physics, D* Vol. 42, No. 2, (Jan), pp. 025102.

Hwang, SJ, Jeng, SC, & Hsieh IM (2010). Nanoparticle-doped polyimide for controlling the pretilt angle of liquid crystals devices. *Optics Express,* Vol. 18, No, 16, (Aug), pp. 16507-16512.

Hwang, SJ, Chen, TA, Lin, KR, & Jeng, SC (2012). Ultraviolet light treated polyimide alignment layers for polarization-independent liquid crystal Fresnel lenses. *Applied Physics B,* Vol. 107, No. 1, (Apr),pp.151–155.

Ichimura, K (2000). Photoalignment of liquid-crystal systems. *Chemical Reviews,* Vol. 100, No. 5, (May), pp. 1847-1874.

Janning, JL (1972). Thin film surface orientation for liquid crystals. *Applied physics letters,* Vol. 21, No. 4, (May), pp. 173.

Jeng, SC, Kuo, CW, Wang, HL & Liao, CC (2007). Nanoparticles-induced vertical alignment in liquid crystal cell. *Applied physics letters,* Vol. 91, No. 6, (Aug), pp. 061112.

Jeng, SC, Hwang, SJ, Horng, JS & Lin, KR (2010). Electrically switchable liquid crystal Fresnel lens using UV-modified alignment film. *Optics Express,* Vol. 18, No, 25, (Dec), pp. 26325-26331.

Komitov, L (2008). Tuning the alignment of liquid crystals by means of nano-structured surfaces. *Journal of the Society for Information Display,* Vol. 16, No. 9, (Sep), pp. 919-925.

Li, YW, Ho, JYL, Yeung, FSY & Kwok, HS (2008). Simultaneous determination of large pretilt angles and cell gap in liquid crystal displays. *Journal of Display Technology,* Vol. 4, No. 1, (Mar), pp. 13–17.

Lee, YJ, Gwag, JS, Kim, YK, Jo, SI, Kang, SG, Park, YR & Kim, JH (2009). Control of liquid crystal pretilt angle by anchoring competition of the stacked alignment layers. *Applied physics letters,* Vol. 94, No. 4, (Jan), pp. 041113.

Lu, J, Deshpande, SV, Gulari, E, Kanickia, J & Warren, WL (1996). Ultraviolet light induced changes in polyimide liquid-crystal alignment films. *Journal of Applied Physics,* Vol. 80, No., (Jul), pp. 5028-5034.

Miyachi, K, Kobayashi, K, Yamada, YY & Mizushima, S (2010). The World's First Photo Alignment LCD Technology Applied to Generation Ten Factory. *SID Symposium Digest of Technical Papers,* Vol. 41, No. 1, (May), pp. 579-582.

Nastishin, YA, Polak, RD, Shiyanovskii, SV, Bodnar, VH & Lavrentovich, OD (1999). Nematic polar anchoring strength measured by electric field techniques. *Journal of Applied Physics,* Vol. 86, No. 8, (Oct), pp. 4199-4213.

Nie, X, Lin, YH, Wu, TX, Wang, H, Ge, Z & Wu, ST (2005). Polar anchoring energy measurement of vertically aligned liquid-crystal cells. *Journal of Applied Physics,* Vol. 98, No. 1, (Jul), pp. 013516.

Nishikawa, M (2000). Design of Polyimides for Liquid Crystal Alignment Films. *Polymers for Advanced Technologies*, Vol. 11, No. 8-12, (Nov), pp. 404–412.

Paek, SH, Duming, CJ, Lee, KW & Lien, A (1998). A mechanistic picture of the effects of rubbing on polyimide surfaces and liquid crystal pretilt angles. *Journal of Applied Physics,* Vol. 83, No. 3, (Sep), pp. 1270-1280.

Scheffer, TJ & Nehring, J (1977). Accurate determination of liquid-crystal tilt bias angle. *Journal of Applied Physics,* Vol. 48, No. 5, (May), pp. 1783–1792.

Tsuda, Y (2011). Polyimides Bearing Long-Chain Alkyl Groups and Their Application for Liquid Crystal Alignment Layer and Printed Electronics, *Features of Liquid Crystal Display Materials and Processes,* Natalia V. Kamanina, pp. 4-24, InTech, ISBN: 978-953-307-899-1.

Uchida, T, Ohgawara, M & Wada, M (1980). Liquid Crystal Orientation on the Surface of Obliquely-Evaporated Silicon Monoxide with Homeotropic Surface Treatment. *Japanese Journal of Applied Physics,* Vol. 19, No. 11, (Nov), pp. 2127-2136

Vaughn, KE, Sousa, M, Kang, D and Rosenblatt, C (2007). Continuous control of liquid crystal pretilt angle from homeotropic to planar. *Applied physics letters,* Vol. 90, No. 19, (May), pp. 194102.

Wu, WY, Wang, CC & Fuh, AYG (2008). Controlling pre-tilt angles of liquid crystal using mixed polyimide alignment layer. *Optics Express*, Vol. 16, No. 21, (Oct), pp. 17131-17137.

Xiao, S, Nguyen, M, Gong, X, Gao, Y, Wu, H, Moses, D & Heeger, AJ (2003). Stabilization of Semiconducting Polymers with Silsesquioxane. *Advanced Functional Materials,* Vol. 13, No. 1, (Jan), pp. 25–29.

Yei, DR, Kuo, SW, Su, YC & Chang, FC (2004). Enhanced thermal properties of PS nanocomposites formed from inorganic POSS-treated montmorillonite. *Polymer,* Vol. 45, No. 8, (Apr), pp. 2633-2640.

Yeung, FS & Kwok, HS (2006).Fast-response no-bias-bend liquid crystal displays using nanostructured surfaces, *Applied physics letters*, Vol. 88, No. 6, (Feb), pp. 063505.

Yeung, FS, Ho, JY, Li, YW, Xie, FC, Tsui, OK, Sheng, P & Kwok, HS (2006). Variable liquid crystal pretilt angles by nanostructured surfaces. *Applied physics letters,* Vol. 88, No. 8, (Jan), pp. 051910.

Yoshida, H & Koike, Y (1997). Inclined homeotropic alignment by irradiation of unpolarized UV light. *Japanese Journal of Applied Physics,* Vol. 36, No. 4A, (Feb), pp. L428-L431.

Yu, XJ & Kwok, HS (2004). Bistable bend-splay liquid crystal display. *Applied physics letters,* Vol. 85, No. 17, (Aug), pp.3711.

Chapter 6

Fabrication of Polyimide Porous Nanostructures for Low-k Materials

Takayuki Ishizaka and Hitoshi Kasai

Additional information is available at the end of the chapter

http://dx.doi.org/10.5772/53458

1. Introduction

The processing speeds of microchips continue to increase as integration densities increase in the semiconductor industry. When the device size is decreased, the so-called resistance–capacitance (RC) delay and the crosstalk noise between metal interconnects offset any gain in chip performance. To avoid these undesirable phenomena, lower dielectric insulating layers must be employed. In the near future, it will be need to develop dielectric materials with ultralow dielectric constants (ultralow-*k*: $k < 2.0$), and a replacement dielectric for carbon-doped silicon dioxide (SiOC) will be required [1]. In addition, high-thermal stability, good adhesion to metals and chemical stability are also important requirements for the interlayer dielectrics. Polyimides (PIs) are among the most promising candidates for use as next-generation interlayer dielectrics because of satisfying all above-mentioned requirements. Indeed, PIs have been widely employed in the fields of microelectronics applications such as substrate of flexible printed circuitry boards, insulating layers in multilevel very-large-scale integrated (VLSI) circuits and buffer coatings in electronic packages. However, the dielectric constant (k) is about 2.4~3.0 even in fluorinated PIs [2-7], which are insufficient for the requirement of ultralow-*k* materials ($k < 2.0$).

According to the Maxwell-Garnett model [8], it is well known that porous structures in PI films could substantially reduce dielectric constants, because the dielectric constant of air is unity. Therefore, many studies on porous PI films have been reported. The porous PI films have been generally prepared by pyrolysis of thermally labile polymer units in phase-separated PI composite films [9-12]. Namely, so-called nanofoams were produced in a PI film, resulting in providing a low dielectric constant (k = 2.56). Their pore sizes are also easily controlled by a component ratio of the copolymer. On the other hand, we have proposed the deposition of PI nanoparticles (NPs) onto a substrate as a novel alternative approach toward the preparation of low-*k* ($k < 2.5$) PI films, that is, by introducing air voids between the PI NPs (Fig. 1a). To obtain ultralow-*k* PI films, we have focused on incorporating pores into the individual PI NPs and the subsequent assembly of multilayered films of porous PI NPs (Fig. 1b). This strategy must be a relatively simple and effective means for introducing air voids uniformly into ultralow-*k* PI films.

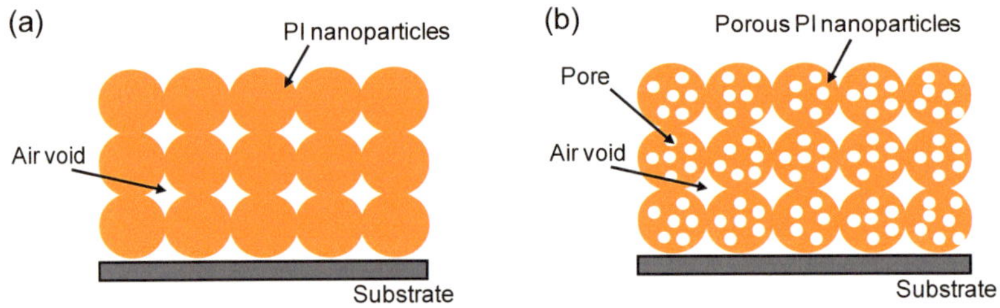

Figure 1. Schematic illustration of our strategy to reduce the dielectric constant (*k*).

In this chapter, we represent fabrication of PI NPs and morphological controlled PI NPs, *i.e.*, variously-sized NPs [13], soccer-ball-like NPs [14], cage-like microparticles [14], golf-ball-like NPs [15-17] and hollow NPs [18], using our technique, the *reprecipitation method* [19]. Furthermore, fabrication of multilayered films of porous PI NPs and their dielectric property are described [18,20].

2. Polyimide nanoparticles

Polymer fine particles are very useful for chromatographic packings; functional coatings; inks and toners; additives for foods, medicines and cosmetics; supports for catalysts; drug delivery systems; optics; and so forth. Polymer fine particles are usually manufactured during polymer synthesis from a monomer, for example, in suspension, emulsion and dispersion polymerizations. The preparations of fine particles of vinyl polymers, *e.g.* polystyrene (PS) and poly(methyl methacrylate) (PMMA), have been well-established by above-mentioned methods, and circumstantially described elsewhere. However, there are a few reports on fine particles of non-vinyl polymers including PI. Lin *et al.* [21] performed the fabrication of submicrometer PI spheres by a cooling a *N*-methyl-2-pyrrolidone (NMP) solution of PI (not PAA), which were given in high temperatures (90°C). Chai *et al.* [22] and Xiong *et al.* [23] obtained PI spheres by the dropwise addition of the precipitant (water or ethanol) into PI solutions. These methods are useful for only soluble PIs because PIs are generally insoluble in common solvents. On the other hand, precursor polymer of PI, poly(amic acid) (PAA), which is generally synthesized by polymerization of tetracarboxylic acid dianhydride and diamine, dissolves in some organic solvents such as NMP, and is easily converted into PI by the imidization treatment, namely, heat treatment or addition of cyclodehydration reagents (Fig. 2). Therefore, PI particles are often fabricated from PAA solutions or PAA particles. Nagata *et al.* [24], Asao *et al.* [25] and Basset *et al.* [26] have reported on poor-solubility PI microparticles prepared by the precipitation polymerization method, *i.e.*, the thermal imidization of PAA dissolved in NMP, which acts as a good solvent for PAA but a poor solvent for PI. Okamura *et al.* [27] precipitated uniform PAA particles of submicron scale by the precipitation polymerization between pyromeritic acid dianhydride and 4,4'-diaminodiphenyl ether in acetone, followed by imidizing these PAA particles. Recently, a few studies on fabrication of polymeric particles composed of PI and other materials (*e.g.* PS , silica) have been also reported [28-31]. On the other hand, we have reported on the reprecipitation method, which is a simple and facile fabrication technique for polymer particles and organic nanocrystals [19,32-34]. We have successfully fabricated PI NPs via the reprecipitation method [13-18]. Furthermore, novel composite PI NPs have been also fabricated by this method [35,36].

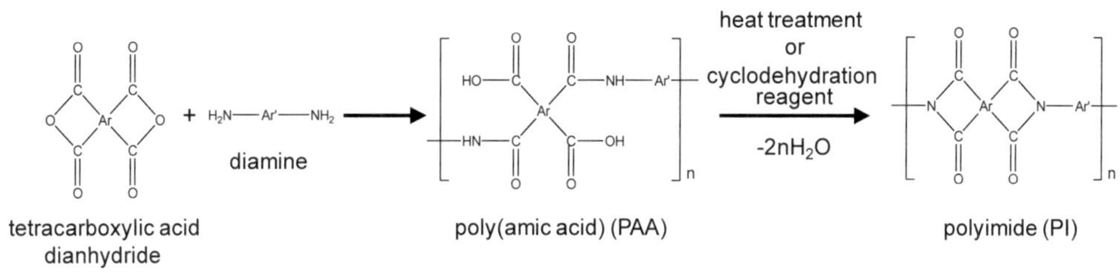

Figure 2. Synthesis route to polyimide.

2.1. Fabrication of polyimide nanoparticles via the reprecipitation method

We have successfully fabricated NPs of various PIs shown in Fig. 3. The fabrication procedure is described as follows. Because PIs are generally insoluble in common organic solvents, we used PAA solutions as starting materials for fabrication of PI NPs. Figure 4 shows the schematic representation of PI NPs via the reprecipitation method. PAA NPs were firstly fabricated by the reprecipitation method and then were converted to PI NPs through the "two-steps imidization", *i.e.* chemical imidization of PAA NPs in the dispersion medium with cyclodehydration reagent, followed by thermal imidization. In a typical experiment, a dispersion liquid of PAA NPs was obtained by injecting NMP solution of PAA (100 μl) into a vigorously stirred poor solvent (10 ml, containing 0.1 wt% of dispersing agent, Acrydic A-1381 (Dainippon ink and chemicals, Inc.)). In the present case, carbon disulfide (CS_2), cyclohexane and their mixture were used as the poor solvent. Chemical imidization of PAA NPs was performed by adding 100 μl of pyridine/acetic anhydride mixture (1:1) into the PAA NPs dispersion liquid and stirring for 2 h. After the chemical imidization, the PI NPs were centrifuged and dried *in vacuo*. Finally, PI NPs were cured at 270°C for 1 h.

PAA NPs and PI NPs obtained through the two-steps imidization were all spherical shape as shown in Fig. 5. Moreover, DLS measurement showed that the mean particle sizes and their distributions almost consisted before and after the chemical imidization [13]. These results indicate that no aggregation occurred during the imidization and we can control size of PI NPs by size-controlling PAA NPs.

A conversion to PI was estimated from IR spectra shown in Fig. 6. After the two-steps imidization (c), the absorption band around 1550 cm^{-1} (amide II: CNH vibration) and 1690 cm-1 (Amide I: C=O stretching vibration) disappeared, while the absorption band of imide IV (bending vibration of cyclic C=O), imide II (C-N stretching vibration), and imide I (stretching vibration of cyclic C=O) newly appeared at 720 cm^{-1}, 1380 cm^{-1} and 1720 cm^{-1}, respectively. A conversion to PI was determined by the following equation [25,37].

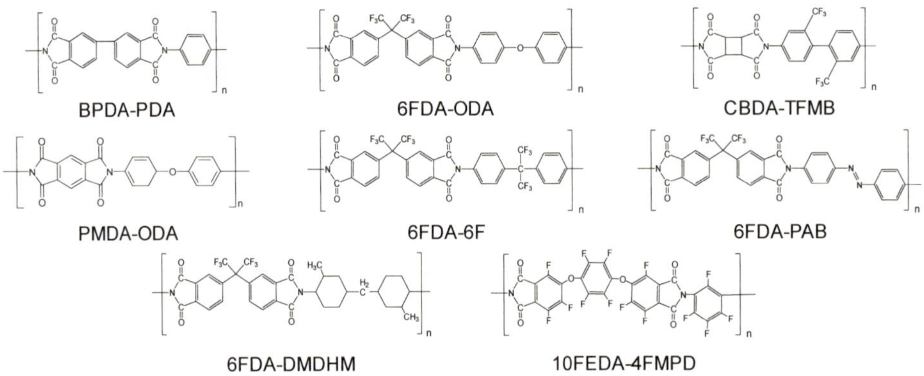

Figure 3. Structure of polyimides used in our study.

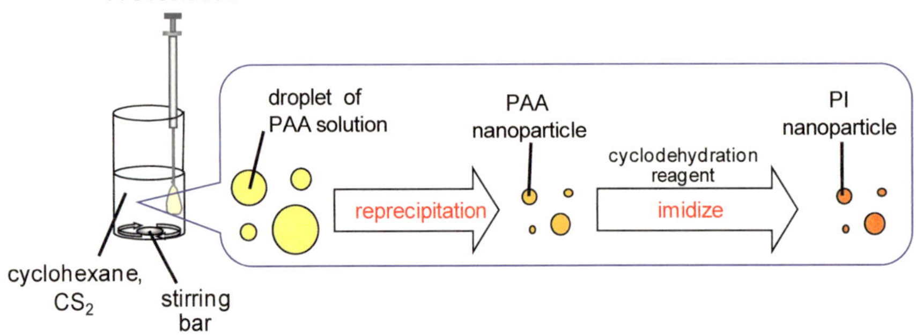

Figure 4. Schematic representation of PI nanoparticles via the reprecipitation method.

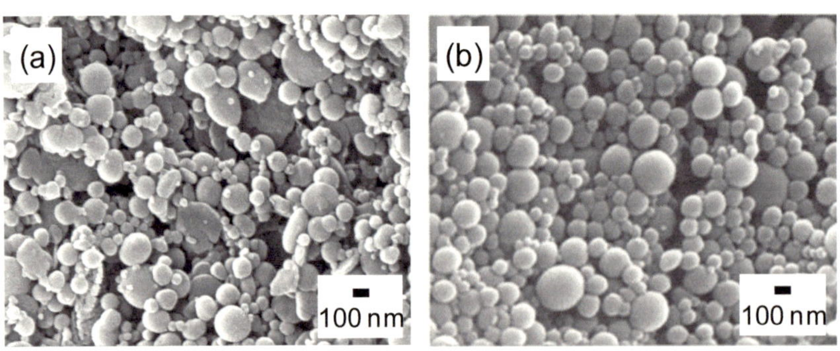

Figure 5. SEM images of 6FDA-ODA-type (a) PAA and (b) PI nanoparticles [13].

$$Conv.(\%) = \frac{\left(D_{1380\text{cm}^{-1}} / D_{1500\text{cm}^{-1}}\right)_{\text{NPs}}}{\left(D_{1380\text{cm}^{-1}} / D_{1500\text{cm}^{-1}}\right)_{\text{Bulk}}} \times 100, \tag{1}$$

where D is the optical density of each absorption bands. The absorption band at 1500 cm^{-1} (C-C stretching vibration of *p*-substituted benzene) was selected as an internal standard because its intensity remains unchanged before and after imidization. A completely-imidized sample, bulk PI, was prepared by curing a PAA cast film at 200°C for 1 h and successively at 350°C for 1 h in nitrogen atmosphere. According to the above equation, the conversion of the PI NPs was 73 % after the chemical imidization, and then the conversion of almost 100 % was found to be achieved by the subsequent curing at 270°C for 1 h. These results indicate that the imidization have been carried out quantitatively without changing their morphology.

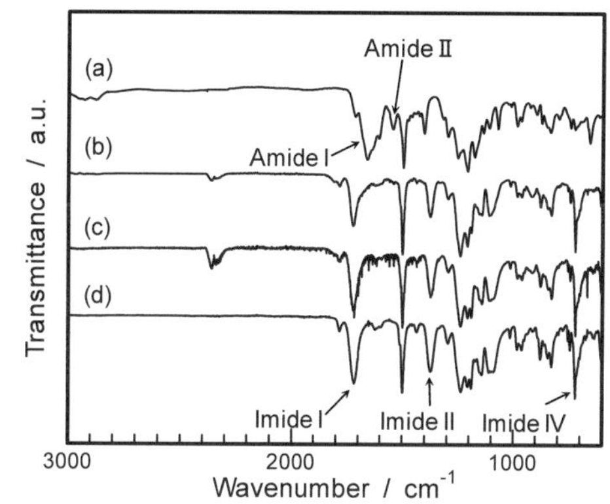

Figure 6. IR specta of (a) PAA nanoparticles, (b) nanoparticles after chemical imidization, (c) nanoparticles after two-steps imidization and (d) bulk polyimide.

2.2. Size-control of polyimide nanoparticles

As mentioned above, size-control of PI NPs will be performed by controlling size of PAA NPs. Size of PAA NPs can be decided by reprecipitation conditions such as concentration of a PAA solution (C), temperature of a poor solvent (T_{poor}) and species of a poor solvent (*e.g.* CS_2 and cyclohexane). Relationships between sizes of final PI NPs and reprecipitation conditions are shown in Fig. 7. As shown in Fig. 7a, a mean particle size increased with increase in C. This tendency is same to that in other organic compound systems previously reported [33]. To understand this phenomenon, we have to consider on the fabrication mechanism of PAA NPs using the reprecipitation method. In the reprecipitation process, first, fine droplets are dispersed in cyclohexane after injection of PAA solution. Next, NMP is removed from droplets by rapid dissolution into cyclohexane, and then PAA NPs form because PAA is insoluble in cyclohexane-rich solvent. When the concentration of the injected solution is higher, higher-concentrated droplets are produced without changing the droplet size, providing lager-sized NPs. On the other hand, a mean particle size decreased with increase in T_{poor} up to 20°C, and same size was given above 20°C (Fig. 7b). This result is explained as follows. NMP diffuses from droplets into cyclohexane slowly because of its high viscosity and poor miscibility with cyclohexane. Hence, droplets grow by agglutination between them before completion of the reprecipitation process, resulting in lager-sized particles are obtained. Under the higher temperature condition, however, the reprecipitation process is performed in a shorter time because the diffusion rate becomes drastically higher. Thus, decrease in the particle size caused by a shorter time period for growth of droplets at higher poor solvent temperature. In addition, a mean particle size also decreased with increasing a volume rate (Φ_{CS2}) of CS_2 in a cyclohexane/CS_2 mixture (Fig. 7c). PAA NPs with a size of *ca.* 50 nm were obtained under the condition using pure CS_2 as a poor solvent. This result is explained by considering miscibility between NMP and poor solvents (cyclohexane and CS_2). A difference in solubility parameters between two kinds of solvents is often used for evaluating their miscibility, and the smaller difference means higher miscibility. The solubility parameters (δ) of NMP, CS_2, and cyclohexane were calculated by the Hansen method [38] and the values of 22.9, 20.5 and 16.8 $MPa^{1/2}$ were obtained, respectively. The differences in solubility parameters indicate that NMP/CS_2 system is higher miscibility than NMP/cyclohexane system. In the case of use of CS_2/cyclohexane mixture as a poor solvent (Fig. 7c), increase in Φ_{CS2} suggests improvement of the miscibility between NMP and poor solvents due to increase in high-miscible CS_2 for NMP. The improvement results in same effect to the temperature dependence, *i.e.*, a shorter time period for droplets growth and a decrease in a size of PAA NP. Thus, we could successfully obtain size-controlled PI NPs via the two-step imidization process. Figure 8 shows SEM images of PI NPs, which were size-controlled in the range of 20 ~ 500 nm by fabricating under various reprecipitation conditions. All PI NPs were almost spherical, free of aggregations, and quantitatively imidized.

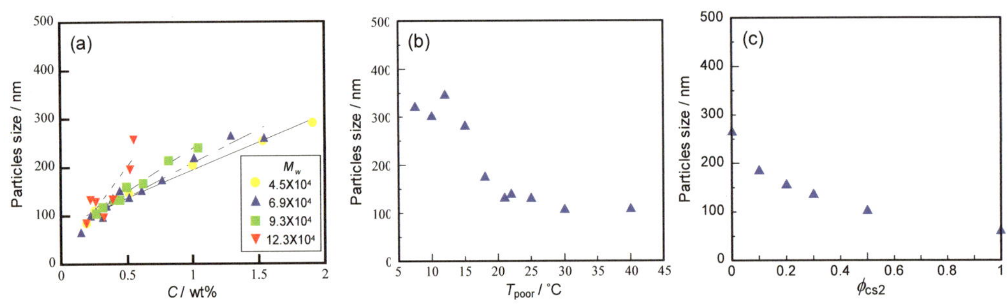

Figure 7. Relationships between size of PI nanoparticles and reprecipitation conditions [13]. Dependence of (a) concentration of PAA solution under T_{poor} = 22°C, Φ_{CS2}=0; (b) temperature of poor solvent under C=0.5 wt%, Φ_{CS2}=0, M_w=69,000; (c) volume fraction of CS_2 in poor solvent (mixture of CS_2 and cyclohexane) under C=1.5 wt% and T_{poor}=22°C, M_w=69,000.

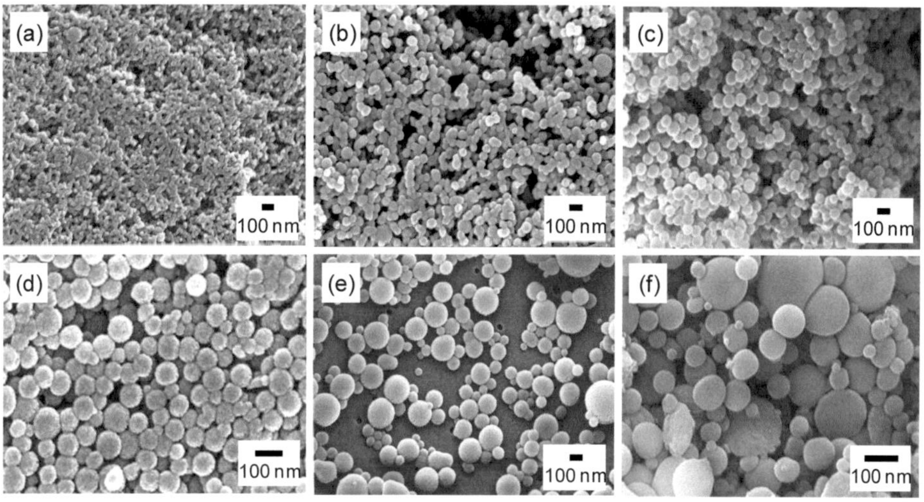

Figure 8. SEM images of 6FDA-ODA PI nanoparticles fabricated under various conditions [13]. Mean particles size: (a) 20 nm, (b) 50 nm, (c) 80 nm, (d) 100 nm, (e) 250 nm, (f) 500 nm.

3. Porous polyimide nanoparticles

Porous NPs are promising materials for low dielectric fillers, adsorbents, drug delivery systems, catalyst carriers and optics. Some studies on fabrication of porous polymer particles have been reported [39-44]. Okubo *et al.* [39] have reported fabrication of micron-sized PMMA/PS composite particles having 50~150 nm-sized dents on their surface. They achieved preparation of these particles by the seeded dispersion polymerization of styrene with PMMA seed particles in a methanol/water medium in the presence of decalin droplets. Griffiths *et al.* [40] have reported porous PS particles fabricated by the surfactant-free polymerization of styrene under the presence of water-soluble natural polymers. Kao *et al.* [41] fabricated porous poly(styrene-divinylbenzen) (PS-DVB) micoparticles by the method of multistep swelling and polymerization involving the use of polymeric porogens as follows. Monodispersed polystyrene seed particles were first prepared by the dispersion polymerization. These seeds were swollen with styrene monomer and then polymerized. Subsequently, styrene and DVB were both absorbed into the particles obtained in the previous step, and then copolymerized within the enlarged particles. In the final step, polystyrene in the particles was eluted by toluene. Unsal *et al.* [42] have also reported poly(glycidyl methacrylate-ethylene dimethacrylate) porous particles prepared by the method of swelling and seeded polymerization using PS seed particles. Koushik *et al.* [43] obtained porous deslorelin-poly(lactide-co-glycolide) microparticles by a supercritical CO_2 treatment of these sphere particles. Albrecht *et al.* [44] have reported microporous particles of poly(ether imide) prepared by a spraying/coagulation process as follows. A NMP solution of the polymer was sprayed through a hollow needle to form droplets. The droplets fell through an air gap in a water coagulation bath, which induced a phase inversion and initiated a fixation of a particle shape, and finally, the microparticles owning micoropores inside were produced. As above-described, most of porous particles sized in the micrometer range, and few studies had been reported on porous polyimide particles [24,29]. Recently, we have successfully fabricated various types of porous polyimide NPs using the reprecipitation method [14-18].

3.1. Fabrication procedure of porous PI nanoparticles via the reprecipitation method

We have focused on incorporating pores into every single PI NPs by using microphase separation of porogen within the particles, and found that LiCl [14], poly(acrylic acid) (PA) [15,16], poly(sodium 4-styrenesulfonate) (PSS) [17], poly(methyl methacrylate) (PMMA) [18] and polyvinylpyrrolidone (PVP) [18], act as useful porogens for fabrication of porous PI NPs. The reprecipitation method was also employed in order to obtain porous PI NPs as follows. A NMP solution of PAA was prepared, and NMP solution of porogen was added subsequently. And then, PAA NPs were fabricated by injecting the polymer blend solution into cyclohexane, and two-step-imidized subsequently.

3.2. Cage-like polyimide microparticles

Interestingly, as shown in Fig. 9a and b, we could obtain some doughnut-like and hollow polyimide NPs without porogens under specific reprecipitation conditions (use of high concentrated PAA solution (e.g. 1.5 wt%) and low temperature cyclohexane (e.g. 7°C)), under which particles are mainly fabricated into the sizes larger than 1000 nm. Furthermore, we sometimes produced intriguing and uniquely shaped objects under otherwise identical conditions. For example, Figure 9c displays 300-nm-diameter polymeric soccer ball-shaped particles that exhibit both pentagonal and hexagonal facets. The line widths of the facet-frames, which are composed of polymer, were ca. 50 nm. We are unaware of any such organic and/or polymeric nanomaterials having been reported previously. We suspect that these strangely shaped NPs were fabricated through phase separation of the polymer in droplets of the solution, as indicated in Fig. 10. First, immediately after injecting the solution, the droplets were produced in cyclohexane (Fig. 10a). As N-methyl-2-pyrrolidinone (NMP) molecules in the droplets gradually dissolved in cyclohexane, the droplet size reduced. As the polymer began to precipitate at the surface of the droplet, capsule structures having sizes greater than 1 μm were formed (Fig. 10b). At this point, if all of the NMP molecules completely diffused from these capsule structures, then hollow structured nanomaterials, such as those displayed in Fig. 10c, were obtained. If, however, some of the NMP molecules remained at the surface layer, so that the polymer molecules had the flexibility to move, shrinkage of the polymer structure in Fig.10b would lead to particles having their most stable structure as that presented in Fig. 10d. Unfortunately, further research will be necessary to improve the reproducibility of our results; indeed, we successfully obtained soccer ball-shaped particles such as those in Fig. 10d less than 5% of the time. Nevertheless, we believe that the ability to prepare these new types of nanosized polymer particles should be of interest to many researchers in this field.

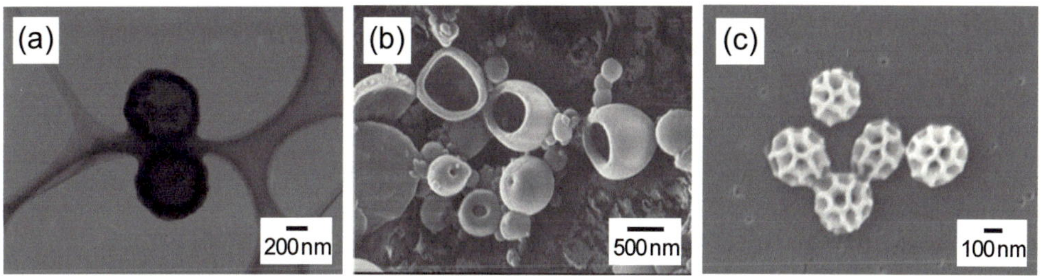

Figure 9. (a) TEM and (b), (c) SEM images of PI nanostructures fabricated by using high concentrated PAA solution and low temperature cyclohexane [14].

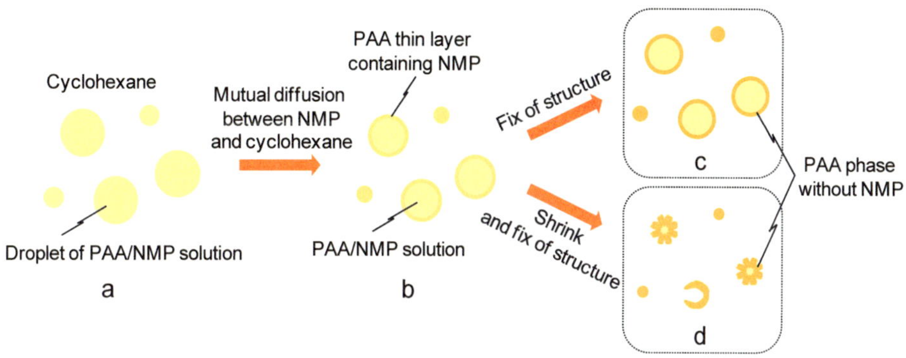

Figure 10. Schematic illustration of the process of fabricating the polyimide nanostructures.

Here, we fine-tuned the reprecipitation method to improve the reproducibility of forming porous polyimide nanostructured materials by using LiCl as a porogen. In short, a solution of poly(amic acid) (PAA) including LiCl was rapidly injected into vigorously stirred cyclohexane at room temperature. Figure 11 displays SEM images of the porous PI particles obtained after imidization and removal of the salt through washing with water. Increasing the weight ratio of LiCl to PAA in the injected solution led to an increase in the number of holes in the porous particles. The diameters of the holes on the surfaces of the particles were almost all ca. 100 nm. Figure 12 presents TEM images of the porous PI particles. We observed that spherical holes existed within the particles, which could be divided roughly into two types of porous materials. When the particle size was 700 nm or more, we observed hollow-type particles, each having a single huge hole in their center (Fig. 12a); for particle sizes below 700 nm, we observed a number of holes of almost equal size within the particles (Fig. 12b).

We have suggested a mechanism for generating these porous particles as follows. Initially, NMP droplets containing PAA and LiCl were generated in the poor solvent immediately after injection.

Next, reprecipitation of the PAA began as the NMP molecules gradually dissolved in the cyclohexane. At this stage, porous particles featuring some holes were formed in the droplets through phase separation of the polymer units. At the same time, LiCl and NMP species became concentrated within the holes. And then, NMP and LiCl eluted perfectly from the polymer particle and into the cyclohexane. Indeed, we did not detect any LiCl in the final products. Finally, the porous PI particles were then obtained through subsequent chemical imidization.

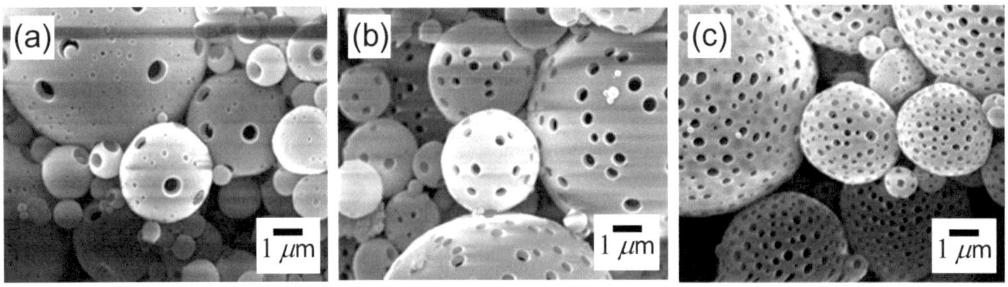

Figure 11. SEM images of porous PI particles fabricated using LiCl as a porogen [14]. Content of LiCl (relative to PAA) in the NMP solution: (a) 5, (b) 10, (c) 20 wt%.

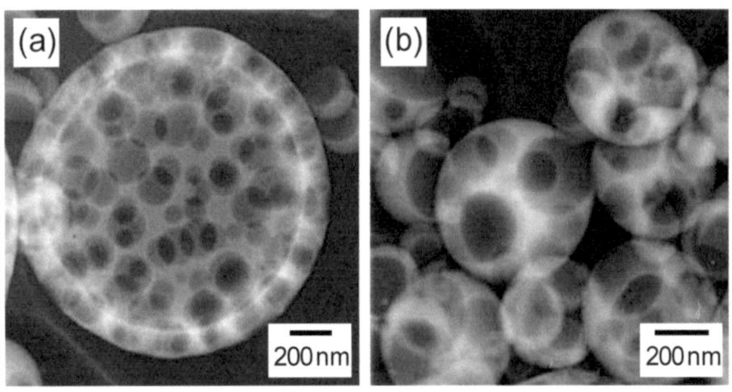

Figure 12. TEM images of the porous PI particles fabricated using LiCl as a porogen [14]. (a) Large (>700 nm) hollow-type porous particles containing a single huge hole; (b) small (<700 nm) porous particles containing multiple and equally sized holes.

3.3. Golf-ball-like polyimide nanoparticles

To fabricate porous PI NPs, the first polymer evaluated as a porogen was poly(vinyl alcohol) (PVA, M_w=500). Although both PVA and PAA (6FDA-ODA, M_w=69,000) dissolved in NMP, only some of the resulting PI NPs possessed nanopores (Fig. 13a and b). A fraction of porous PI NPs was not, however, affected by a content of PVA added. The second polymer evaluated as a porogen was poly(acrylic acid) (PA, M_w=2,000). Porous PI NPs were successfully fabricated as indicated in Fig.13c–e. The diameters of pores were in the range of 20~70 nm and the surface morphology remained almost constant when the PA content was greater than 40 wt %. Thus, PA acted as a progen for fabrication of porous PI NPs, and we have also obtained those consisted of other types of PI (10FEDA-4FMPD, BPDA-PDA and CBDA-TFMB) (Fig. 13f–h)

To consider the mechanism of formation for the porous structure, we observed SEM images of the porous PAA and/or PI NPs at each stages of their fabrication (Fig. 14). It is apparent that nanopores were already formed on the surface of the PAA NPs prior to their imidization. In addition, the surface morphology was similar before and after performing

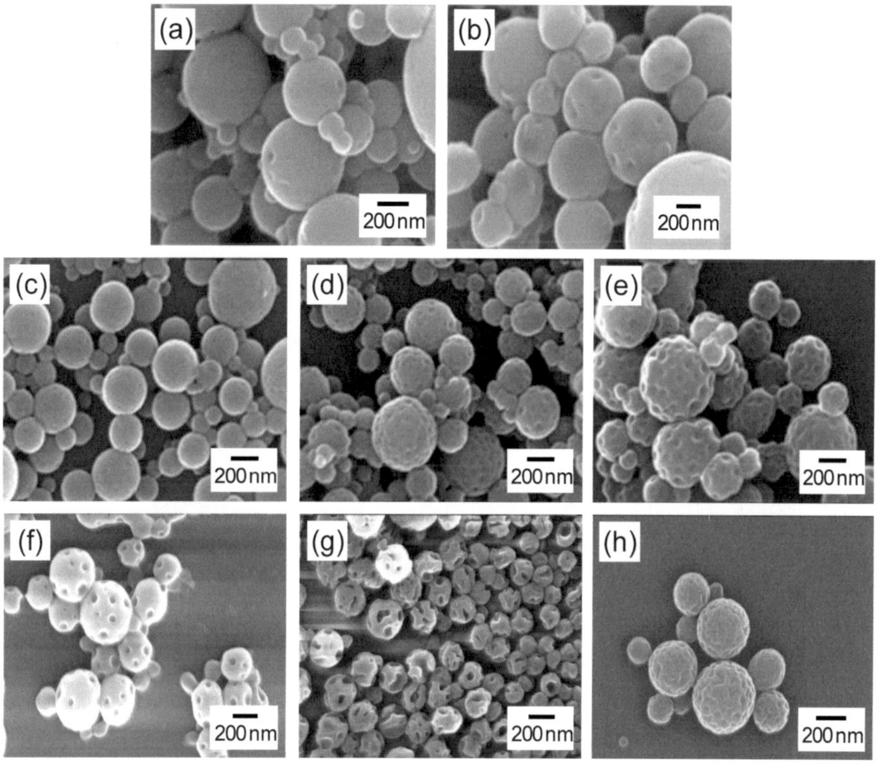

Figure 13. SEM images of porous PI nanoparticles fabricated by using various blend polymer solutions [15,16]. (a) PVA (20 wt% to PAA)/6FDA-ODA, (b) PVA (50 wt%)/6FDA-ODA, (c) PA (20 wt%)/6FDA-ODA, (d) PA (40 wt%)/6FDA-ODA, (e) PA (60 wt%)/6FDA-ODA, (f) PA (40 wt%)/10FEDA-4FMPD, (g) PA (20 wt%)/BPDA-PDA, (h) PA (40 wt%)/CBDA-TFMB.

the two-step imidization process. These findings have led us to propose a plausible mechanism of formation for the porous NPs (Fig. 15). At first, fine droplets consisting of mixture of NMP, PAA, and PA were generated in cyclohexane immediately after injection. According to the solubility parameters, cyclohexane is a much poorer solvent for PA than for PAA. As the cyclohexane and NMP diffused into one another, PAA and PA began to precipitate and microphase-separate in the surface layer of the fine droplets. At this stage, the content of cyclohexane increased gradually within each fine droplet, with NMP remaining as the main component, but with a reduced solvent power. Thus, PA-rich microdomains, which consist mainly of PA and NMP, formed within the fine droplets, because PA is more miscible with NMP. Because these internal PA-rich microdomains did not possess such a clear interface, they might have then diffused toward the surface layer. Finally, the discontinuous and isolated PA-rich microdomains that laid in the surface layer were eliminated, leading to the formation of micropores. Undoubtedly, the microphase separation giving these porous surface nanostructure is a much more complicated process than this mechanism described above; it must also depend on the molar mass of PA, the interfacial tension, the viscosity of the solvents, the mutual diffusion, and the compatibility between PAA and the porogen. With regard to this mechanism, the selection of the porogen is very important aspect of the successful formation of the porous NPs. A suitable porogen must be compatible with PAA to some extent; a suitable choice can be determined qualitatively by comparing the difference between the solubility parameters (δ) of PAA and the porogen [45]. The solubility parameter of PA (δ_{PA}= 23.6 MPa$^{1/2}$) is closer to that of PAA ($\delta_{FDA\text{-}ODA}$=20.8 MPa$^{1/2}$) than it is to that of PVA (δ_{PA}=30.5 MPa$^{1/2}$), which implies that PA is more compatible with PAA. Therefore, we obtained better results when using PA as the porogen.

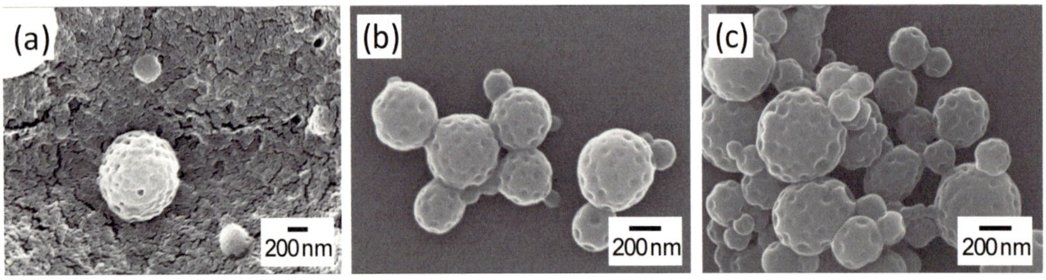

Figure 14. SEM images of porous PAA and/or PI nanoparticles at each stage of fabrication [15]. (a) immediately after injection, (b) after chemical imidization, (c) after thermal imidization.

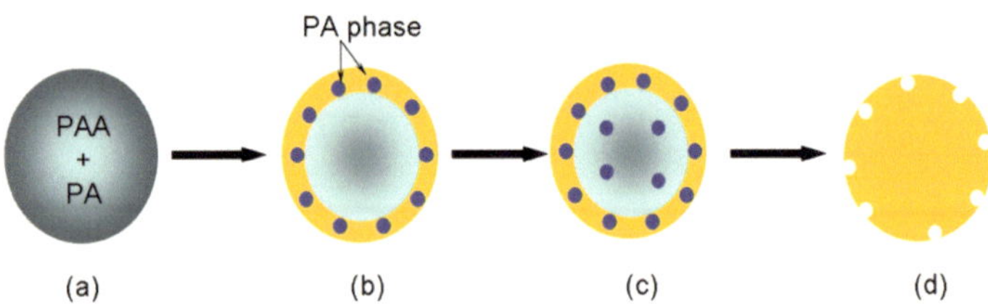

Figure 15. Schematic representation of the mechanism of formation of porous nanoparticles [15]. (a) A fine droplet of NMP, PAA, and PA formed immediately after injection; (b, c) possible intermediate states, in which the yellow regions represent zones in which NMP and cyclohexane exchange through mutual diffusion processes; (d) the resulting porous PAA NP.

It would be necessary to eliminate residual PA from the porous PI NPs if they were to be used in device applications. Figure 16 displays the results of thermogravimetric analysis (TGA) measurements. TGA curve i is that of ordinary PI NPs that lack porous structures; they decomposed thermally at a temperature above 550°C. On the other hand, PA began to decompose at ca. 200°C (TGA curve iv). The TGA curve ii is that of the chemically imidized porous PI NPs, which also decomposed gradually at ca. 200°C as a result of the removal of its residual PA and solvents. The thermally imidized porous PI NPs (TGA curve iii) yielded their 5% weight-loss temperature at 400°C; they were thermally stable up to 300°C, almost identical to the behavior of the ordinary PI NPs (TGA curve i). This result suggests that almost no PA remained within the thermally imidized porous PI NPs. Similar conclusions were drawn from the IR spectra of the various NPs. The spectrum of the thermally imidized porous PI NPs was virtually identical to that of the nonporous PI NPs. This finding suggests that PA was almost completely eliminated from the porous PI NPs during thermal imidization, consistent with the results of TGA.

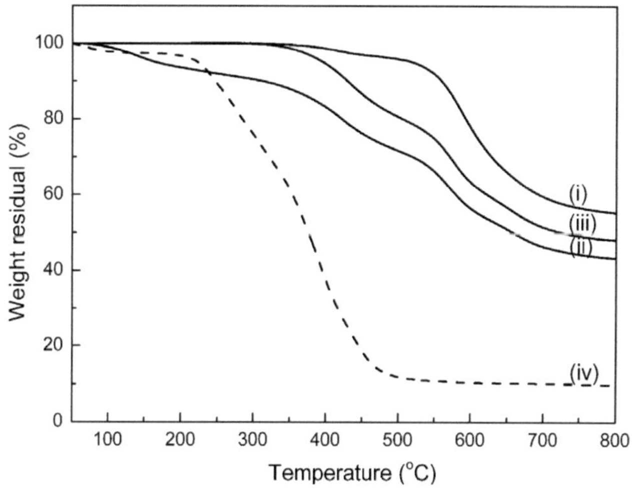

Figure 16. TGA (weight loss) curves recorded under N_2 [15]: (i) PI nanoparticles, (ii) chemically imidized porous PI nanoparticles, (iii) thermally imidized porous PI nanoparticles, (iv) PA powder. PA (40 wt %) was added to samples (ii) and (iii).

3.4. Hollow polyimide nanoparticles

By using other porogen (poly(methyl methacrylate) (PMMA) or polyvinylpyrrolidone (PVP)), hollow/multi-hollow PI NPs could be fabricated via the reprecipitation method. Figure 17 reveals that multi-hollow PI NPs were formed when we used PMMA or PVP as a porogen. Only a few small hollow cores (diameters: < 20 nm) appeared inside the PI NPs after thermal imidization when the PMMA content was 20 wt% (Fig. 17a). Upon increasing the PMMA content, larger hollow cores (diameters: ca. 100 nm), surrounded by several smaller cores, were generated; *i.e.*, the volume of the hollow structures increased accordingly (Fig. 17a–c). In contrast, we obtained only larger hollow pores ranging in a diameter from 100 to 200 nm when using PVP as a porogen (Fig. 17d, e). The number of hollow cores increased upon increasing the PVP content up to 40 wt%, but, due to high viscosity, agglomeration of the PAA NPs occurred at higher PVP contents. Clearly, the choice of a suitable porogen is an important feature affecting the generation of hollow structures. Suitable porogens can be determined qualitatively by comparing their solubility parameter with that of PAA. The solubility parameter of PAA (6FDA-ODA) ($\delta_{6FDA\text{-}ODA}$ = 20.8 MPa$^{1/2}$) is closer to that of PMMA (δ_{PMMA} = 19.7 MPa$^{1/2}$) than that of PA (δ_{PA} = 23.6 MPa$^{1/2}$), which acts as the porogen for fabrication of golf-ball-like particles (see section 3.3), indicating that PMMA is more compatible with PAA. Such a relatively high compatibility makes the PMMA phase penetrate deeper and remain within the PAA NPs, resulting in the formation of multi-hollow structures after the removal of the PMMA phase, rather than merely superficial pores. When we employed PVP (δ_{PVP} = 19.9) as an even more compatible porogen for PAA (6FDA-ODA), the PVP phases penetrated further to generate several large cores within the individual PI NPs. These findings indicate quite dramatically that the compatibility of the porogen and the PAA influences the number of hollow cores within the resulting PI NPs.

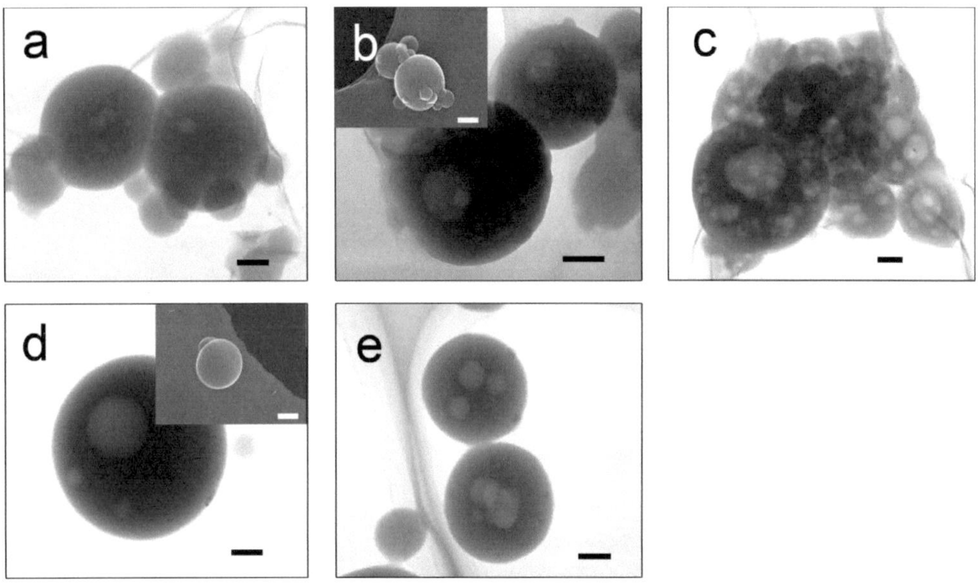

Figure 17. TEM images (scale bars: 100 nm) of multi-hollow PI(6FDA-ODA) NPs prepared using (a) PMMA 20 wt %, (b) PMMA 40 wt %, (c) PMMA 80 wt %, (d) PVP 20 wt %, and (e) PVP 40 wt % [18]. The insets display corresponding SEM images (scale bars: 200 nm).

The selection of a suitable porogen for a specific PAA allows the formation of single hollow core within each PI NP. Our approach is also applicable to other PAAs—as long as they can dissolve in the solvent. The second PI that we evaluated was PI(CBDA-TFMB) ($\delta_{CBDA-TFMB}$ = 20.0). Interestingly, the TEM images in Fig. 18 reveal only one type of hollow morphology formed—namely, a single hollow core—when using either PMMA or PVP as the porogen. Because these porogens are more compatible with PAA(CBDA-TFMB) than with PAA(6FDA-ODA), the porogen phase penetrated deeper into the center of each PI NP and integrated into a single domain, which ultimately formed the single hollow core. Moreover, the volume fraction of the hollow cores increased upon increasing the content of added PVP/PMMA. At 80 wt%, we prepared hollow PI NPs possessing core diameters in the range of 200~400 nm—the highest degree of porosity obtained in this present study.

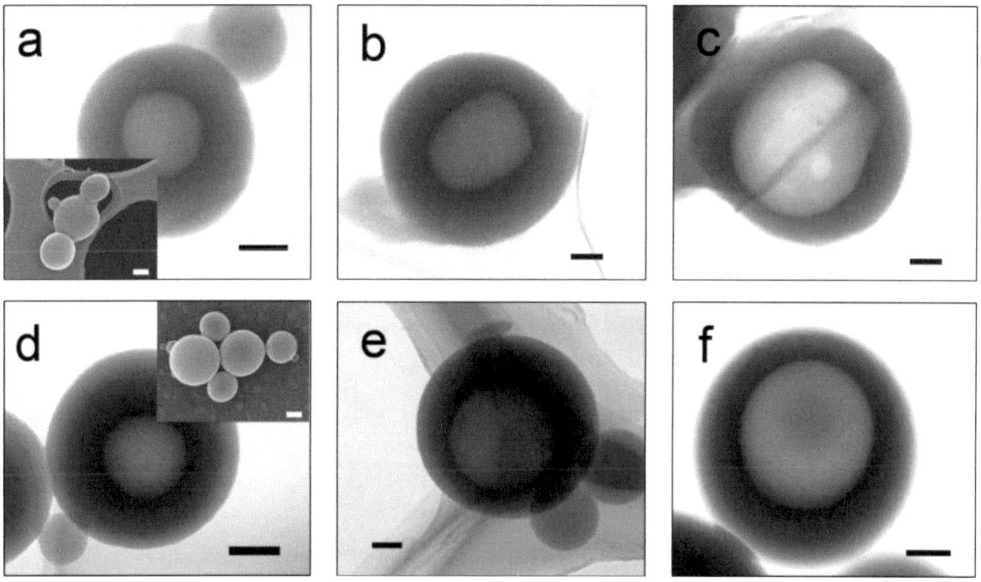

Figure 18. TEM images (scale bars: 100 nm) of hollow PI(CBDA-TFMB) NPs prepared using PMMA or PVP as the porogen at contents of (a) PMMA 20 wt %, (b) PMMA 40 wt %, (c) PMMA 80 wt %, (d) PVP 20 wt %, (e) PVP 40 wt %, and (f) PVP 80 wt % [18]. The insets display corresponding SEM images (scale bars: 200 nm).

polymer	Solubility parameter ($MPa^{1/2}$)
PAA(6FDA-ODA)	20.8
PAA(CBDA-TFMB)	20.0
poly(methyl methacrylate) (PMMA)	19.7
poly(acrylic acid) (PA)	23.6
polyvinylpyrrolidone (PVP)	19.9

Table 1. Solubility Parameters calculated using the Hoftyzer and van Krevelen method [46].

According to TGA and IR spectra of the resulting hollow PI NPs, little PVP remained within the thermally imidized hollow PI NPs, *i.e.*, the hollow PI NPs formed through thermal decomposition of the microphase-separated porogen within the composite PI/porogen NPs. This result is caused by high pyrolysis temperature of PVP (drastic weight loss from 400 to 470°C). On the other hand, PMMA was dramatically decomposed from 350 to 420°C. Therefore, no PMMA remained within the hollow PI NPs derived from the composite PI/PMMA NPs due to its lower pyrolysis temperature compared with that of PVP.

4. Low-*k* porous polyimide films

Porous PI films have been generally prepared by pyrolysis of thermally labile polymer units in phase-separated PI composite films, for example, derived from poly(propylene oxide)-,

PMMA- and poly(α-methyl-styrene)-PI block copolymer [9,10]; blend solution containing polyurethane and PAA [11]; and polyacrylamide- and PMMA-grafted PI [12]. They have achieved 10~30% decrease in a dielectric constant of polyimide films, and low-k porous films ($k \approx 2.2$) have been provided. Also, other methods, such as using water droplets and LiCl crystal as templates, and extracting a porogen polymer from composite PI film by supercritical CO_2, have been known [47-49]. On the other hand, we have obtained ultralow-k porous films by depositing porous PI NPs onto substrates [18,20].

4.1. Preparation of multilayered films of polyimide nanoparticles

We employed the electrophoretic deposition method [50] to assemble the obtained hollow PI NPs into multilayered films. The typical experiment procedure is described as follows. DC voltages were applied to indium tin oxide (ITO) electrodes immersed in a face-to-face arrangement in a suspension of PI NPs. The PI NPs, which were minus-charged in cyclohexane, were electrophoretically deposited on the anodes, and multilayered films were quickly prepared within 1 minute. The films were dried and then cured at 270 °C for 1 h to perform pyrolysis of a porogen. The electrophoretic deposition process was repeated to fill some cracks formed through coalescence of the NPs during evaporation of dispersion medium. And then, PAA solution was spin-coated onto the resulting film. After thermal imidization, a multilayer PI NPs film having a smooth surface was obtained.

Uniform films of PI NPs were easily obtained by this method as shown in Fig. 19. Furthermore, multilayered films with different thickness were prepared by changing dispersion concentration of PI NPs. As shown in Fig. 20a, film thickness increased with increase in the dispersion concentration. On the other hand, applied voltage also influenced the film thickness dramatically (Fig. 20b). Under using the dispersion liquid with solid concentration of 0.2 wt%, films became thicker as the applied voltage increased from 50 to 2000 V/cm. However, no further increase of the film thickness was observed when the applied voltage was higher than 2000 V/cm. After the electrophoretic deposition process under the applied voltage above 2000 V/cm, a transparent liquid left in sample tubes. This result indicates 100% consumption of PI NPs. Thus, the film thickness was controlled in the range of several hundred nm ~ several μm by varing dispersion concentration and applied voltage.

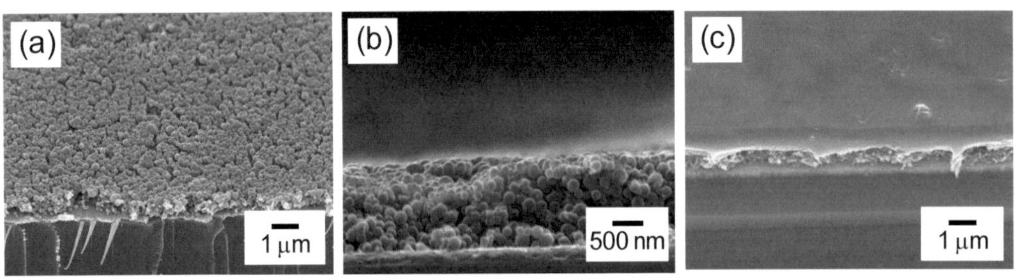

Figure 19. SEM images of multilayered films of PI nanoparticles on ITO substrates. (a) after 1st electrophoretic deposition, (b) cross section, (c) after further spin coat.

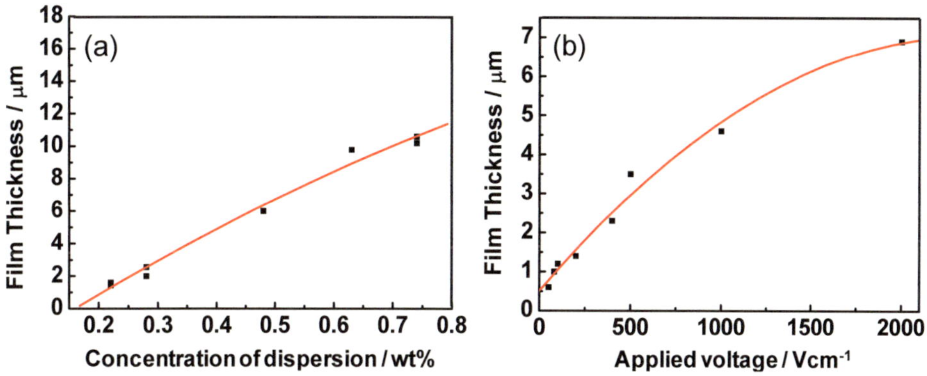

Figure 20. Dependence of (a) concentration of PI nanoparticles dispersion, (b) applied voltage, on thickness of layered film of PI nanoparlticles.

4.2. Dielectric constant of multilayered films assembled by various type of PI nanoparticles

We evaluated a dielectric constant of layered films of PI NPs prepared. Capacitance, C_P, of films was measured by a LCR meter in air and at room temperature, and a dielectric constant, k, was calculated by the equation (2).

$$k = \frac{tC_P}{\varepsilon_0 S}, \tag{2}$$

Here, C_P is a capacitance measured, t is a thickness of a film, ε_{r0} is the electric constant and S is an electrode area. The dielectric constants of various films, *i.e.*, cast films (non particles films), layered films of nonporous NPs and layered films of porous NPs, were summarized in Table 2. The dielectric constants of the cast films consisted of BPDA-PDA, 6FDA-ODA and CBDA-TFMB were 3.17, 2.76 and 2.52, respectively. The dielectric constants of multilayer films of nonporous PI NPs were somewhat lower, resulting from generation of air voids between NPs. The porosity of the porous films was 15~22%, calculated by using the Maxwell-Garnett model, the equation (3) [8].

$$\frac{k_{\text{porous}} - 1}{k_{\text{porous}} + 2} = \left(1 - \frac{P}{100}\right)\frac{k_{\text{solid}} - 1}{k_{\text{solid}} + 2}, \tag{3}$$

where k_{solid} is a dielectric constant of a nonporous film, and P is a porosity of a porous film. It was even lower than that of rhombohedral packing of spherical particles (porosity=26% [51]). Polydispersity of PI NPs may result in a denser packing arrangement. Moreover, the dielectric constants decreased significantly upon increasing the degree of porosity of the PI films (porosity =31~33%), *i.e.*, using hollow PI NPs instead of nonporous PI NPs. Amazingly, the dielectric constant was ca. 1.9 for the multilayer films formed from the hollow PI(CBDA-TFMB) NPs. This dielectric constant is even lower than that of a cast film of poly(tetrafluoroethylene), which has the lowest dielectric constant among polymeric insulators. Thus, the introduction of air voids between and within the PI NPs reduced the dielectric constant significantly. In addition, the electrophoretic deposition appears to be an effective means of generating multilayer ultralow-*k* films from hollow PI NPs.

PI structures	Film types	Film thickness (μm)	Dielectric constant (*k*) @1MHz	Porosity (%)
BPDA-PDA	cast	2.14	3.17	–
	nonporous NPs	2.62	2.67	15
	golf-ball-like NPs	5.81	2.27	31
6FDA-ODA	cast	1.39	2.76	–
	nonporous NPs	3.18	2.36	17
	hollow NPs	1.90	2.00	32
CBDA-TFMB	cast	1.39	2.52	–
	nonporous NPs	4.40	2.07	22
	hollow NPs	2.50	1.87	33

Table 2. Dielectric constants and porosities of various type of PI films.

5. Conclusion

We proposed the depositing of porous polyimide nanoparticles (PI NPs) onto a substrate as the novel strategy, that is, introducing air voids within and between the PI NPs, toward the preparation of ultralow-k ($k < 2.0$) PI films. We presented fabrication of porous PI NPs and porous films, which are assembled by the porous NPs, for low-*k* materials on the basis of this strategy. In particular, Size-controlled NPs were successfully fabricated by using our technique, the *reprecipitation method,* followed by the two-step imidization. Moreover, addition of various progens, such as LiCl, poly(acrylic acid) (PA), poly(methyl methacrylate) (PMMA) and polyvinylpyrrolidone (PVP), to starting solutions of poly(amic acid) (PAA) gave cage-like, golf-ball-like, multi-hollow and hollow NPs, respectively. Also, multilayered films of porous PI NPs were successfully prepared by the electrophoretic deposition method within 1 minute. We obtained a particularly low dielectric constant (k = ca. 1.9) for a multilayer film assembled by one such type of hollow PI NPs. Thus, our strategy is an effective means for preparing ultralow-*k* PI films featuring unique porous structures, which are alternative candidates for use in dielectric interlayer applications.

Author details

Takayuki Ishizaka*

Research Center for Compact System,
AIST (National Institute of Advanced Industrial Science and Technology), Sendai, Japan

Hitoshi Kasai*

Institute of Multidisciplinary Research for Advanced Materials, Tohoku University, Sendai, Japan

* Corresponding Authors

Acknowledgement

Authors would like to thank Dr. M. Suzuki, Mr. H. Mitsui, Dr. G. Zhao, Prof. H. Oikawa, Prof. H. Nakanishi, Tohoku University; Prof. M. Hasegawa, Toho University and Prof. T. Furukawa, Tokyo University of Science for useful assistances and supports in this study.

6. References

[1] International Technology Roadmap for Semiconductors (ITRS), 2010 Update, Interconnect. Available: http://www.itrs.net/Links/2010ITRS/Home2010.htm. Accessed 2012 Sep 27.

[2] Anderson M R, Davis R M, Taylor C D, Parker M, Clark S, Marciu D, Miller M (2001) Thin Polyimide Films Prepared by Ionic Self-Assembly. Langmuir 17: 8380–8385.

[3] Eichstadt A E, Ward T C, Bagwell M D, Farr I V, Dunson D L, McGrath J E (2002) Synthesis and Characterization of Amorphous Partially Aliphatic Polyimide Copolymers Based on Bisphenol-A Dianhydride. Macromolecules 35: 7561–7568.

[4] Ando S, Matsuura T, Sasaki S (1992) Perfluorinated Polyimide Synthesis. Macromolecules 25: 5858–5860.

[5] Feiring A E, Auman B C, Wonchoba E R (1993) Synthesis and Properties of Fluorinated Polyimides from Novel 2,2'-bis(fluoroalkoxy)benzidines. Macromolecules 26: 2779–2784.

[6] Hasegawa M, Horiuchi M, Wada Y (2007) Polyimides Containing Trans-1,4-cyclohexane Unit (II). Low-K and Low-CTE Semi- and Wholly Cycloaliphatic Polyimides. High Perform. Polym. 19: 175–193.

[7] Hasegawa M (2001) Semi-Aromatic Polyimides with Low Dielectric Constant and Low CTE. High Perform. Polym. 13: S93–S106.

[8] Garnett J C M (1904) Colours in Metal Glasses and in Metallic Films. Philos. Trans. R. Soc. London Ser. A 203: 385-420.

[9] Hedrick J L, Miller R D, Hawker C J, Carther K R, Volksen W, Yoon D Y, Trollsås M (1998) Templating Nanoporosity in Thin-film Dielectric Insulators. Adv. Mater. 10: 1049-1053.

[10] Carter K R, DiPetro R A, Sanchez M I, Swanson S A (2001) Nanoporous Polyimides Derived from Highly Fluorinated Polyimide/Poly(propylene Oxide) Copolymers. Chem. Mater. 13: 213-221.

[11] Krishnan P S G, Cheng C Z, Cheng Y S, Cheng J W C (2003) Preparation of Nanoporous Polyimide Films from Poly(urethane-imide) by Thermal Treatment. Macromol. Mater. Eng. 288: 730-736.

[12] Wang W-C, Vora R H, Kang E-T, Neoh K-G, Ong C-K, Chen L-F (2004) Nanoporous Ultra-low-κ Films Prepared from Fluorinated Polyimide with Grafted Poly(acrylic acid) Side Chains. Adv. Mater. 16: 54-57.

[13] Suzuki M, Kasai H, Ishizaka T, Miura H, Okada S, Oikawa H, Nihira T, Fukuro H, Nakanishi H (2007) Fabrication of Size-controlled Polyimide Nanoparticles. J. Nanosci. Nanotechnol. 7: 2748-2752.

[14] Kasai H, Mitsui H, Zhao G, Ishizaka T, Suzuki M, Oikawa H, Nakanishi H (2008) Fabrication of Porous Nanoscale Polyimide Structures. Chem. Lett. 37: 1056-1057.

[15] Zhao G, Ishizaka T, Kasai H, Oikawa H, Nakanishi H (2007) Fabrication of Unique Porous Polyimide Nanoparticles Using Reprecipitation Method. Chem. Mater. 19: 1901-1905.

[16] Zhao G, Ishizaka T, Kasai H, Nakanishi H, Oikawa H (2007) Introducing Porosity into Polyimide Nanoparticles. J. Nanosci. Nanotechnol. 8: 1-5.

[17] Zhao G, Ishizaka T, Kasai H, Hasegawa M, Nakanishi H, Oikawa H (2009) Using a Polyelectrolyte to Fabricate Porous Polyimide Nanoparticles with Crater-like Pores. Polym. Adv. Technol. 20:43-47.

[18] Zhao G, Ishizaka T, Kasai H, Hasegawa M, Furukawa T, Nakanishi H, Oikawa H (2009) Ultralow-dielectric-constant Films Prepared from Hollow Polyimide Nanoparticles Possessing Controllable Core Sizes. Chem. Mater. 21: 419-424.

[19] Kasai H, Nalwa H S, Oikawa H, Okada S, Matsuda H, Minami N, Kakuta A, Ono K, Mukoh A, Naknishi H (1992) A Novel Preparation Method of Organic Microcrystals. Jpn. J. Appl. Phys. 31: L1132-L1134.

[20] Zhao G, Ishizaka T, Kasai H, Oikawa H, Nakanishi H (2007) Preparation of Multilayered Film of Polyimide Nanoparticles for Low-k Applications. Mol. Cryst. Liq. Cryst. 464: 31-38.

[21] Lin T, Stickney K W, Rogers M, Riffle J S, McGrath J E, Marand H, Yu T H, Davis R M (1993) Preparation of Sub-micron Polyimide Particles by Precipitation from Solution. Polymer 34: 772-777.

[22] Chai Z, Zheng X, Sun X (2003) Preparation of Polymer Microspheres from Solutions. J. Polym. Sci. Part B: Polym. Phys. 41: 159-165.

[23] Xiong J Y, Liu X Y, Chen S B, Chung T S (2004) Surfactant Free Fabrication of Polyimide Nanoparticles. Appl. Phys. Lett. 85: 5733-5735.

[24] Nagata Y, Ohnishi Y, Kajiyama T (1996) Highly Crystalline Polyimide Particles. Polym. J. 28: 980-985.

[25] Asao K, Ohnishi H, Morita H (2000) Preparation of Polyimide Particles by Precipitation Polymerization. Kobunshi Ronbunshu *in Japanese* 57: 271-276.

[26] Basset F, Lefrant A, Pascal T, Gallot B, Sillion B (1998) Crystalline Polyimide Particles Generated via Thermal Imidization in a Heterogeneous Medium. Polym. Adv. Technol. 9: 202-209.

[27] Okamura A, Fujimoto K, Kawaguchi H, Nishizawa H, Hirai O (1996) Preprints of 9th Polymeric Microspheres Symp.: 167 p.

[28] Omi S, Matsuda A, Imamura K, Nagai M, Ma G H (1999) Synthesis of Monodisperse Polymeric Microspheres Including Polyimide Prepolymer by Using SPG Emulsification Technique. Colloids surf. A 153: 373-381.

[29] Watanabe S, Ueno K, Murata M, Masuda Y (2006) Preparation of Polystyrene-Polyimide Particles by Dispersion Polymerization of Styrene Using Poly(amic acid) as a Stabilizer. Polym. J. 38: 471-476.

[30] Watanabe S, Ueno K, Kudoh K, Murata M, Masuda Y (2000) Preparation of Core-shell Polystyrene-polyimide Particles by Dispersion Polymerization of Styrene Using Poly(amic acid) as a Stabilizer. Macromol. Rapid Commun. 21: 1323-1326.

[31] Kim T H, Ki C D, Cho H, Chang T, Chang J Y (2005) Facile Preparation of Core-Shell Type Molecularly Imprinted Particles: Molecular Imprinting into Aromatic Polyimide Coated on Silica Spheres. Macromolecules 38, 6423-6428.

[32] Kasai H, Kamatani H, Yoshikawa Y, Okada S, Oikawa H, Watanabe A, Ito O, Nakanishi H (1997) Crystal Size Dependence of Emission from Perylene Microcrystals. Chem. Lett. 11: 1181-1182.

[33] Kasai H, Nalwa H S, Okada S, Oikawa H, Nakanishi H (2000) Fabrication and Spectroscopic Characterization of Organic Nanocrystals. In: Nalwa H S, editor. Handbook of Nanostructured Materials and Nanotechnology, San Diego:Academic Press. Chap. 8.

[34] Kasai H, Okazaki S, Hanada T, Okada S, Oikawa H, Adschiri T, Arai K, Yase K, Nakanishi H (2000) Preparation of C60 Microcrystals Using High-Temperature and High-Pressure Liquid Crystallization Method. Chem. Lett. 12: 1392-1393.

[35] Ishizaka T, Kasai H, Nakanishi H (2004) Intensity Controllable Luminescence of Eu^{3+}-doped Polyimide Nanoparticles by UV-irradiation and Thermal Treatment. Jpn. J. Appl. Phys. 43: L516-518.

[36] Ishizaka T, Kasai H, Nakanishi H (2008) Fabrication of Eu-complex/Polyimide Compostie Nanoparticles. J. Mater. Sci. 44: 166-169.

[37] Nishino T, Kotera M, Inayoshi N, Miki N, Nakamae K (2000) Residual Stress and Microstructures of Aromatic Polyimide with Different Imidization Processes. Polymer 41:6913-6918.

[38] Barton A F M (1975) Solubility Parameters. Chem. Rev. 75: 731-753.

[39] Okubo M, Takekoh R, Suzuki A (2002) Preparation of Micron-sized, Monodisperse Poly(methyl methacrylate)/Polystyrene Composite Particles Having a Large Number of Dents on Their Surfaces by Seeded Dispersion Polymerization in the Presence of Decalin. Colloid Polym. Sci. 280: 1057-1061.

[40] Griffiths P C, Wellappili C, Hemsley A R, Stephens R (2004) Ultra-porous Hollow Particles. Colloid Poly. Sci. 282: 1155-1159.

[41] Kao C-Y, Lo T-C, Lee W-C (2003) Influence of Polyvinylpyrrolidone on the Hydrophobic Properties of Partially Porous Poly(Styrene–Divinylbenzene) Particles for Biological Applications. J. Appl. Polym. Sci. 87: 1818-1824.

[42] Unsal E, Irmak T, Durusoy E, Tuncel M, Tuncel A (2006) Monodisperse Porous Polymer Particles with Polyionic Ligands for Ion Exchange Separation of Proteins. Anal. Chim. Acta 570: 240-248.

[43] Koushik K, Kompella U B (2004) Preparation of Large Porous Deslorelin-PLGA Microparticles with Reduced Residual Solvent and Cellular Uptake Using a Supercritical CO_2 Process. Pharm. Res. 21: 524-535.

[44] Albrecht W, Lützow K, Weigel T, Groth T, Schossig M, Lendlein A (2006) Development of Highly Porous Microparticles from Poly(ether imide) Prepared by a Spraying/Coagulation Process. J. Membr. Sci. 273: 106-115.

[45] Coleman M M, Serman C J, Bhagwagar D E, Painter P C (1990) A practical Guide to Polymer Miscibility. Polymer 31:1187-1203.

[46] Krevelen D W V, Nijenhuis K T (2009) Cohesive Properties and Solubility. In: Properties of Polymers, 4th Edition. Amsterdam: Elsevier Science. pp. 189-228.

[47] Yabu H, Tanaka M, Ijiro K, Shimomura M (2003) Preparation of Honeycomb-patterned Polyimide Films by Self-organization. Langmuir 19: 6297-6300.

[48] Niyogi S, Adhikari B (2002) Preparation and Characterization of a Polyimide Membrane. Eur. Polym. J. 38: 1237-1243.

[49] Mochizuki A, Fukuoka T, Kanada M, Kinjou N, Yamamoto T (2002) Development of Photosensitive Porous Polyimide with Low Dielectric Constant. J. Photopolym. Sci. Technol. 15: 159-166.

[50] Trau M, Saville D A, Aksay I A (1996) Field-induced Layering of Colloidal Crystals. Science. 272: 706-709.

[51] Graton L C, Fraser H J (1935) Systematic Packing of Spheres with Particular Reference to Porosity and Permeability. J. Geol. 43: 785.

Chapter 7

Novel Polyimide Materials Produced by Electrospinning

Guangming Gong and Juntao Wu*

Additional information is available at the end of the chapter

http://dx.doi.org/10.5772/53459

1. Introduction

1.1. Electrospinning

Due to one-dimentional nanostructures' unique properties and their intriguing applications, there has being a great demanding on the technologies which are able to produce such structures. Among the applicable strategies, electrospinning seems to be the simplest approach to generate continues nanofibers with ultrathin and uniform diameters (Li & Xia, 2004). Electrospinning is a spin technology which involves the use of a high voltage, usually direct-current, to trigger the formation of a liquid jet (Li & Xia, 2004; Greiner & Wendorff, 2007).

1.1.1. The Mechanism of Electrospinning

Briefly, the mechanism of electrospinning is often to be explained as the collaborative effects of electrostatic repulsion by the accumulated charges on the surface of polymer solution and the Coulombic force exerted by the external electric field (Li & Xia, 2004; Greiner & Wendorff, 2007). A illustrative figure may help the understanding of it. As can be seen in Figure 1, polymer solutions are added in a syringe with a metal spinneret connected on its tip. Driven by a syringe pump or gravity, the solution starts to flow out. When a high voltage (usually in the range from 1 to 30 kv), the pendent drop of the solution will be highly electrified. During this process, the liquid drop will be shaped into a cone-like object, as known as "Taylor Cone". And the charges on the surface will repulse each other, overcome the surface tension of the solution until reaching a certain threshold. Then an ejection of solution is formed. Driven by the mighty electric field, the liquid jet undergoes a mighty stretching force, elongating the jet, thinning the diameter to a certain range. While during this period, the solvent evaporates intensively. Thus, the electrospun polymer fibers are formed. And a typical SEM photo of the electrospun fiber is illustrated in the insert of

Figure 1. Electrospinning is able to produce fibers whose diameters ranged from tens of nanos to a few microns. And a great amount of polymers are reported to be successfully spun into fibers via this technology. Based on the experimental data and electrohydrodynamic theories, several groups have built mathematical model to describe behaviors of electrospinning. Further information is not explained here, if interested, one can refer to these literatures: (Reneker, Yarin & Fong, et al., 2000; Yarin, Koombhongse & Reneker, 2001; Hohman, Shin & Rutledge, et al., 2001, etc.)

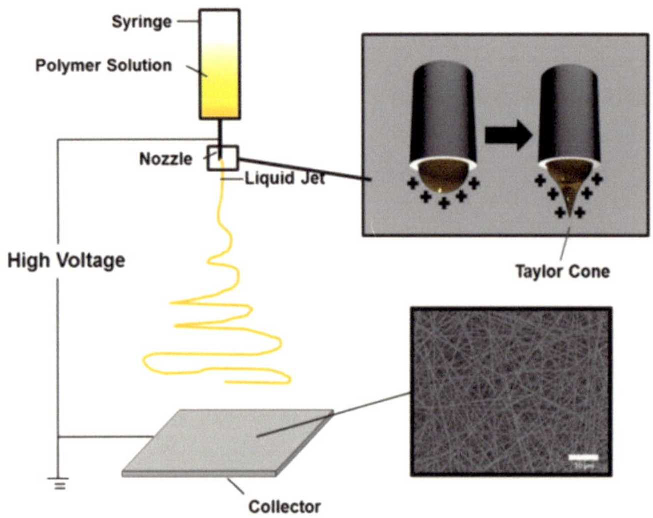

Figure 1. An illustration of a typical electrospinning set and the formation of the liquid jet (in enlargement). The inserted is an SEM picture of nowoven PI mat prepared in our laboratory.

1.1.2. A Brief History of Electrospinning

The first patent that describe the operation of electrospinning appeared in 1934 (Formalas, 1934), when Formalas disclosed an apparatus for producing polymer filaments by taking the advantage of electrostatic repulsions between surface charges. It has been almost 80 years since the patent but electrospinning existed only as a theory in people's mind until 1990s. (Li & Xia, 2004) During that period, there were only few reports on ES. Its value was not fully attended so that no further development was made. But in the beginning of 1990s, by the efforts of several research groups, especially the ones leaded by Prof. D. H. Reneker and Prof. G.C. Rutledge, people demonstrated that ES was able to produce a great amount of polymer fibers. A large campaign of polymer ES researches was waged. (Wikipedia, Online) The timely demonstrations popularized the term "electrospining" in literatures that we see today.

1.1.3. Development and Applications

The diameters of polymer fibers produced by electrospinning were able to reach an incredible small range. Compared to traditional polymer fibers, ES fibers have greater specific areas and aspect ratio (Li & Xia, 2004; Greiner & Wendorff, 2007). Electrospinning is a method to produce not only pure fiber structures, but also many other morphologies. By tuning electrospinning parameters (e.g. electric field strength, viscosity of spinning solution, solvents, etc.) and/or choosing different ES methods, controllable morphologies of ES products can be achieved. Owing to the world wide collaborative efforts, so far, people are able to electrospin pearl-necklace-like beaded fiber structures (Greiner & Wendorff, 2007), highly porous fiber structures (Bognitzki et al., 2001), grafted fiber structure (Hou H, Reneker D H, 2004; Chang Z, 2011), hollow interior micro tubes (Li, Wang & Xia, 2004; Zhao, Cao & Jiang, 2007), wire-in-tube structures (Greiner & Wendorff, 2007), What's more, twisted fiber bundles, golfball-like micro particles and multi-chambered hollow spheres (Chen et al., 2008) are also able to be generated via this technique. These fibrous structures and particles obtained by electrospinning have some unmatchable properties, like the

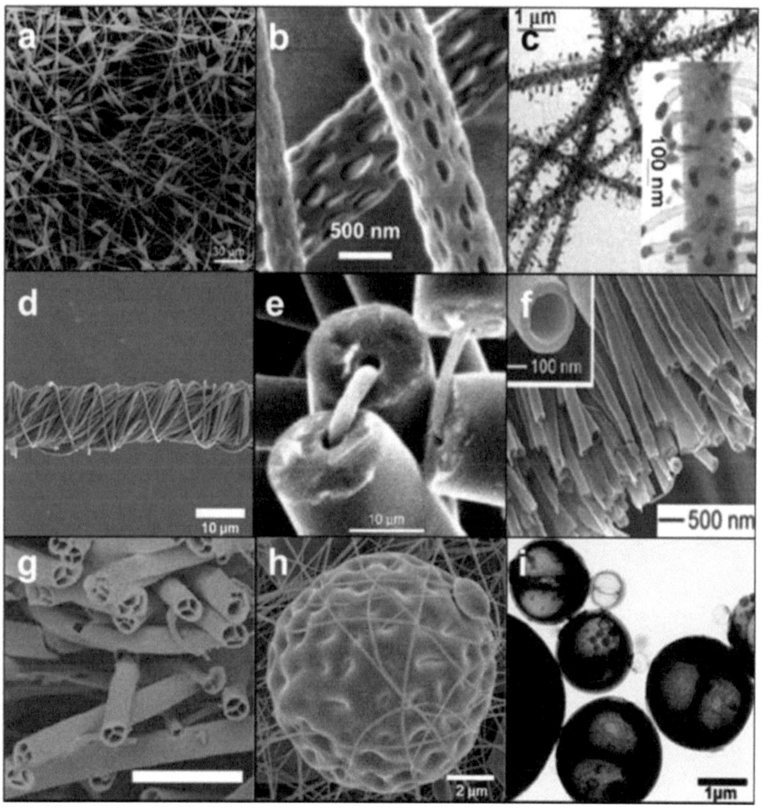

Figure 2. Morphologies of electrospun products: a) pear-necklace like fibers. Copyright © 2007 WILEY-VCH, reused with permission; b) highly porous fibers. Copyright © 2001 WILEY-VCH, reused with permission; c) grafted fibers. Copyright © 2004 WILEY-VCH, reused with permission; d) a bundle of electrospun PI micro rope (Gong & Wu, unpublished work); e) wire-in-tube fibers. Copyright © 2007 WILEY-VCH, reused with permission; f); aligned micro tubes. Copyright © 2004, American Chemical Society, reused with permission; g) multi-channel tubes. Copyright © 2007, American Chemical Society, reused with permission; h) golfball-like spheres (Gong & Wu, unpublished work); i)multi-chamber spheres. Copyright © 2008, American Chemical Society, reused with permission

extremely long length, high surface area, complex pore structure, alignment on the molecule level, etc. Such properties enable the electrospun materials possessing the potential values in applications and research fields, e.g. templates, filter and textile, catalysis and enzyme carriers, nanofiber reinforcement, medical applications, and surfaces with special wettability, etc. Due to the space constraints, detailed instances are not discussed here. The further introductions and examples, if one is interested, can be reffered to these literatures: (Li & Xia, 2004; Greiner & Wendorff, 2007; Lu, Wang & Wei, 2009)

1.2. Polyimide (Ding, 2011)

Polyimide (PI) is a polymeric material which contains imide rings in their molecule backbones. PI is mainly obtained through a two-step method: fragrant diamine and dianhydride undergo a condensational polymerization and then a thermal/chemical imidization, to afford PI, illustrated in Figure 3. The representative PI is Kapton®, developed and commercialized by Dupont™ in 1960 and 1965, respectively. This golden film presents good mechanical properties and high thermal stability. After 50 years, Kapton® is still a leading material in thermal tolerable materials. After Kapton®'s success, a series of fragrant polyimides were developed and accepted by the market. PI's wide popularity is due to its comprehensive performances: wide applicable temperature range, good mechanical, electrical properties and bio-compatability. Besides, PI's broad molecular designing window also guarantees its rapid development: by modifying molecular structures of diamine and dianhydride, different kinds of soluble and thermal shapeable PI were prepared. After a long term of research and developing state, PI is now extensively employed in the aerospace and aviation industries and in the microelectronic and electric fields as advanced packaging and insulating materials. With the rapid development of these advanced industry, novel PI materials with high performances and new functions are usually required.

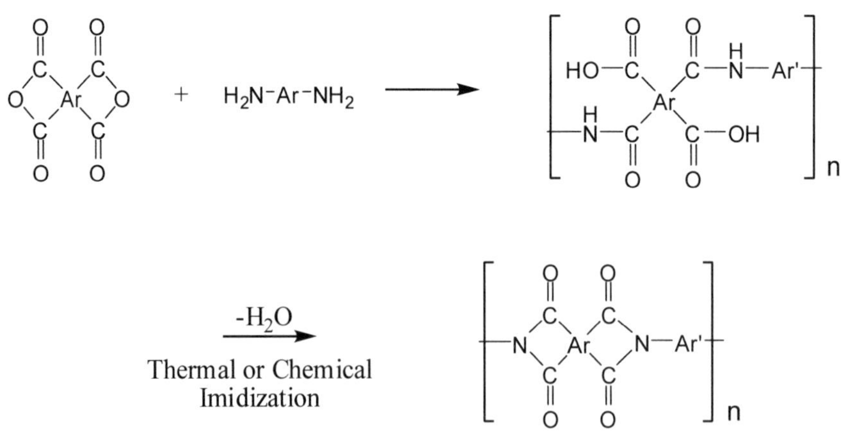

Figure 3. A two-step PI synthetic route

1.3. Electrospun polyimide materials

In 1996, Reneker referred that PI nano fibers are obtainable via electrospinning (Reneker & Chun, 1996). In 2003, Nah et al. described in details the method to produce PI nano fibers (Nah, Han & Lee, 2003), declared the successful preparation of ultra-thin PI fibers. The morphology of the PI fibers is shown in Figure 4. The successful preparation was a milestone that it was when a polymer with comprehensive high performances encountered a versatile and effective method for fiber producing. Their combination afforded a series of results with both scientific amd industrial values. Herein, this chapter summarized and categorized the works of electrospun PI materials since 2003. In the meantime, some outlooks of this research field were also given.

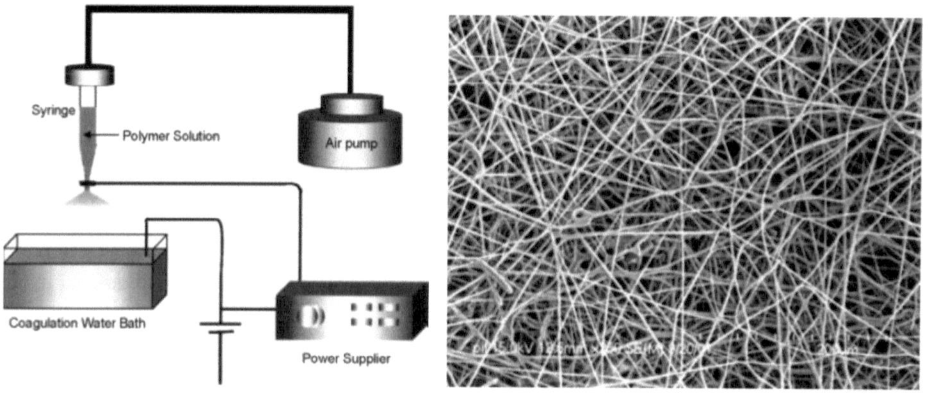

Figure 4. Nah et al. obtained PI fibers by electrospinning for the first time, left: the electrospinning set; right: SEM of PI mat. Copyright © 2003 WILEY-VCH, reused with permission

2. Carbon fiber precursors: the preparations of electrospun nano PI fibers

Carbon fibers produced from traditional fibers possess high tensile strength and modulus of elasticity. People wish to be able to produce thinner carbon fibers from electrospun polymers after the electrospinning technique was born. Due to the ultra-thin diameter of the electrospun fibers, compared to the ordinary art of carbon fibers, the ultra-thin carbon fibers produced from electrospun fibers possess much higher specific area. Such property is in favor of super capacitors or the carrier of the catalyst (Li & Xia, 2004; Dong, 2009). The researchers also found that, interestingly, carbon fibers produced from electrospun PI fibers possess higher conductivity than some ordinary materials using the same method. As a result, ressearchers attempted to get carbonized products with higher performances from the electrospun PI fibers. Some have achieved initial results and they are introduced as follow.

In 2003, Yang et al. first reported the method of carbon fiber preparation by carbonizing the electrospun PI fibers (Yang et al, 2003). They discussed the parameters for the PI electrospinning, pointing out the proper concentrations and viscosity of the solutions, as well as the voltage and electric field. Their job provided precious experience for the follow-up works. In their work, 4,4'-oxydianiline (ODA) and pyromellitic dianhydride (PMDA) underwent a condensational polymerization to form polyamic acid (PAA), the precursor of PI. Certain amount of PAA with certain concentration (by weight) was added into the syringe to be electrospun into fiber samples with diameters ranged from 2 to 3μm. After thermal imidization, the diameters shrank to 1~2μm. After the final carbonizing, carbon fibers were obtained. Yang et al. discovered that the conductivity of the samples increased as the carbonizing temperature rose. The conductivity reached 2.5S/cm after carbonized under 1000°C, significantly higher than the one of which produced from PAN fibers (1.96S/cm) treated under the same conditions and procedures. An SEM photo of the carbonized mat is shown in Figure 5.

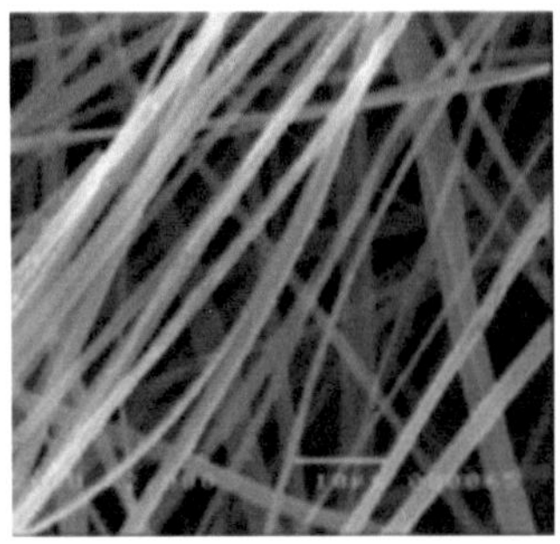

Figure 5. SEM of PI nonwoven mat after carbonization. Copyright © 2003 Elsevier, reused with permission

In 2004, Kim et al. reported the preparations of carbon fibrous electrode with high performance via electrospinning of PI (Kim et al, 2004). The nonwoven mat they prepared was made into electrode after carbonizing at 2200°C, the conductivity was as high as 5.26S/cm, detected by the four-probe method. The specific capacity could reach 175F/g at most. Cyclic Voltammetry and Alternating Current Impedance Spectra both demonstrated its excellent performance. Due to the super thin diameters of each single fiber, a huge amount of micro and nano gaps were introduced into the body of the electrode. As a result, the specific area of the electrode is incredibly high, $1453M^2/g$ (detected by BET test), which is unmatched by ordinary electrode. The high porosity, specific area and conductivity of the ultra-thin carbon fiber produced from electrospun PI, brought about high specific capacity and high reaction reversibility to such kind of electrodes, enabling its popularity in the preparations of high performance electrodes, super capacitor and energy storing.

Chung et al. reported a new way to produce carbon fibers in 2005 (Chung et al, 2005). The soluble PI, Matrimid® 5218, was directly electrospun into nonwoven mat and then carbonized. In the electrospinning process, the authors added a certain amount of $[CH_3COCH{=}C(O^-)CH_3]_3Fe$ as additives. In the following characterizations, XRD, Raman Spectra, SEM, TGA all demonstrated that $[CH_3COCH{=}C(O^-)CH_3]_3Fe$ might promote the carbon yield, enlarge the crystal dimension, increase the thermal stability at the same time.

In 2007, Xuyen et al. for the first time discussed how the electrospinning parameters would affect the PI fibers quantitatively (Xuyen et al., 2007), part of the data is illustrated in Figure 6. Further they discussed the relationship between the diameters of PI fibers and the

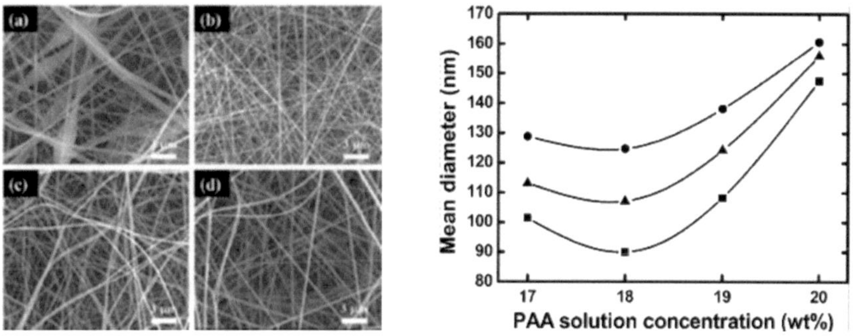

Figure 6. The study of parameters on the PAA electrospun diameters. Left: fiber morphology affected by the amount of TEA while electrospinning, a) 0wt%, b) 1wt%, c) 3wt%, d) 5wt%; right: fiber diameters affected by concentrations of PAA solution. Copyright © 2007, American Chemical Society, reused with permission

conductivity of their carbonized products. The parameters they referred included the amount of catalyst (triethyl aminde, TEA), the electric field and concentrations of the solution (wt%). This research revealed that: in some range, fiber diameter decreased when the amount of TEA increased; the fiber diameter increased while the concentration increased; the critical electric field was quadratic to the molar mass. According to the authors' explanations, firstly, TEA promoted the polymerization of ODA and PMDA, leading to a high viscosity during the synthesis, which stabilized the Taylor Cone, inhibited the variation of the fiber diameter during electrospinning; Secondly, at the same feeding rate, while the concentration increased, the amount of PAA increased in the electrospun sample, leading to the expansion of the fiber diameters; At last, due to the quadratic relationship (Jones &Richards, 1999) between surface tension of the PAA solution and the molecule weight, while the critical electric field was directly linked to the surface tension (Taylor, 1969), as a result, the critical electric field and molecule weight could be described as follows:

$$E_c^{\,2} = 563 \bullet [1 - (\frac{57178}{M_n})^{1.8}]$$

where E_c is the critical electric field and M_n is the number-average molar mass. Experimental data fitted the equation well. During the research, by controlling the diameter of the fibers and the pressure during the carbonization, Xuyen et al. discovered that the conductivity increased when the diameter decreased and the pressure increased. The maximum value they obtained was as high as 16 S/cm, and the fiber diameter was only 80 nm. Compared to the formal reported ones (Yang et al, 2003; Kim et al, 2004), the fiber diameter was sharply reduced while the conductivity was greatly enhanced, demonstrating the great advantages of electrospinning in carbon fiber preparations.

3. PI nanocomposites prepared via electrospinning

During the beginning of electrospinning process, the solution jet bears huge shear force and stretching force, so that the fiber diameter reaches the nano range. Then three factors will limit the agglomerations of the nano particles in electrospun nanocomposites. The three factors are: the constrain of the fiber diameter, mighty electric force and the surface tension of the polymer solution as well. In addition, during the end of the process, due to the fast evaporation of the solvents, the polymer fiber will be solidified rapidly as if the liquid was "frozen". Once "frozen", the reagglomerations of the nano particles will be greatly prohibited. These factors enable electrospinning a versatile and effective method for the dispersion of nano particles in the polymer body (Behler et al., 2009). There have been a great amount of reports on this technique about the dispersions of functional nano particles into polymer fibers, involving the functions of catalyst, thermal conductivity, light adsorption and bio characteristics, etc. Detailed examples are not discussed here.

What's more, PI is a intrinsically tough resin with comprehensive high performance as we introduced above. By electrospinning, we may be able to disperse a variety of functional particles into its body in order to prepare PI materials with high performance and certain functionalities. Such idea is worth of great scientific and industrial values because it might realize the functionalization of PI and in the meantime, expand PI's applications. Several research groups are carrying out such works and a few results are mentioned as follow.

Zhang et al. reported the successful preparation of nano Ag particles/PI fiber composite in 2007 (Zhang et al., 2007). Due to the excellent optical, electrical, catalytic and anti-micro-organism properties, nano Ag particles are widely applied in composite preparations. The Ag/PI composite that Zhang et al. prepared are potentially applicable in optics and catalysts.

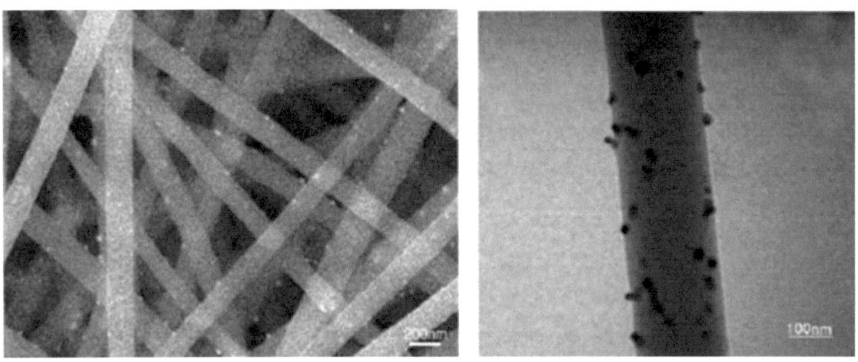

Figure 7. SEM (left) and TEM (right) of nano Ag/PI fiber composites. Copyright © 2007 Elsevier, reused with permission

In 2009, Chen et al. reported the successful preparation of multi-wall carbon nanotubes (MWCNTs)/PI composite via electrospinning (Chen et al., 2009). The aim of the research was to add MWCNTs into PI fiber systems to enhance the mechanial properties of the composite. The key problem is to disperse the MWCNTs homogeneously. However, MWCNTs are not compatable with the PAA solution. To increase the compatibility of the nano filament with the polymer, MWCNTs were treated by high concentrations of nitrate, in order to introduce oxygen-contained groups onto their surfaces. Then after in-situ polymerization, the MWCNTs/PAA was then electrospun into nano fibers. During the ES process, a special set was used to introduce alignment into the fiber belt. After the final thermal imidization, the mechanical properties of the aligned fibrous MWCNTs/PI composite were examined. Compared to the neat PI fiber belt, the mechanical performances of the composite were obviously more advantageous: the yield strength, tensile strength and tensile modulus were 200.9 MPa, 239.7 MPa, 2.56 GPa, respectively. And the elongations at break could reach as high as 90.5%. At the same time, the thermal stability of the composite was greatly enhanced. The TEM demonstrated the well dispersion of the filament in PI matrix and the filament were well aligned along the fiber direction. The alignment of these filament might contributed to the shearing and stretching force of the mighty electric field.

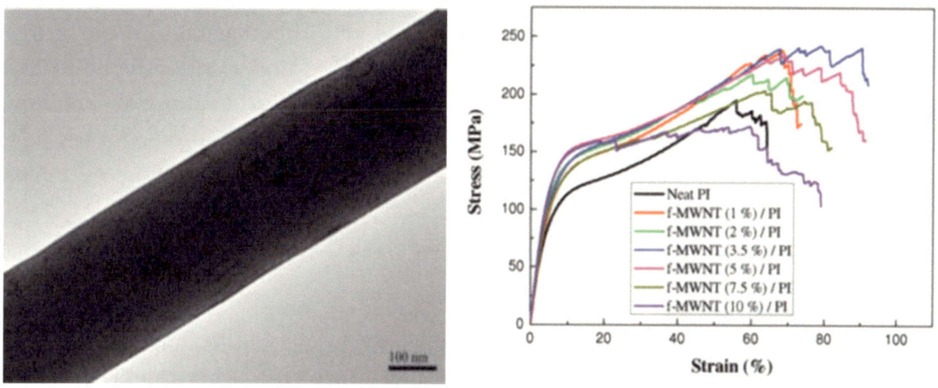

Figure 8. TEM of nano-carbon tubes/PI fiber composites (left) and its stress-strain curve (right). Copyright © 2007, American Chemical Society, reused with permission

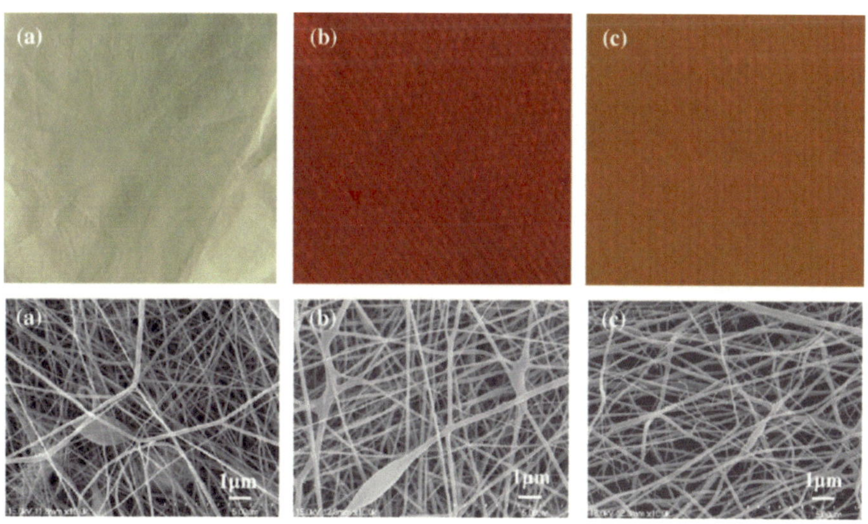

Figure 9. Images of PI-(hemicyanine dye) hybrid. The upper row are digital pictures of a): Pure PAA nonwoven mat, b):PAA-(hemicyanine dye) hybrid nonwoven mat, c): the hybrid mat after imidization; the lower row are the corresponding SEM images. Copyright © 2009 Elsevier, reused with permission

In 2010, the methodology of preparations of PI composite with magnetic effect was reported by Zhu et al (Zhu et al., 2010). In their work, Fe-FeO nano particles with core-shell structures were doped into soluble PI, Matrimid 5218®. The inorganic particles were well compatible with the PI matrix without further treatment. The research found out that the ES has no major negative effects on the saturation magnetization of the particles, while after the ES process, the coercivity was increased from 62.3 Oe (neat particles) to 188.2 Oe (composite), which meant the composite's anti-magnetic ability was improved. Zhu et al. pointed out that such improvement was triggered by the uniform dispersion of the inorganic particles which would lead to the weakening of the dipolar interactions between each and every one.

Moreover, there were other reports on the electrospun PI composite: Qin et al. reported the fluorescent performance of electrospun PI web mixed with hemicyanine dye (Figure 9), pointed out that the nano fiber structure were helpful to the fluorescence (Qin et al., 2009); Cheng et al. reported the preparation of nonwoven PI/silica hybrid fabrics by combining electrospinning and controlled in situ sol–gel techniques, their product had better mechanical and thermal properties than ordinary PI electrospun fabrics (Cheng et al., 2009).

4. Novel light and tough electrospun PI fabrics

Due to PI's excellent mechanical properties, it is widely applied in military industry, engineering, aeronautices and astronautics. The most intriguing point is that electrospinning is able to improve PI's mechanical properties. The molecule chains will be greatly constrained within the nanofibers. As a result, the crystallization area is increased, leading to the enhancement of the nanofiber's mechanical property. Now that nano PI fibers are achievable via electrospinning and such fibrous PI materials with lighter weight and excellent performances are intriguing, we consider this research of great importance and several scientific outcome are shown below.

In 2006, Huang et al. prepared PI nonwoven mat with considerable mechanical properties (Huang et al., 2006). 3,3',4,4'- biphenyl dianhydride (BPDA) and P-phenylenediamine (PDA) were polymerized into PAA with high molar mass. After ES and imidization, the mat's tensile strength and Young's modulus were detected as 210 MPa and 2.5 GPa. Compared to the formerly reported mats (Cheng et al., 2009; Kim et al., 2004; Yang et al., 2003), this product exhibited higher mechanical performances. In this work, they mentioned the high molar mass guaranteed the high performance. In the same year, they reported a new way (Huang et al., 2006) to fabricate PI mat with higher mechanical properties. In that work, a rotating wheel with a 8 mm wide edge was utilized to collect the electrospun PI fibers. The fibers in the as-prepared PI fiber belt were highly aligned and along the align direction, the tensile strength and Young's modulus reached as high as 664 MPa and 15.3 GPa, respectively.

In 2008, for the first time, the mechanical property of a single BPDA-PDA nano fiber was reported (Chen et al., 2008). The authors used a square frame to collect a few strands of PI fibers by waving across through the space between the spineret and collector rapidly several times. Then special care were taken to mount one single fiber into the micro tensile testing machine. In the described details, the mechanical properties of both PAA and PI single fibers with diameters aroung 300nm were characterized: the data of PAA fiber was 766±41MPa in tensile strength, 13±0.4GPa in tensile modulus, ~43% in elongation at break; the data for the PI fiber was 1.7±0.12GPa in tensile strength, 76±12GPa in tensile modulus, ~3% in elongation at break. According to the authors, the excellent mechanical performance stemmed from the high alignment of the macromolecules in the fiber structure, which was confirmed by Wide-angle XRD.

In this series of works, the PI matrix was synthesized from BPDA and PDA. In such kind of polymer chains, there are few flexible groups, which will make the material hard and brittle. This explains why those PI fiber belt possess low elongation at break and unsatisfactory toughness and flexibility. To improve the toughness and flexibility of the electrospun PI materials, in 2008, Chen et al. added 4,4'-oxydianiline (ODA) into the former system (BPDA+PDA), to form block copolymers (Chen et al., 2008). Similar to the methods that have been discussed before (Huang et al., 2006), the copolymers were also electrospun into aligned fiber belt but with ameliorated elongation properties. Due to the introduction of flexible groups, the elongation at break of the copolymer could reach ~20%, and the tensile strength and modulus were 1103±61MPa, 6.2±0.7GPa. Such a PI belt with 7.5 mm wide and 1.5 μm thick could bear a 10 kg load, illustrated in Figure 10.

In 2009, they reported a PI fibrous material with even higher stretching ability (Cheng et al., 2010): the elongation at break could reach as high as 200%. This PI was synthesized from BPDA and 2-Bis[4-(4-Aminophenoxy)Phenyl] Hexafluoropropane (6FBAPP). 6FBAPP contains more flexible "-O-" groups, so such PI materials are more pliable but not strong enough. The BPDA-6FBAPP exhibited 308±14 MPa in tensile strength and 2.08 ±0.25 GPa in tensile modulus. From the results of the series of researches, in summary, the molecule

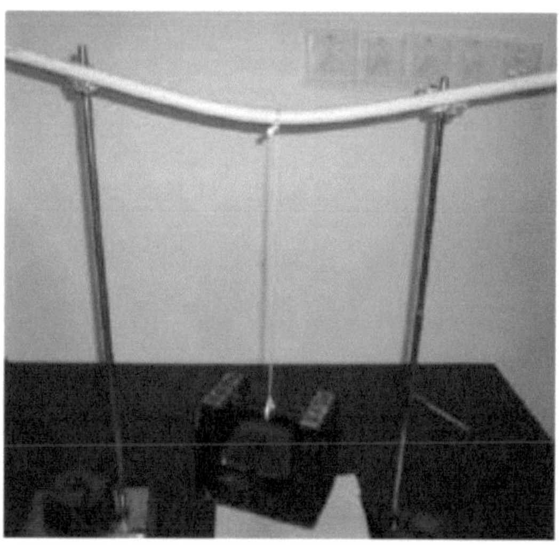

Figure 10. A photograph of the electrospun copolyimide nanofiber belt hanging a weight of 10 kg; the belt has 7.5 mm width and 12 μm thickness. Copyright © 2008 IOP Publishing Ltd, reused with permission

structures greatly affect the macroscopic property of electrospun PI materials. The molecule structures are summerized in Table 1. In addition, the molar mass and the aligment of fibers are also vital keys to such kind of material.

Dianhydride	Diamine	Notes
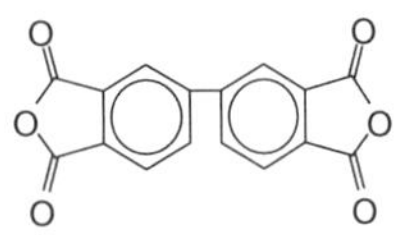 BPDA	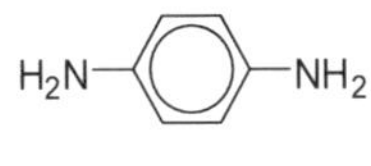 PDA	Diamine used in [52–54] the material is rigid due to lack of flexible groups in molecule backbones
	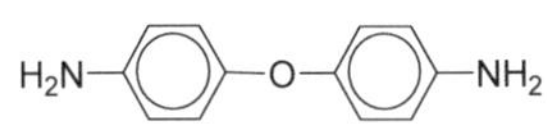 **ODA**	Diamine used in [55] The as-prepared PI is tough and flexible, owing to the existence of flexible groups in the backbone
	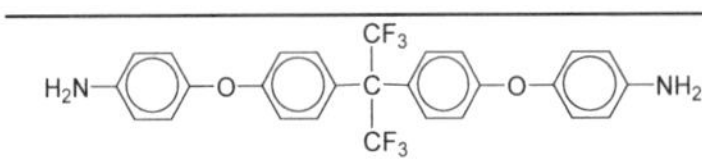 **6FBAPP**	Diamine used in [56] Flexibility of the PI materials is improved due to additional flexible groups in the backbone

Table 1. A list of molecule structure of aromatic anhydrides and amines referred in Section 4

The PI fabrics discussed above were only aligned in one certain direction, in other words, the mechanical property could only be presented anisotropically. So far, constrained by the development of patterning technology of electrospinning, there have been no reports on multi-directional aligned PI fabrics with isotropic high mechanical performances. However, in 2008, Carnell et al. reported a novel way to fabricate perpendicularly aligned PI fabrics (Carnell et al., 2008). They applied a needle-like auxillary electrode opposite to the spinneret, and used a metal roller to collect the fibers. The set for electrospinning the aligned fibers is shown in Figure 11. In their work, when negative high voltage was applied on the auxiliary electrode, because of the guide and confinement of the concentrated electric field lines, the instability of the spin jet was greatly suppressed, leading to fine alignment in one certain direction. What's more, by re-place the collector, the aligned fibers can be vertically

Figure 11. The set for electrospinning the aligned fibers. Copyright © 2008, American Chemical Society, reused with permission

staggered (shown in Figure 12), to form fabrics with double-directional alignment. Such technique could improve the continuity and yield of the aligned fibers, and especially it expanded the dimensions for the alignment of electrospun nano fibers.

In 2010, Lee et al. reported a method to optimize the fiber morphology and mechanical property (Lee et al., 2010). During the ES process of co-Poly-(amide)-imide (PAI), the fibers were soaked in glycerine. And during the imidization, the fibers were stretched. After these procedures, the PAI fibers presented smoother sectional shapes, better alignment and as a result, as we have discussed above, the mechanical property should be improved but not discussed in that article. In fact, the stretching imidization is a critical step for enhancing the mechanical performance of PI material. So the stretching imidization is also expected to be one effective way to improve the characteristics of the electrospun PI fibers.

Figure 12. Horizontally and vertically aligned PI nonwoven mat prepared through the way that reported by *Carnell et al.* Copyright © 2008, American Chemical Society, reused with permission

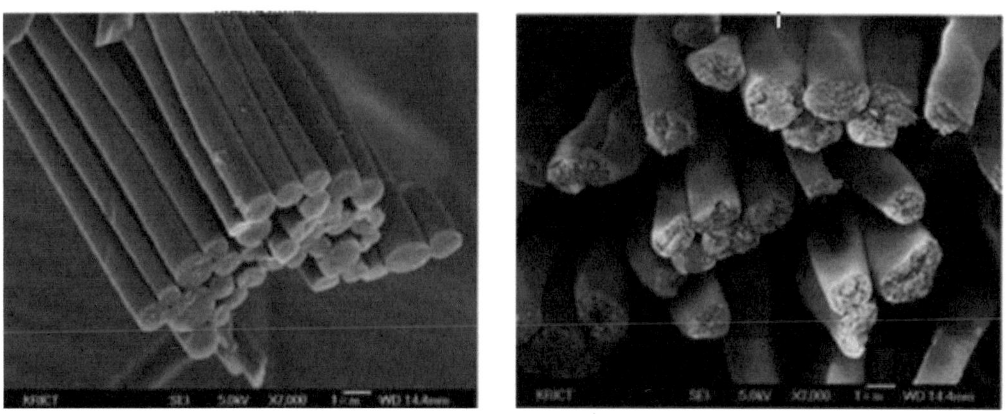

Figure 13. SEM photos of the cross-sectional surfaces of continuous copolymer nanofiber bundles: imidized under tension after glycerol pre-treatment. Copyright © 2010 WILEY-VCH, reused with permission

5. Others

In recent years, reports on other functional electrospun PI fibers thrived. Herein we have chosen some typical works to discuss. In 2007, Liu et al. reported a new way to produce PI material with ultralow dielectric constant (*k*) via electrospinning (Liu et al., 2007). The traditional *k* of a piece of PI film, e.g. Kapton®, is about 3.5. However, the electropsun PI mat's *k* could reach as low as 1.5 according to Liu et al. The theory could be explained as this: during the process of ES, due to the nano size of the fiber diameters, a great amount of mesoscopic gaps would be introduced into the body of the PI mat. As the fiber diameter decreased and/or the aspect ratio increased, the amount of gaps increased, making the inner space of the mat more and more similar to pure air, leading to the decrease of *k* of the PI mat.

In 2008, Lv et al. reported that they used two kinds of diamines: diaminotetraphenylporphyrin and ODA, copolymerized with PMDA to form porphyrin modified PAA (Lv et al., 2008). Then it was electrospun and imidized into fluorescent nonwoven mat. The characterizations, including UV-vis spectra, fluorescence spectra, ^{1}H NMR spectra, TGA, demonstrated that the concentration of porphyrin, the nanostructures were the main factors which would influence the quantum yield and the fluorescent property. The higher the concentration is and the thinner the fiber diameters are, the higher the quantum yield is and the stronger the fluorescence is. In addition, the existence of heavy metal ions, e.g. Hg^{2+}, will lead to fluorescence quenching which can be detected by naked eye. So that this PI mat can be used as the sensor of heavy metal ions.

In 2009, Chang et al. utilized the technique of "click chemistry" to successfully link alkynyl-terminated polymethyl methacrylate (PMMA) onto the surfaces of the electrospun PI nanofibers (Chang et al., 2009). The route was illustrated in Figure 14. The extremely high specific areas of the nanofibers were the key to the success and Chang's work was the first attempt to graft other polymers onto the PI surface, provided a novel idea to the surface modification of PI.

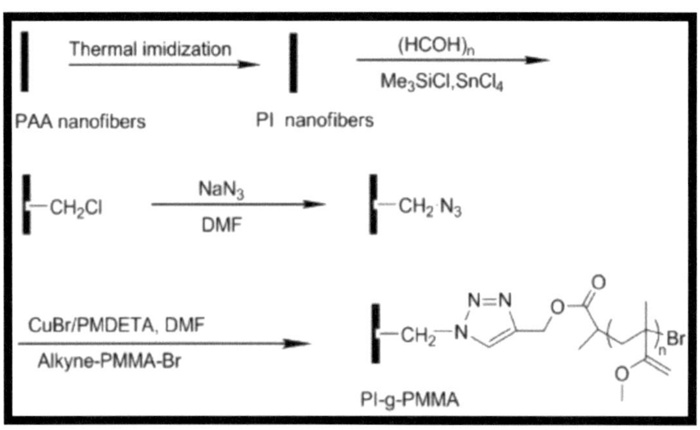

Figure 14. Synthetic route for the "Click Chemistry" for preparing the grafted PI fibers. Copyright © 2009, American Chemical Society, reused with permission

6. Conclusion

In the past few years, researchers pave their way through the field of the preparations of electrospun PI materials. The combination of the facility and effectiveness of electrospinning and the high performance of PI give birth to a series of novel PI materials with hierarchical constructions and multi-functionalization. In the coming researches of PI materials, the electrospinning way is worthy of noting and will attract more and more research interest. And we hope that novel PI materials produced by electrospinning can be used in the field of civil engineering, electrics and aero&astronautics in the near future.

Author details

Guangming Gong and Juntao Wu*
Key Laboratory of Bio-Inspired Smart Interfacial Science and Technology of Ministry of Education,School of Chemistry and Environment, Beihang University, Beijing 100191, PR China

Acknowledgement

This work is financially supported by the National Natural Science Foundation of China (No. 51003004), the National Research Fund for Fundamental Key Projects (No. 2010CB934700, 2012CB933200), the Scientific Research Foundation for the Returned Overseas Chinese Scholars, State Education Ministry and the Fundamental Research Funds for the Central Universities.

7. References

Behler, K.; Stravato, A.; Mochalin, V.; Korneva, G.; Yushin, G. & Gogotsi, Y. (2009). Nanodiamond-Polymer Composite Fibers and Coatings. *ACS Nano*, Vol. 3 No. 2, (Feb., 2009), pp. 363-369, ISSN: 1936-0851

Bognitzki M, Czado W, Frese T, Schaper A, Hellwig M, Steinhart M, Greiner A, and Wendorff J H, (2001). Nanostructured Fibers via Electrospinning, *Adv. Mater.*, Vol. 13, No. 1, (Jan, 2001), pp. 70-72, ISSN 1521-4095

Carnell, L. S.; Siochi, E. J.; Holloway, N. M.; Stephens, R. M.; Rhim, C.; Niklason, L. E. & Clark, R. L. (2008). Aligned Mats from Electrospun Single Fibers. *Macromolecules*, Vol. *41* No. 14, (June,2008), pp. 5345–5349, ISSN: 0024-9297

Chang Z, (2011). "Firecracker-shaped" ZnO/polyimide hybrid nanofibers via electrospinning and hydrothermal process. *Chem. Commun.*, Vol. 47, (Feb, 2011), pp. 4427–4429, ISSN: 1364-548X

Chang, Z.; Xu, Y.; Zhao, X.; Zhang, Q. & Chen, D. (2009). Grafting Poly(methyl methacrylate) onto Polyimide Nanofibers via "Click" Reaction. Vol. 1, No. 12, (Dec. 2009), pp. 2804–2811, ISSN: 1944-8252

* Corresponding Author

Chen H, Song Y, Zhao Y & Jiang L, (2008). One-Step Multicomponent Encapsulation by Compound-Fluidic Electrospray. *J. Am. Chem. Soc.*, Vol. 130, No. 25, (May, 2008), pp. 7800-7801, ISSN: 0002-7863

Chen, D.; Liu, T.; Zhou X.; Tjiu, W. C. & Hou, H. (2009). Electrospinning Fabrication of High Strength and Toughness Polyimide Nanofiber Membranes Containing Multiwalled Carbon Nanotubes. *J. Phys. Chem. B*, Vol. 113, No. 29, (July, 2009) pp. 9741–9748, ISSN: 1520-6106

Chen, F.; Peng, X.; Li, T.; Chen, S.; Wu, X.; Reneker, D. H. & Hou, H. (2008). Mechanical characterization of single high-strength electrospun polyimide nanofibres. *J. Phys. D: Appl. Phys.*, Vol. 41, (Jan. 2008), pp. 025308, ISSN: 0022-3727

Chen, S.; Hu, P.; Greiner, A.; Cheng C.; Cheng, H.; Chen, F. & Hou, H. (2008). Electrospun nanofiber belts made from high performance copolyimide. *Nanotechnology*, Vol. 19, (Nov. 2007), pp. 015604, ISSN: 0957-4484

Cheng, C.; Chen, J.; Chen, F.; Hu, P.; Wu, X.; Reneker, D. H. & Hou, H. (2009). High-Strength and High-Toughness Polyimide Nanofibers: Synthesis and Characterization. *J. Appl. Polym. Sci.*,Vol. 116, No. 3, (Jan. 2010), pp. 1581–1586, ISSN: 0021-8995

Cheng, S.; Shen, D.; Zhu, X.; Tian, X.; Zhou, D. & Fan L. (2009). Preparation of nonwoven polyimide/silica hybrid nanofiberous fabricsby combining electrospinning and controlled in situ so-gel techniques. *European Polymer Journal*, Vol. 45, (July, 2009), pp. 2767-2778, ISSN: 0014-3057

Chung, G. S.; Jo, S. M. & Kim, B. C., Properties of Carbon Nanofibers Prepared from Electrospun Polyimide, *J. Appl. Polym. Sci.*, Vol. 97, No. 1, (Apr., 2005), pp. 165-170, ISSN: 0887-624X

Ding, M., (2011). *Polyimide: monomers, polymerization and materials*, Science Press, ISBN: 978-7-03-031080-4, Beijing, China

Dong, J. H., (2009). Frontier and Progress in Polymer Science (2nd Ed.), Science Press., ISBN: 978-7-03-023197-0, Beijing, China

Formalas A, US patent, 1975504, 1934

Gong G, Wu J. Unpublished work

Greiner A & Wendorff J H. (2007). Electrospinning: A Fascinating Method for the Preparation of Ultrathin Fibers. *Angew. Chem. Int. Ed.*, (June, 2007). Vol. 46, pp. 5670-5703, ISSN 1521-3773

Hohman M M, Shin M & Rutledge G C, (2001). Electrospinning and electrically forced jets. II. Applications, *Phys. Fluidics.*, (May, 2001). Vol. 13, No. 8, pp. 2201-2217, ISSN 1070-6631

Hou H, Reneker D H, (2004). Carbon Nanotubes on Carbon Nanofibers: A Novel Structure Based on Electrospun Polymer Nanofibers. *Adv. Mater.*,Vol. 16, No. 1, (Jan, 2004), pp. 69-73, ISSN 1521-4095

Huang, C.; Chen, S.; Reneker, D. H.; Lai, C. & Hou, H. (2006). High-Strength Mats from Electrospun Poly(*p*-Phenylene Biphenyltetracarboximide) Nanofibers. *Adv. Mater.*, Vol. 18, No. 5, (Mar. 2006), pp. 668-671, ISSN: 1521-4095

Huang, C.; Wang, S.; Zhang, H.; Li, T.; Chen, S.; Lai, C. & Hou, H. (2006). High strength electrospun polymer nanofibers made from BPDA-PDA polyimide. *European Polymer Journal*, Vol. 42, (Dec. 2005), pp. 1099-1104, ISSN: 0014-3057

Jones, R. L. & Richards, R. W. (1999). Polymers at Surface and Interfaces, Cambridge University Press, ISBN-13: 978-0521479653, Cambridge

Kim, C.; Choi Y.; Lee, W. & Yang, K., (2004), Supercapacitor performances of activated carbon fiber webs prepared by electrospinning of PMDA-ODA poly(amic acid) solutions. *Electrochimica Acta.* Vol. 50, (Aug., 2004), pp. 883–887, ISSN: 0013-4686

Lee, S. H.; Kim, S. Y.; Youn, J. R.; Seong D. G.; Jee, S. Y.; Choi, J. I. & Lee, J. R. (2010). Processing of continuous poly(amide-imide) nanofibers by electrospinning. *Polym. Int.*, Vol., 59, No. 2, (Feb. 2010) pp. 212–217, ISSN: 1097-0126

Li D & Xia Y. (2004). Elctrospinning of Nanofibers: Reinventing the Wheel?. *Adv. Mater.*, Vol. 16, No. 14, (July, 2004), pp.1151-1170, ISSN 1521-4095

Li D, Wang Y, Xia Y, (2004). Direct Fabrication of Composite and Ceramic Hollow Nanofibers by Electrospinning, *Nano Lett.*, Vol. 4, No. 5, (Mar., 2004), pp. 933-938, ISSN: 1530-6984

Liu, J.; Min, Y.; Chen, J.; Zhou, H. & Wang, C. (2007). Preparation of the Ultra-Low Dielectric Constant Polyimide Fiber Membranes Enabled by Electrospinning. *Macromol. Rapid Commun.*, Vol. 28, No. 2, (Jan., 2007), pp. 215–219, ISSN: 1521-3927

Lu X, Wang C & Wei Y, (2009). One-Dimensional Composite Nanomaterials: Synthesis by Electrospinning and Their Applications, *Small*, Vol. 5, No. 21, (Nov, 2009), pp. 2349-2370, ISSN: 1613-6810

Lv, Y.; Wu, J.; Wan, L. & Xu, Z. (2008). Novel Porphyrinated Polyimide Nanofibers by Electrospinning. *J. Phys. Chem. C.* Vol. 112, No. 29, (Jun., 2008), pp. 10609-10615, ISSN: 0022-3654

Nah, C., Han, S. H., Lee, M. H., Kim, J. S. & Lee, D. S., (2003). Characteristics of polyimide ultrafine fibers prepared through electrospinning. *Polym Int.*, Vol. 52, No. 3, (Feb, 2003), pp. 429–432, ISSN: 0959-8103

Qin, C.; Wang, J.; Cheng, S.; Wang, X.; Dai, L. & Chen G. (2009). Fluorescent performance of electrospun polyimide web mixed with hemicyanine dye. *Materials Letters*. Vol. 63, No. 15, (June, 2009), pp. 1239-1241, ISSN: 0167-577X

Reneker D H, Yarin A L, Fong H & Koombhhongse S, (2000). Bending instability of electrically charged liquid jets of polymer solutions in electrospinning. *J. Appl. Phys.*, (Jan, 2000). Vol. 87, No. 9, pp. 4531-4548, ISSN 0021-8979

Reneker, D H & Chun, I., (1996). Nanometre diameter fibres of polymer, produced by electrospinning, *Nanotechnology*, Vol. 7, No. 3, (n.d.), ISSN: 0957-4484

Taylor, G. I. (1969). Electrically Driven Jets. *Proc. Roy. Soc. A*, ISBN: 1471-2946, London, Dec., 1969

Wikipedia Online. (n.d.). Electrospinning, Available from: http://en.wikipedia.org/wiki/Electrospinning#History

Xuyen, N. T.; Ra, E. J.; Geng, H.; Kim, K. K.; An, K. K. & Lee, Y. H. (2007). Enhancement of Conductivity by Diameter Control of Polyimide-Based Electrospun Carbon Nanofibers. *J. Phys. Chem. B*, Vol. 111, No. 39, (Aug., 2007), pp. 11350-11353, ISSN: 1520-6106

Yang, K. S.; Edie D. D.; Lim, D. Y.; Kim, Y.M. & Choi Y.O., (2003), Preparation of carbon fiber web from electrostatic spinning of PMDA-ODA poly(amic acid) solution, *Carbon*, Vol. 41, No. 11, (Apr., 2003), pp. 2039–2046, ISSN: 0008-6223

Yarin A L, Koombhhongse S & Reneker D H, (2001). Taylor cone and jetting from liquid droplets in electrospinning of nanofibers, *J. Appl. Phys.*, (Aug, 2001). Vol. 90, No. 9, pp. 4836-4847, ISSN 0021-8979

Zhang, Q.; Wu D.; Qi S.; Wu Z.; Yang X. & Jin R. (2007). Preparation of ultra-fine polyimide fibers containing silver nanoparticles via in situ technique. *Materials Letters*. Vol. 61 No. 19-20, (Jan. 2007), pp. 4027-4030, ISSN: 0167-577X

Zhao Y, Cao X & Jiang L, (2007). Bio-mimic Multichannel Microtubes by a Facile Method. *J. Am. Chem. Soc.*, Vol. 129, No. 4, (Jan, 2007), pp. 764-765, ISSN: 0002-7863

Zhu, J.; Wei, S.; Chen, X.; Karki, A. B.; Rutman, D.; Young, D. P. & Guo, Z. (2010). Electrospun Polyimide Nanocomposite Fibers Reinforced with Core-Shell Fe-FeO Nanoparticles. *J. Phys. Chem. B*, Vol. 114, No. 19, (Apr., 2010), pp. 8844–8850, ISSN: 1520-6106

Chapter 8

Polyimide-Coated Fiber Bragg Grating Sensors for Humidity Measurements

Lutang Wang, Nian Fang and Zhaoming Huang

Additional information is available at the end of the chapter

http://dx.doi.org/10.5772/53551

1. Introduction

Humidity is one of important environmental parameters, which refers to the water vapor content in the air. As one of the fundamental abiotic factors, humidity determines an environment where animals and plants can thrive, also influences the human life. There are three items commonly used to describe humidity, which are absolute humidity (AH), relative humidity (RH) and specific humidity (SH) [33]. In daily life and industrial measurement fields, relative humidity is one of the important and most commonly used parameters.

With the developments of sciences and technologies, in the recent years, humidity measurements and controls are becoming more and more important and required greatly in a wide range of areas, such as in environmental monitoring for meteorological services, in agriculture for seed storage, in industrial fields for chemical, biomedical, food, and electronic processing, and in civil engineering for bridge and building constructions, as well as in daily life for air conditioner in living rooms, hospitals, museums and libraries [24, 33]. Besides the humidity measurements, moisture measurements also are very important, such as in high voltage engineering for measurements of water content in transformer oils. At present, there are various types of humidity sensors available for humidity or moisture measurements, which can be categorized into capacity, resistive, gravimetric and mechanical types based on the sensing principles used [5, 11, 30, 34]. Most of the humidity sensors need a layer of humidity sensitive material, such as polymers or hydrogels. It is clear that no one humidity sensing technology can cover for all applications, however the sensors with a wide detection range, linear response, small hysteresis and fast exchange with water vapor will offer the best potential for use in a broad range of applications.

Compared to the conventional mechanical and electrical/electronic humidity sensors, fiber-optic based humidity sensors [6, 14], including silica glass fiber and polymer optical fiber (POF) based sensors [28, 46], demonstrate many unique advantages, such as small size and low weight, immunity to electromagnetic interference, corrosion resistance, potentials for remote operation and distributed sensing. In fiber-optic sensor family, the fiber grating based sensors are playing the important role in wide range of industrial measurement fields [12, 15]. The fiber grating sensors with many intrinsic characteristics as mentioned above can further offer the potentials of multiple-parameter sensing, embedding into other structures [37], and multiplexing in a single mode fiber to form an all-fiber sensor network [26]. In the past several decades, a lot of types of fiber grating sensors have been proposed and developed for measurements of strain, temperature and pressure as well as acceleration/vibration in structural health monitoring [4, 8, 24, 38]. These types of sensors can be roughly classified into Bragg grating, chirped grating and long period grating (LPG) as well as in-line grating based interferometer types. In the family of fiber grating based sensors, however, the uniform fiber Bragg grating (FBG) based sensors have been accepted widely, owing to their simplicity in structure as well as in fabrication.

Fiber-optic based humidity sensors can be realized by means of FBG technology and the polymer coating process. Polyimide is a class of thermally stable polymers that are often based on stiff aromatic backbones, which demonstrates many excellent properties–mechanical and tensile strength, heat resistance and adhesiveness to various substrates as well as unique hygroscopic property. When the polyimide resin is coated on the FBG sensor to form the moisture sensitive layer, this polyimide coating will undergo a volume expansion as the water molecules migrate into it, which directly induces a strain imposed on the FBG and in turn results in a proportional shift of the Bragg wavelength of the FBG sensor. Based on this sensing mechanism, the humidity measurements are carried out by directly reading this wavelength shift through spectrum analysis methods [9, 16, 31, 36, 42–45].

One important topic in sensor applications is how to extend the sensor system ability to achieve the simultaneous detections of multiple physical parameters within a single measurement process [1, 23, 26, 27, 29, 37]. In many applications, for example, in the pump system, the vibration and temperature are two important parameters that often are used for evaluating working states of the pump system, especially of bearings, shafts and gear box subsystems. In the high-voltage transformer case, the vibration, temperature and moisture in the insulation oil are three important parameters that should be detected simultaneously and analyzed real time. A transformer in a good state will produce low levels of inherent vibration. Signatures in vibration signals with respect of intensity and frequency reflect well the working state of the transformer and indicate if there is part loosing occurring. The frequency response of a sensor system for transformer condition monitoring, often is required to be capable of detecting vibration signal in a wider frequency range, such as ranging from 5 Hz to 1000 Hz. A rapidly increasing in the temperature implies some short-turned events happening inside the transformer's coils. The moisture in the insulation oil inside the transformer easily induces the partial discharge occurring, which also should be real-time monitored. The fiber laser sensor system equipped with multiple polyimide-coated FBG sensors working in a multi-wavelength lasing mode can provide this diversity to integrate several sensing functions in a sensor system configuration.

The aim of this chapter is to give an introduction to the polyimide as an excellent moisture sensitive coating material used to form the fiber-optic humidity sensors. The remainder of this chapter will be arranged as follows: a brief review on the structure, optical properties and sensing principles of polyimide-coated FBG humidity sensor as well as preliminary experimental results for investigating the sensor properties are given in Section 2. In Section 3, we will demonstrate a prototype of semiconductor optical amplifier (SOA) based fiber laser sensor system used for multiple physical parameter measurements and present some experimental results on the simultaneous measurements of vibration, temperature and humidity, based on our multi-function FBG sensor and interrogation techniques developed recently. Finally a conclusion for summarizing our work is given to close this chapter.

2. FBG humidity sensors

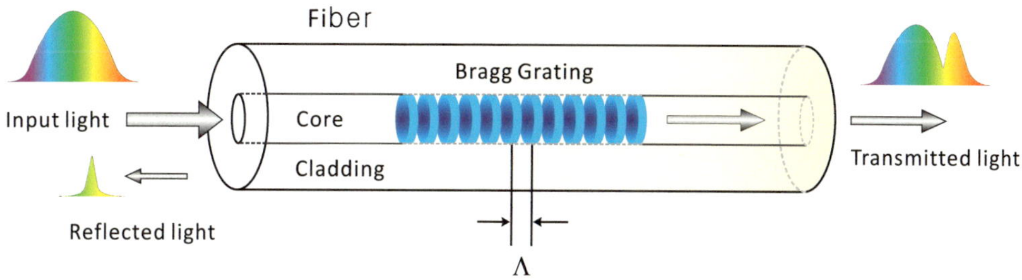

Figure 1. Structure and principle of a FBG

2.1. Theory

A uniform FBG sensor contains a periodically varied refractive index, distributed along the core of the optical fiber within a selected length, as illustrated in Figure 1. When a photosensitive optical fiber is irradiated by an ultraviolet (UV) laser beam with a wavelength band within 230~250 nm, through a phase mask or by holographic recording process, a periodic refractive index variation along the fiber core is permanently recorded and a fiber Bragg grating is produced in this way. As an important optical property, the FBG shows the wavelength selectivity in its reflection spectrum or transmission spectrum. When a broad-band light beam propagating along the optical fiber will interact with each grating plane where the only the part of light beam with a specific wavelength met with the Bragg condition will be reflected and propagate in opposite direction, and other parts of light beam will pass through this grating without an obvious optical loss. According the mode coupling theory, this Bragg condition can be expressed as

$$\lambda_B = 2n_{eff}\Lambda \tag{1}$$

where λ_B indicates the Bragg wavelength or called the center wavelength of the Bragg grating; n_{eff} denotes the effective refractive index of fiber core; Λ is the pitch length of grating plane or called the grating period. From this equation, it is clear that any external perturbation will alternate λ_B through modifying n_{eff} and/or Λ.

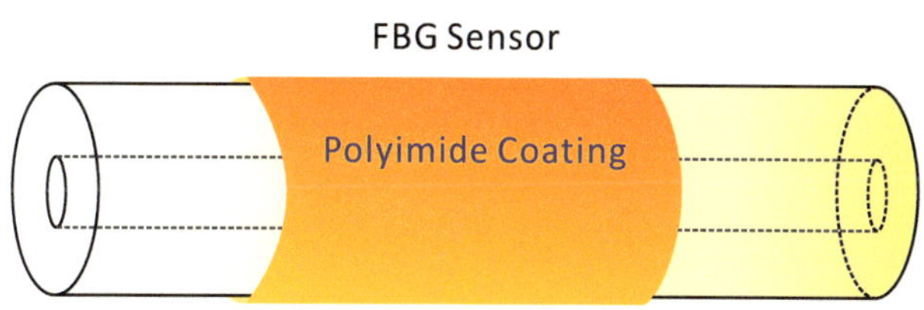

Figure 2. Illustration of a FBG sensor with polyimide coating

The FBG humidity sensor is fabricated by coating suitable hygroscopic film on the surface of fiber cladding covering the Bragg grating section, as illustrated in Figure 2. The polyimide resin can be selected as ideal coating material for making FBG humidity sensor, because of its linear response in volume swelling with respect to the humidity level. After a grating is inscribed in the fiber, the fiber should be taken an annealing treatment first, and then dipped into the polyimide solution for 5∼10 minutes and dried in a drying cabinet for a short thermal treatment at 150 °C, in order to get a uniform polyimide film on the surface of the fiber. This process will be repeated several times in order to obtain the desired film thickness. As a final step, the coated sensor is put into an oven for curing at 180 °C for about 60 minutes, which will stabilize the adhesion of the coating to the fiber. The thickness of the polyimide film determines the sensing sensitivity and the response time. In addition, the experimental results indicated that a multiple layer coating can significantly improve the sensor sensitivity to moisture. However the stable time after the sensor is placed into a humidity environment will become longer than that of the sensor with single polyimide layer. The fabricated FBG humidity sensor should be well packaged with an appropriately designed casing before it can be practically used in the field for humidity measurements.

As indicated by other researchers, the Bragg wavelength λ_B is sensitive to ambient temperature and strain imposed on the fiber, so this property can be utilized for sensing applications. When an axial strain ε is imposed on the FBG, the grating period Λ will alter due to a change of fiber physical length, and the effective refractive index of the fiber core n_{eff} also will alter through the photoelastic effects in the fiber core. The shift of Bragg wavelength induced by ε, therefore, can be obtained with a total differential operation of Equation 1, given as

$$\Delta\lambda_{B,S} = \lambda_B \left(\frac{\Delta\Lambda_S}{\Lambda} + \frac{\Delta n_{eff,S}}{n_{eff}} \right) = \lambda_B (1 - P_e)\varepsilon \tag{2}$$

where $\Delta\Lambda_S/\Lambda = \varepsilon$, $\Delta n_{eff,S}/n_{eff} = -P_e\varepsilon$ denotes the change of effective refractive index induced by through the photoelastic effects, and P_e is a photoelastic constant of the fiber [25], expressed as

$$P_e = \frac{n_{eff}^2}{2} \left[P_{12} - \mu(P_{11} + P_{12}) \right] \tag{3}$$

where μ is the Poisson ratio; P_{11} and P_{12} are two elasto-optic tensor coefficients (Pockel's coefficients) along the fiber axis.

For a temperature change ΔT, the grating period Λ will alter with ΔT through a thermal expansion effect in the fiber, and the effective refractive index of the fiber core n_{eff} also will alter through a thermo-optic effect. Therefore, in same way, the shift of Bragg wavelength induced by the temperature change ΔT is given as

$$\Delta\lambda_{B,T} = \lambda_B \left(\frac{\Delta\Lambda_T}{\Lambda} + \frac{\Delta n_{eff,T}}{n_{eff}} \right) = \lambda_B (\alpha_T + \zeta)\Delta T \tag{4}$$

where α_T is the thermal expansion coefficient of the fiber, expressed as

$$\alpha_T = \frac{1}{\Lambda}\frac{\partial\Lambda}{\partial T} \tag{5}$$

and ζ denotes the thermo-optic coefficient

$$\zeta = \frac{1}{n_{eff}}\frac{\partial n_{eff}}{\partial T} \tag{6}$$

With Equation 2 and Equation 4, the shift of Bragg wavelength of the FBG sensor under the external perturbations from strain and temperature changes, expressed as

$$\Delta\lambda_B = \Delta\lambda_{B,S} + \Delta\lambda_{B,T} = \lambda_B\left[(1-P_e)\varepsilon + (\alpha_T+\zeta)\Delta T\right] \tag{7}$$

For the polyimide-coated FBG humidity sensor, when ambient temperature and the humidity level change, the Bragg wavelength will shift through strains induced by thermal longitudinal expansion and hygroscopic longitudinal expansion of the polyimide film. The strain imposed on the FBG due to thermal expansion of the polyimide film therefore can be expressed as

$$\varepsilon_T = (\alpha_{RH} - \alpha_T)\Delta T \tag{8}$$

where α_{RH} is the thermal expansion coefficient of hygroscopic material (polyimide film). When the polyimide film absorbs or desorbs the moisture, its volume changes proportionally to the moisture quantity absorbed within a unsaturation range. This volume change causes directly an axial strain on the fiber, called humidity-induced strain [2], which is defined as

$$\varepsilon_M = \int_{RH_1}^{RH_2} \beta(RH,T)dRH \tag{9}$$

were β is humidity-induced hygroscopic longitudinal expansion coefficient of polyimide film; T is temperature; RH is relative humidity. The saturation absorption of polyimide film decides the maximum change range of humidity-induced strain. If in the interest humidity measurement range, ε_M only is a linear function of relative humidity in a given temperature range, which can be approximately expressed as

$$\varepsilon_M = \bar{\beta}\Delta RH \tag{10}$$

where $\bar{\beta}$ is an average moisture expansion coefficient, and $\Delta RH = RH_2 - RH_1$ is the humidity difference. $\bar{\beta}$ can be determined through experimental measurements.

Finally, the total strains imposed on the FBG humidity sensor can obtained according to a superposition principle, expressed as

$$\varepsilon = \varepsilon_T + \varepsilon_M \tag{11}$$

or

$$\varepsilon = (\alpha_{RH} - \alpha_T)\Delta T + \bar{\beta}\Delta RH \tag{12}$$

Combining Equation 7 and Equation 12, the total shifts of Bragg wavelength of the polyimide-coated FBG humidity sensor, under a conjunct influence from relative humidity and temperature, can be obtained accordingly, expressed as

$$\Delta\lambda_B = \lambda_B\{(1-P_e)[(\alpha_{RH}-\alpha_T)\Delta T + \bar{\beta}\Delta RH] + (\alpha_T+\zeta)\Delta T\} \tag{13}$$

or

$$\begin{aligned}\Delta\lambda_B &= \lambda_B\{(1-P_e)\bar{\beta}\Delta RH + [\alpha_{RH} - P_e(\alpha_{RH}-\alpha_T) + \zeta]\Delta T\} \\ &= \lambda_B(S_{RH}\Delta RH + S_T\Delta T)\end{aligned} \tag{14}$$

here S_{RH} and S_T are sensitivity coefficients of humidity and temperature of the polyimide-coated FBG humidity sensor, respectively [16].

It should be noted that for a polyimide-coated FBG humidity sensor, actually there is a humidity-temperature cross sensitivity [27]. Therefore in an actual humidity sensing application, one should consider to take an effective temperature compensation to reduce this cross-sensitivity influence. This can be achieved by writing another normal FBG with different wavelength in same fiber close to the FBG humidity sensor to detect the temperature surrounding the sensors or by using other thermometers to obtain the actual ambient temperature value, or by utilizing an athermal package for the sensor [24].

2.2. Humidity sensing experiments

In this section, we will demonstrate some fundamental experimental results on measurements of relative humidity and moisture in industrial oils with our fabricated polyimide-coated FBG humidity sensors.

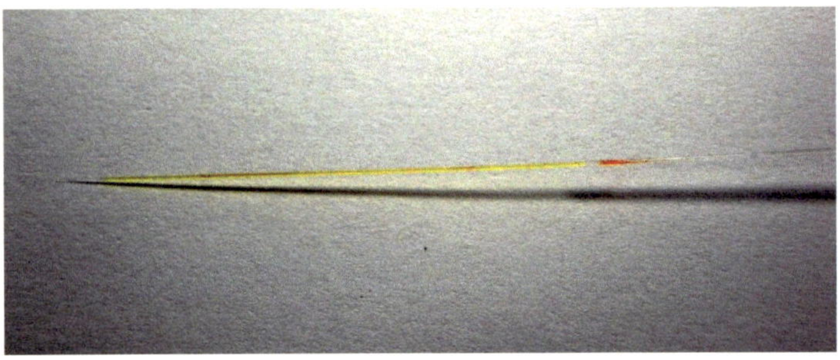

Figure 3. Photograph of fabricated FBG sensor with polyimide coating used for humidity sensing

Five pieces of FBG sensors with same Bragg wavelength were fabricated. Each Bragg grating was written into the core of a single-mode fiber (SMF-28, Corning Inc.) that was hydrogen-loaded beforehand, by using the irradiation method with an ArF Excimer laser (λ = 248 nm) and a phase mask. The sensors then were re-coated with the hygroscopic material (polyimide solution, PI-2560). The coating process adopted is one as described in above section. The dip coating was repeated ten times in order to form a ten-layer polyimide film. The average coating thickness in these sensors was estimated to be 35 $\pm 1{,}0\ \mu m$. Figure 3 is the photograph of one of fabricated polyimide-coated FBG humidity sensors.

The primary experiment for characterization of the polyimide-coated FBG sensor was conducted based on an experimental setup as illustrated schematically in Figure 4. In this work, a sensor with a Bragg wavelength at 1540,58 nm at 25 °C was used. A super-luminescent emitting diodes (SLED) as a broadband light source was used to provide the probe light that was launched into the sensor through an optical circulator. The sensor was placed inside an environmental chamber (ESL-64KAD, Espec, Guangzhou) with a temperature and relative humidity controllable environment. The light reflected from the sensor firstly passed through the optical circulator again and then was fed into an optical spectrum analyzer (OSA, YOKOGAWA AQ6370B) with a fine resolution bandwidth of 0,01 nm, in which the shift of Bragg wavelength of the sensor as the relative humidity changes was observed and recorded.

A group of the reflection spectrums under different relative humidity levels is illustrated in Figure 5. In this experiment, the chamber temperature was fixed at 35 °C and the relative humidity level was adjusted from 30 %RH to 80 %RH in a step of 10 %RH. From these results,

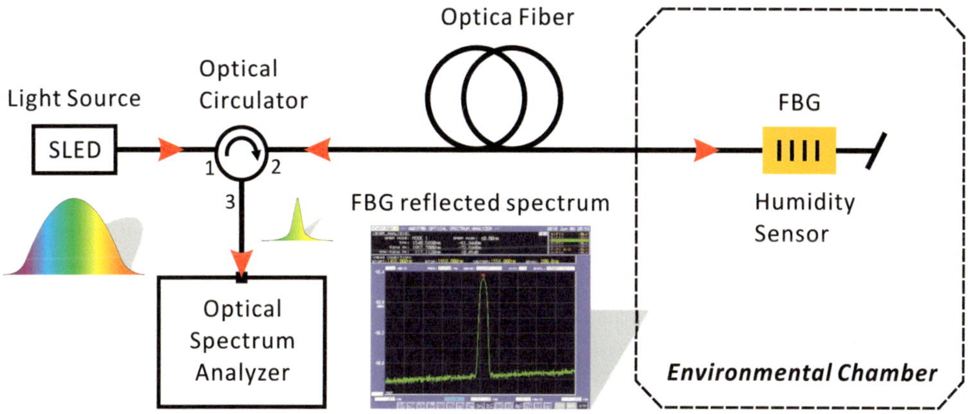

Figure 4. Schematic of experimental setup. Inset is a reflected spectrum of FBG humidity sensor at room condition (25 °C, 60 %RH)

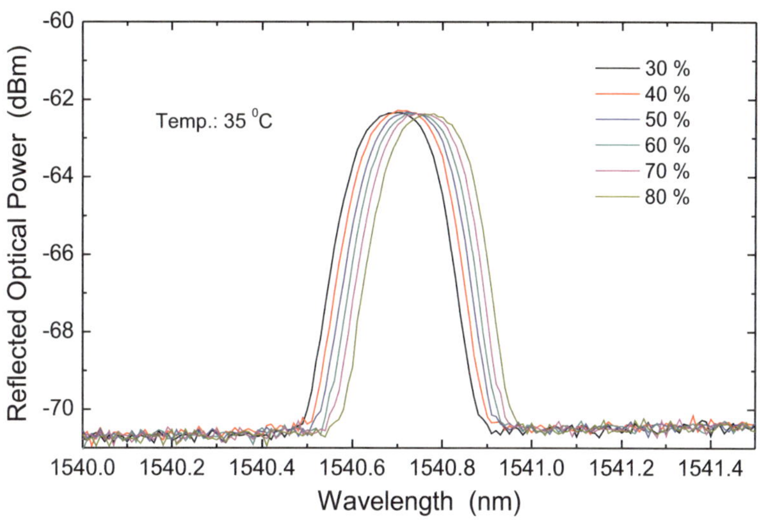

Figure 5. Measured reflection spectrums of sensor under different reflective humidity levels

it is clearly seen that the Bragg wavelength of the polyimide-coated FBG humidity sensor shifts toward the long-wavelength direction (red shift) as the relative humidity level increases.

Figure 6(a) is another group of measured results on Bragg wavelength shifts under different environmental conditions, where the temperature in the environmental chamber was respectively set at 25 °C, 35 °C and 45 °C and the relative humidity level was changed from 30 %RH to 80 %RH in a step of 10 %RH. From these results, a good linear relation between the Bragg wavelength and the relative humidity level can be confirmed. However the Bragg wavelength shifts in high temperature range are apart little bit from this linear relation. This is considered that the polyimide coating on the FBG humidity sensor fabricated is not uniform all over whole FBG section as well as the adhesiveness of polyimide coating to the fiber will decrease in high temperature condition, which brought the FBG humidity sensor a complicated humidity sensing property. Figure 6(b) is a group of measured results on the long-term stability of the sensor during humidity measurements. In this experiment, the

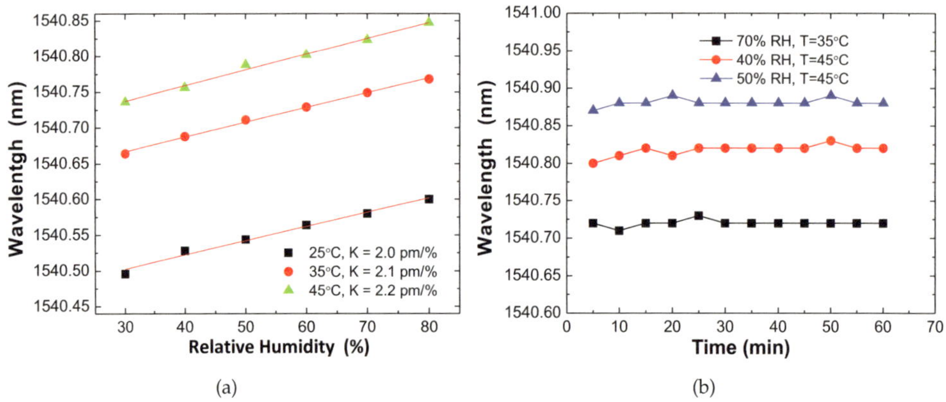

Figure 6. Measured results, (a) Bragg wavelength shifts under different environmental conditions, (b) long-term sensing stability

Bragg wavelengths were recorded within 60 minutes at different temperatures and relative humidity levels. These results show that this fabricated polyimide-coated FBG humidity sensor has a stable sensing property for a long-term relative humidity monitoring.

2.3. Moisture sensing experiments

The monitoring of the moisture in oil and in soil has become very important recently in the industrial field as well as in agriculture, forestry and geography [18, 46]. In the industrial field, the moisture in oil may induce the corrosion on the surfaces of metal parts in the engine and the occurrences of partial discharges in high-voltage electrical transformers. In these cases, compared to conventional electrical sensors, the fiber based moisture sensors play a very important role. Two experiments for measuring the moisture quantity in oil were carried out by using one of our fabricated polyimide-coated FBG humidity sensors. In these experiments, in order to simulate the actual diffusion process of the moisture in oil, we first poured a certain amount of water into the oil and stirred them with a mixer for several minutes, and then inserted the FBG humidity sensor into the oil and waited for 15 minutes before starting a stable measurement. Figure 7 is a set of photographes showing our experimental appliances used in the experiments.

In these experiments, two types of industrial oils were utilized as samples, one was the engine oil as the lubrication oil used in the gear box and the other one was the insulation oil used in high-voltage transformers. Figure 8(a) is a group of experimental data on the measurements of water contents in 20-ml engine oil at two temperatures, 25 °C, 35 °C, respectively. In this experiment, 1-ml pure water was poured each time into the oil. Figure 8(b) is a group of long-term measurement results at different temperatures by adding 3-ml water into 20-ml engine oil. From these results, a stable sensing property of the FBG humidity sensor has been verified.

Figure 9 is the measured data in another experiment on the water contents in the insulation oil, where 100-ml insulation oil was used and 1-ml, total 12-ml, water was added into the insulation oil each time. The experiment was carried out at room temperature (25 °C) and Bragg wavelength of the sensor was measured after a stabilization time of 15 minutes. From

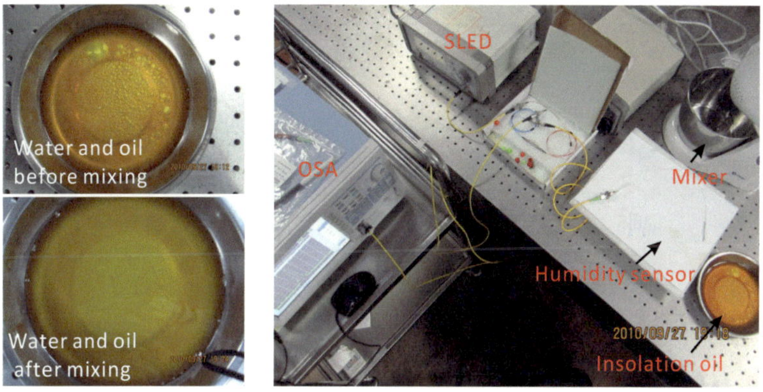

Figure 7. Experimental appliances used in measurements of moisture in oil

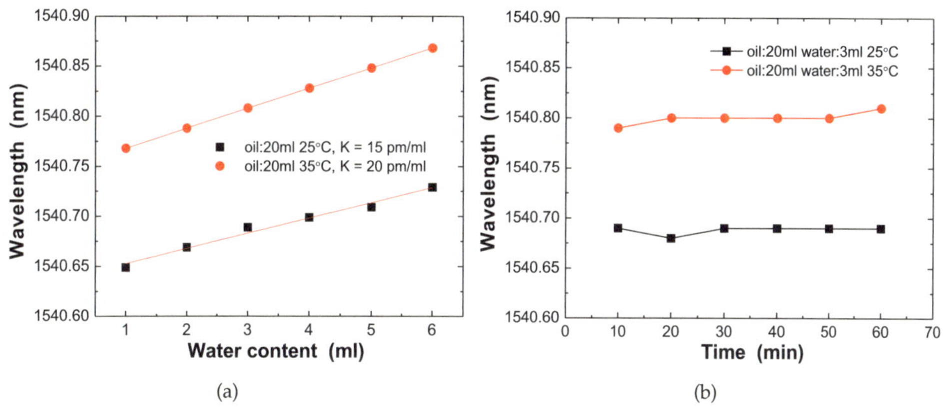

Figure 8. Measured results, (a) Bragg wavelength shifts with water contents in engine oil at two temperatures, (b) long-term sensing stability

the measured results, an increasing shift of the Bragg wavelength following the increase of water contents in the insulation oil can be verified.

From a practical view for moisture monitoring with the FBG humidity sensor, considering a fact that the moisture distribution in oil is not uniform, in order to obtain a credible result, it is important to install multiple sensors at different positions in the oil container and need to take an average operation with the measured values from all sensors installed at different locations.

3. SOA-based FBG fiber laser for multiple parameter simultaneous measurements

3.1. Technical background

Based on the phenomenon of the Bragg wavelength of FBG changing with many physical parameters, FBG sensors are mainly used for measurements of temperature, strain, pressure, and vibration [3, 39]. On the other hand, incorporated with recently developed fiber laser technologies, fiber laser sensors as active sensor systems with many unique advantages, such

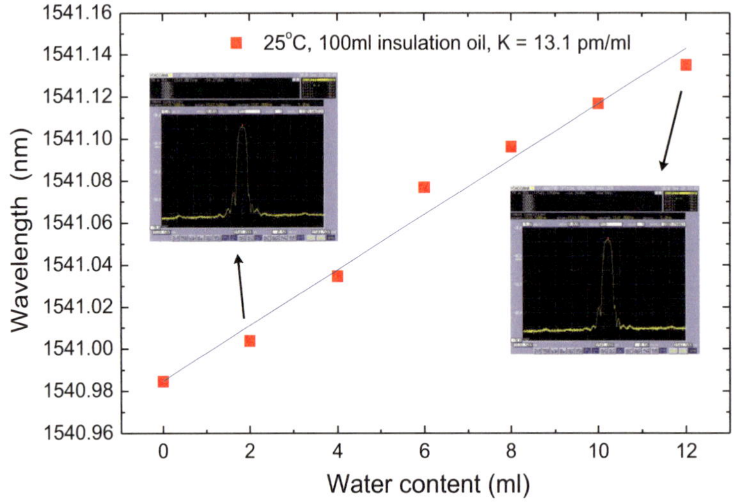

Figure 9. Bragg wavelength shifts with water contents in insulation oil at room temperature. Insets are two reflected spectrums measured at corresponding water contents

as high detection sensitivity, large dynamic range and quickly responding speed, as well as easily achieving remote sensing, recently have been developed well and widely employed in many measurement fields [23, 32, 41]. In addition, it is well known that when the fiber is coated by the polyimide resin, its mechanical strength, especially tensile strength and sensitivity to pressure will have an obvious enhancement, almost up to 30 times over those of the bare fiber [19, 35]. This property is very suitable for FBG sensors used in many specific applications for large-scale measurements, such as for measurements of strain, pressure and acceleration, where high mechanical strengths in FBG sensor itself often are required.

In many types of fiber laser systems, the recently-developed semiconductor optical amplifier (SOA) as a gain medium with an inhomogeneous broadening bandwidth over 40 nm and ultra low polarization dependency has been widely used in the construction of multi-wavelength fiber laser system with a stable lasing output [22]. The FBGs incorporated in this kind of fiber laser system are taken not only as wavelength selective mirrors, but also as sensing components to transfer various physical parameters into the changes of the system property in terms of lasing wavelength and output optical power.

In previous work, we had presented a novel SOA-based fiber laser sensor system with an ability of simultaneously measuring vibration, temperature and humidity [23]. In this proposed system, two polyimide-coated FBG sensors were used as the vibration or temperature sensor and the humidity sensor, respectively. Other two FBGs as wavelength reference elements inside the interrogator were employed, with which the required measurands in terms of temperature and humidity could be deduced from the Bragg wavelength shifts of sensors. The experimental results showed that this proposed fiber laser sensor system had an ability of simultaneously measuring the mechanical vibration in a wide frequency range from 40 Hz to 1 kHz and the temperature increase up to 110 °C as well as the relative humidity level ranging from 25 %RH to 85 %RH. In following parts of this section, we will firstly give a brief description of sensing principles in measurements of vibration, temperature and relative humidity based on this system configuration and interrogation methods, and then illustrate some primary experimental data to demonstrate the system performances.

3.2. System configuration and sensing principles

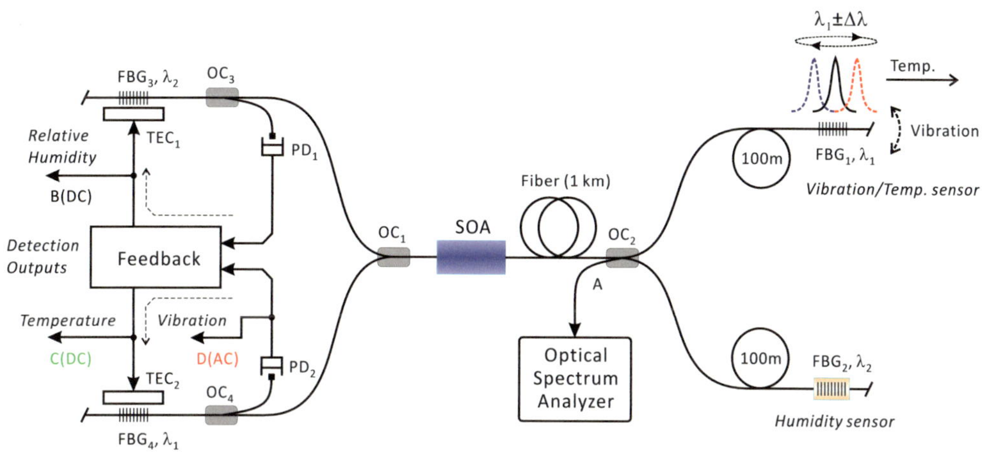

Figure 10. Schematic of proposed SOA-based fiber laser sensor system for multi-parameter sensing, in which FBG_1 and FBG_2 are taken as vibration/temperature and humidity sensors, respectively. FBG_3 and FBG_4 are two wavelength-matched reference FBGs. 1-km long fiber used in sensor system is to verify a remote sensing ability of system

The proposed experimental system is schematically illustrated in Figure 10. In this configuration, a 1550-nm band, polarization insensitive SOA (COVEGA's 1013) was used as a gain medium to generate lasing outputs. Two pairs of FBGs with different Bragg wavelengths used as two wavelength selective mirrors in the fiber laser system were incorporated with a 1-km long and two 100-m long single mode fibers to form two in-line optical cavities working in different wavelength ranges.

One pair of FBGs, FBG_1 and FBG_4 both having same Bragg wavelength at $\lambda = 1537{,}28$ nm were coated by one-layer polyimide film. FBG_1 was taken as a vibration or temperature sensor and FBG_4 as a wavelength reference of FBG_1. Another pair of FBGs, FBG_2 and FBG_3 also had same Bragg wavelength at $\lambda = 1541{,}64$ nm. FBG_2 was coated with a ten-layer polyimide film as a humidity sensor, while FBG_3 coated with single polyimide layer as FBG_4's wavelength reference. All FBGs were 2-mm long in length with 75 % reflectivity and 1-nm linewidth. This sensor system actually is a dual-wavelength fiber laser with continuing-wave (CW) outputs. Two reflected lights from FBG_1 and FBG_2 through two 3-dB fiber optic couplers (OC), OC_3 and OC_4, were fed into two photo-detectors, PD_1 and PD_2, respectively, where they were optoelectronic-detected and converted into corresponding electrical signals.

The interrogation method used in our system to demodulate the reflected signal lights from FBG_1 and FBG_2, was a so-called wavelength matching method [7]. By adjusting the Bragg wavelength of reference FBG through the mechanical or thermal means to match it with that of the sensor. In this way, the reflected light from the FBG sensor can be demodulated. In our system, this wavelength tuning operation was completed automatically by adjusting the surface temperature of a thermoelectric cooler (TEC) on which the reference FBG was attached, with a control voltage. It is clear that the Bragg wavelength shift of the reference FBG linearly depends on the TEC temperature and resultantly on the control voltage of the TEC. By measuring this control voltage, one can actually obtain the magnitude of wavelength shift of corresponding sensor, which is directly related to the required measurand [10, 13]. The TEC was driven by an IC chip and its surface temperature was set through a feedback control driving with an error signal related to a wavelength difference between the sensor and the reference FBGs.

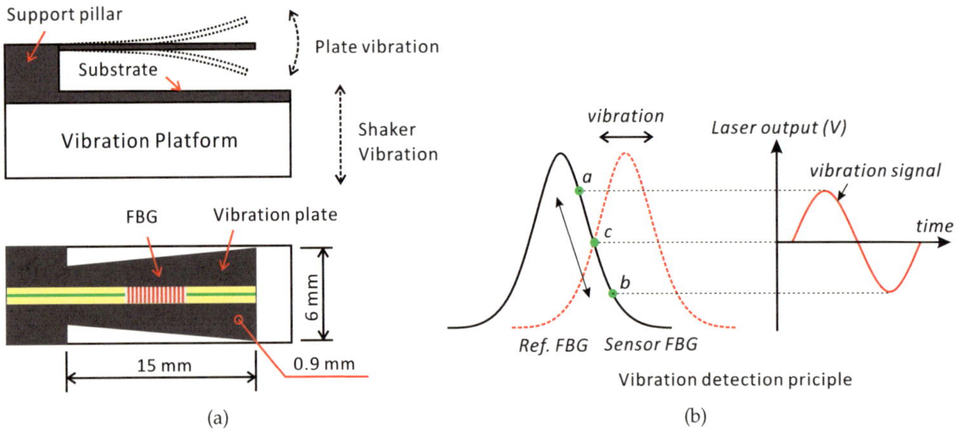

Figure 11. (a) Structure of FBG-based vibration/temperature sensor and (b) operation principle for vibration measurements

For vibration measurements, a vibration/temperature sensor was fabricated with FBG_1, as shown in Figure 11(a), which was constructed based on a mechanical amplifier structure, including a vibration plate mounted on a support pillar, and a substrate. The FBG sensor was adhered on the vibration plate with the epoxy glue. In this way the FBG sensor can be insulated from the effects of surrounding moisture. In sensing principle, when the sensor is mounted on the surface of a vibrating object, such as a shaker, a periodically up-down moving of the vibrating object will induce a periodical deformation of the vibration plate, as shown in Figure 11(a), which in turn generates a time-varying strain on the FBG_1 to periodically alter its Bragg wavelength. The magnitude of the wavelength variation of the sensor is proportional to its mechanical vibration amplitude. In sensor interrogations, when the Bragg wavelength of a reference FBG, such as FBG_4, is offset a little bit cross at the center point of the spectral profile of the sensor, marked as "C" in Figure 11(b), the periodical change in FBG_1's Bragg wavelength can be easily converted into an intensity change in the fiber laser output. Based on this sensing principle, a vibration signal will be directly obtained from the PD_2 detection output as an AC signal.

For temperature measurements with the FBG-based vibration/temperature sensor during the simultaneous measurement of vibration, when the Bragg wavelength of FBG_1 slowly shifts due to an ambient temperature change, the Bragg wavelength of FBG_4 will be automatically adjusted by TEC_2 controller to match it with that of FBG_1. Since the magnitude of control signal voltage to TEC_2 is proportional to the magnitude of Bragg wavelength shift, in turn to temperature changes surrounding FBG_1, it can be utilized directly for temperature measurements.

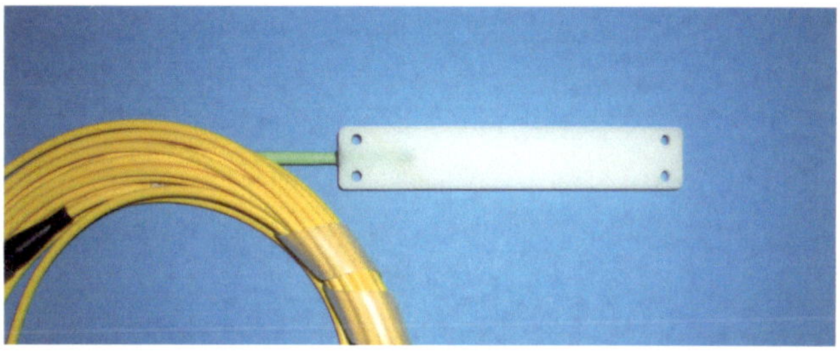

Figure 12. A photograph of packaged FBG humidity sensor

For humidity measurements, a packaged, ten-layer polyimide-coated FBG humidity sensor was used, as illustrated in Figure 12. The fundamental experimental investigations on performances of this sensor in measurements of relative humidity and water contents in oil had been carried out previously. The sensor was packaged in a specific plastic box formed with a 3-D printer, which was air permeable and allowed the moisture easily to penetrate the cover and arrived at the sensor.

The principle used for humidity sensing is similar to that employed for temperature sensing. The change of the relative humidity level surrounding FBG_2 will induce a Bragg wavelength shift of the sensor. As a direct reaction, the Bragg wavelength of FBG_3 will be automatically adjusted by TEC_1 controller to trace and finally match with the Bragg wavelength of the sensor. So the control voltage to TEC_1 can be taken directly as the measurement result of the humidity sensing.

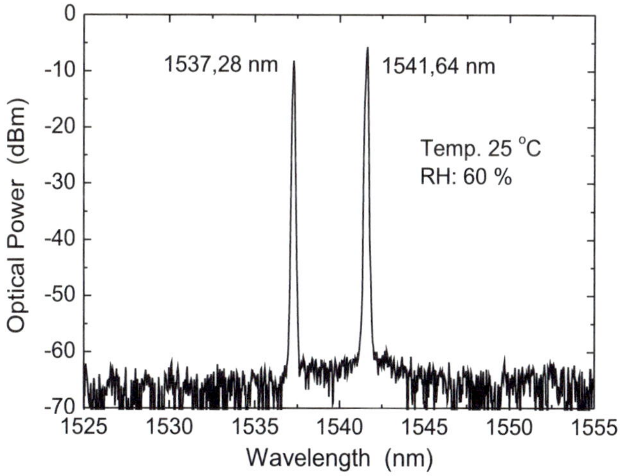

Figure 13. Output spectrum of SOA-based fiber laser sensor system

It should be noted that since the FBG-based humidity sensor has a humidity-temperature cross sensitivity, for a pure humidity measurement, however the temperature effects on the sensor should be taken into account and the temperature compensation should be taken with suitable techniques. Our system was intended to be used in a relatively closed environment for multiple parameters monitoring, for example, in a cabinet of high-voltage equipment, in which the temperature distribution is regarded to be uniform. Therefore the temperature compensation can be real-time completed by using a measured value on the temperature change provided by the FBG-based vibration/temperature sensor.

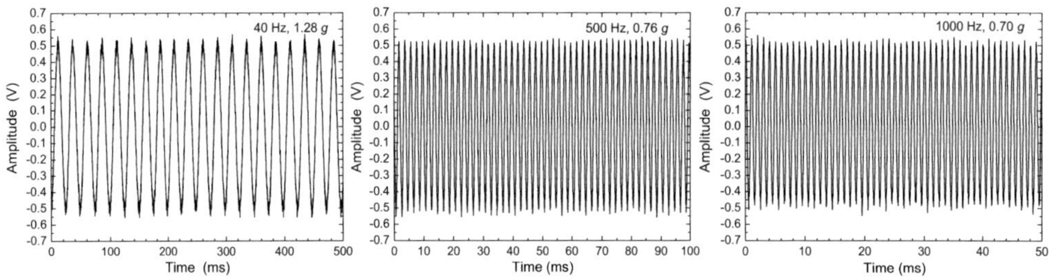

Figure 14. Three measured vibration waveforms at frequency, 40 Hz, 500 Hz and 1000 Hz, respectively

3.3. Experimental results

Several sets of experimental data for demonstrating the performances of proposed fiber laser sensor system on simultaneous measurements of vibration, temperature and humidity are presented here. Figure 13 is the output spectrum of the fiber laser sensor, measured at port "A" of a 3-dB optical coupler, OC_1. From this result, it is clear that this sensor system actually worked in a dual-wavelength lasing mode with two oscillating wavelengths at 1537,28 nm and 1541,64 nm, respectively. Figure 14 is a group of vibration waveforms measured at frequency, 40 Hz, 500 Hz and 1000 Hz, respectively, with the FBG-based vibration/temperature sensor. In this experiment, a shaker (Gilchrist TX Technology, Inc.) was used to generate the required mechanical vibrations with controlled frequency and acceleration values. The shaker had been calibrated at 100 Hz with 1 *g* acceleration. The sensor was placed on the vibration platform of the shaker and the output waveforms at PD_2 were monitored and recorded with a digital oscilloscope. From these waveforms measured at different frequency points, one can clearly observe that this sensor system possesses fair good performances in vibration measurements within a very wide frequency range from 40 Hz to 1000 Hz. We also investigated the frequency response of the sensor by tuning the mechanical vibration frequency of the shaker from 10 Hz to 2000 Hz and measured the average amplitude of each vibration signal at corresponding frequency point. The measured results are plotted in Figure 15, in which the red line with solid markers represents the frequency response curve of the sensor and the black line with circle markers is the acceleration profile of the shaker used. From this result, we further confirm that this FBG-based vibration/temperature sensor has a fairly flattening frequency response property from 40 Hz to 1,1 kHz and its resonance peak appears at near 1,3 kHz.

This sensor system not only can be used for continues vibration measurements, but also can be utilized for impact detections. Figure 16(Left) is a measured shock signal waveform when an impact force was applied on the table where the sensor was mounted on. Figure 16(Right) is the corresponding FFT spectrum of the shock signal. It is seen that the sensor and the designed interrogator have the ability for large-scale dynamic strain measurements within a wide frequency range up to 120 kHz.

We also investigated the system performances in tracing temperature changes. For a comparison, Figure 17 shows a group of measured vibration signal waveforms at 300 Hz with (b) and without (a) TEC_2 temperature control. It is clear that without TEC_2 temperature control, the amplitude of the vibration signal varied along with ambient temperature

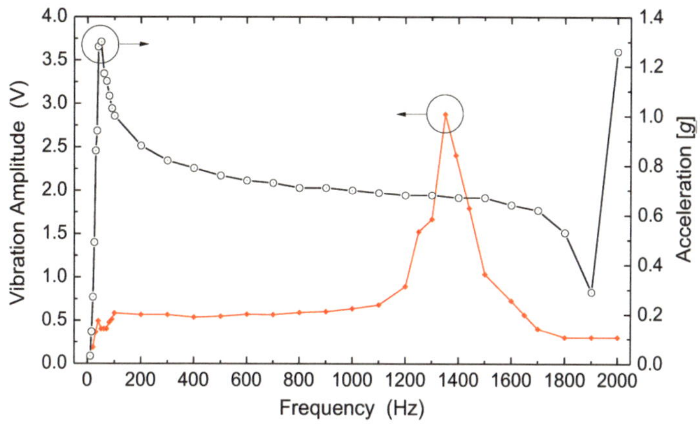

Figure 15. Measured frequency response of FBG vibration/temperature sensor and acceleration profile of shaker used in experiment. Resonance peak of sensor appears at near 1300 Hz

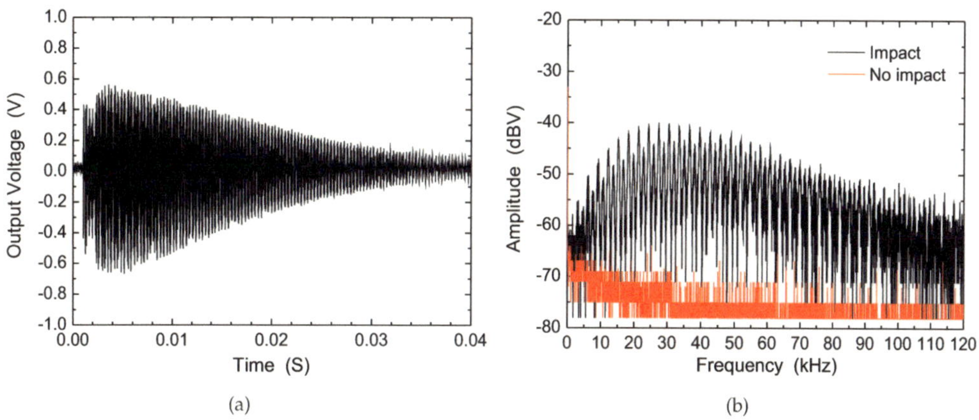

Figure 16. Measured results, (a) detected shock signal waveform, (b) corresponding FFT spectrum

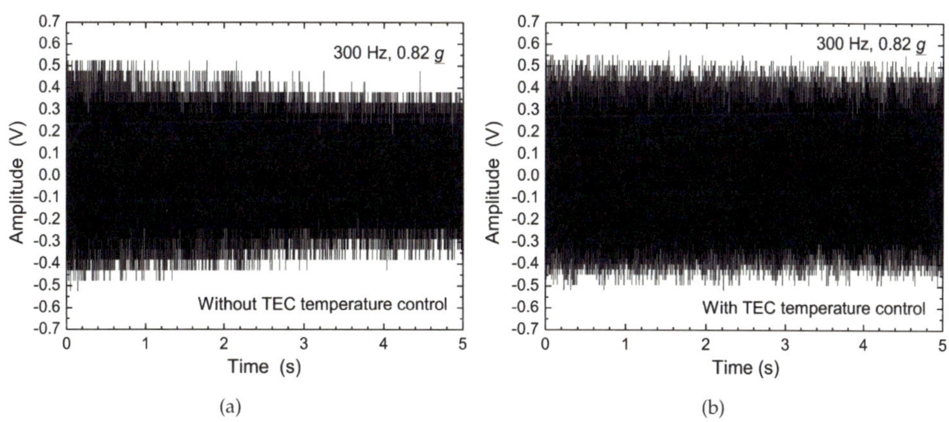

Figure 17. Measured vibration waveforms without (a) and with (b) TEC control

fluctuations. It should be noted that the TEC temperature actually is reversely proportional to its control voltage. Figure 18 is a profile of TEC_2 control voltage. From this we can obtain the information on the temperature variations. Figure 19 is a group of measured results on the amplitude variations of the vibration signal measured with the TEC temperature control. In this experiment, we placed the sensor inside a small size, thermal-insulation box with internal temperature controlled. This box then was mounted on the vibration platform of the shaker, vibrating at 500 Hz and the temperature inside this box was changed from 30 °C to 110 °C within 5 minutes. It is clear that the fluctuations in the vibration amplitude induced by temperature changes can be greatly suppressed into a small region of lower than 10 % with a suitable TEC temperature control process.

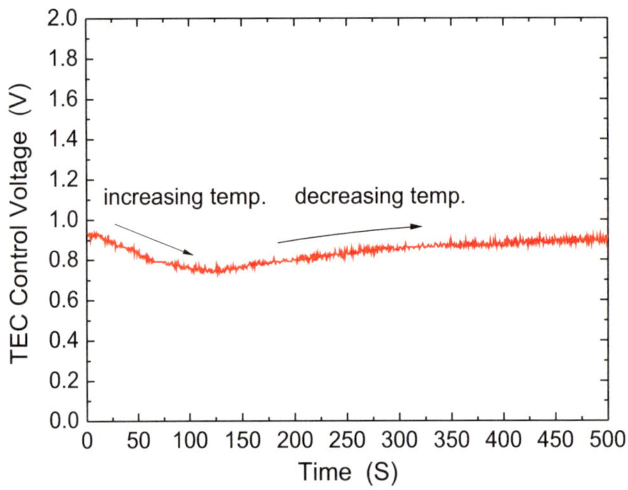

Figure 18. A profile of TEC_2 control voltage when temperature was changed

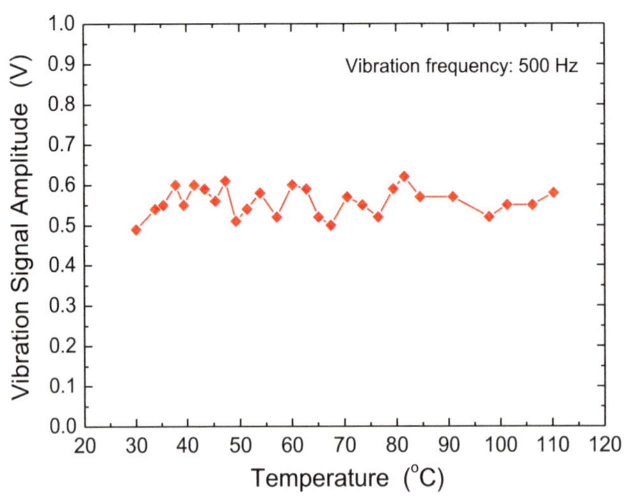

Figure 19. Amplitude fluctuation versus temperature change in a vibration measurement with TEC temperature control

Figure 20 shows an experimental result on temperature measurements with the FBG-based vibration/temperature sensor by reading TEC_2 control voltage. In this experiment, the sensor was placed into the temperature control box as mentioned above. The temperature inside the box was changed gradually form 25 °C to 110 °C within 30 minutes. During this period, the vibration and temperature were simultaneously measured. With this measured result through a linear fitting operation, we obtained a slope value of 0,00914 V/°C as the temperature sensitivity of the sensor.

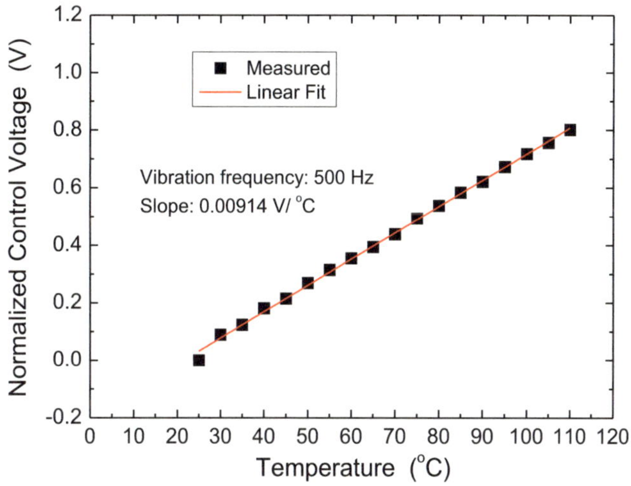

Figure 20. Measured normalized TEC control voltage versus temperature. A slope of 0.00914 V/°C was obtained

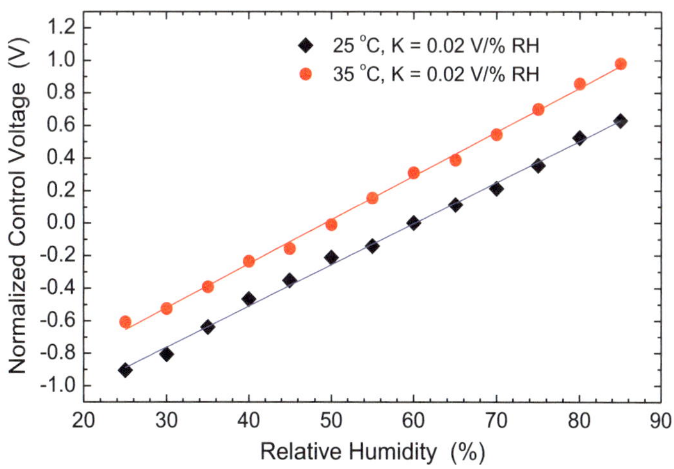

Figure 21. Measured normalized TEC control voltages versus different relative humidity levels at two environmental temperatures, 25 °C and 35 °C, respectively

Figure 21 is two sets of measured results on relative humidity measurements with the polyimide-coated FBG humidity sensor. In this experiment, the sensor was placed inside the environmental chamber, and the relative humidity level in the chamber was changed from 25 %RH to 85 %RH in a step of 5 %RH at two chamber temperatures, 25 °C and 35 °C, respectively. In these results, the linear relationships at different temperatures between the control voltages of TEC_3 and the actual RH values in the chamber can be confirmed. Also it is clear that when the temperature in the chamber is changed the whole measured RH curve will move proportionally to the amount of the temperature change. In addition, the sensitivities of the sensor at 25 °C and 35 °C are almost same, equal to $K = 0.02$ V/%RH.

4. Conclusion

In this chapter, the sensing principle of polyimide-coated FBG humidity sensors has been introduced. Also some primary experimental results for investigations of sensor properties in respect of humidity sensing have been presented. The polyimide resin is an ideal coating material for fabricating the FBG-based humidity sensor. The polyimide coating swelling due to moisture absorption generates a strain imposed on the FBG, which in turn directly results in a linear and reversible Bragg wavelength shift with the relative humidity level over a wide range. This is a basis by using FBG sensors with polyimide coating for humidity sensing.The sensitivity of the sensor is dependent on the thickness of polyimide coating [21]. A thicker polyimide layer may effectively induce larger strains imposed on the FBG to increase the magnitude of Bragg wavelength shift. However this in turn inevitably degrades the time response property of the sensor and increases its susceptibility to surrounding temperature changes. The cross sensitivity of the polyimide-coated FBG humidity sensor to temperature as well as to humidity is an important influence factor in an actual sensing application for humidity measurements, which should be taken into account. Considering the importance of moisture detection and water content monitoring in industrial measurement fields, we tried to use the polyimide-coated FBG humidity sensor to measure the water contents in two types of industrial oils, including the engine oil and the insulation oil. The measured data showed that the fabricated humidity sensor has enough high sensitivity to detect the low-density moisture (< 2 %) in both industrial oils.

A prototype of SOA-based, dual-wavelength fiber laser system capable of simultaneous measurements of multiple physical parameters has been demonstrated. In this system configuration, a one-layer polyimide-coated FBG was used as the vibration/temperature sensor, and another ten-layer polyimide-coated FBG sensor was used for humidity sensing. The system performances have been investigated experimentally in various aspects, showing that this system has an ability of simultaneously sensing vibration, temperature and relative humidity in a relatively closed environment. Further work is to explore the polyimide-coated FBG sensor for other physical parameters sensing [17], and to develop the smart interrogation techniques as well as self-adapting systems [20, 40], which includes to employ the fiber laser system configuration together with wavelength multiplexing and de-multiplexing technology to form a sensor array for large scale, multiple point and multiple parameter measurements.

Acknowledgements

The authors would like to acknowledge the support and guidance provided by Key Laboratory of Specialty Fiber Optics and Optical Access Networks, School of Communication and Information Engineering, Shanghai University, China, Thanks to Prof. Ting-Yun Wang for his continuous support and encouragement during this work.

Author details

Lutang Wang, Nian Fang and Zhaoming Huang
Key Laboratory of Specialty Fiber Optics and Optical Access Networks, School of Communication and Information Engineering, Shanghai University, China

5. References

[1] Bremer, K., Lewis, E., Moss, B., Leen, G., Lochmann, S. & Mueller, I. [2009]. Conception and preliminary evaluation of an optical fibre sensor for simultaneous measurement of pressure and temperature, *Proceedings of SPIE 20th International Conference on Optical Fibre Sensors*, Vol. 7503, SPIE, Edinburgh, United Kingdom, pp. 1–4.

[2] Buchhold, R., Nakladal, A., Gerlach, G., Sahre, K. & Eichhorn, K. [1998]. Mechanical stress in micromachined components caused by humidity-induced in-plane expansion of thin polymer films, *Thin Solid Films* Vol. 312(No. 1-2): 232–239.

[3] Chang, J., Huo, D., Ma, L., Liu, X., Liu, T. & C., W. [2008]. Interrogation a fiber bragg grating vibration sensor by narrow linewidth light, *Proceedings of Optical Fiber Sensors Conference, APOS 2008*, Chengdu, China, pp. 1–4.

[4] Cheng, L. K. [2005]. High-speed dense channel fiber bragg grating sensor array for structural health monitoring, *Proceedings of International Conference on MEMS, NANO and Smart Systems, 2005*, IEEE Computer Society, Banff, Alberta, Canada, pp. 364–365.

[5] Chil-Won Lee, Hee-Woo Rhee, M.-S. G. [2001]. Humidity sensor using epoxy resin containing quaternary ammonium salts, *Sensors and Actuators B: Chemical* Vol. 73(No. 2-3): 124–129.

[6] Claus, R. O., Distler, T., Mecham, J., Davis, B., Arregui, F. J. & Matias, I. R. [2004]. Optical fiber sensor for breathing diagnostics, *Proceedings of SPIE in Optical Fibers and Sensors for Medical Applications IV*, Vol. 5317, SPIE, San Jose, CA, USA, pp. 167–171.

[7] Davis, M. A. & Kersey, A. D. [1995]. Matched-filter interrogation technique for fibre bragg grating arrays, *Electronics Letters* Vol. 31(No. 10): 822–823.

[8] Fan, H., Qian, J., Zhang, Y. & Shen, L. [2007]. A novel fbg sensors network for smart structure vibration test, *Proceedings of the 2007 IEEE International Conference on Robotics and Biomimetics*, IEEE, Sanya, China, pp. 1109–1113.

[9] Fuxin, D., Lutang, W., Nian, F. & Zhaoming, H. [2010]. Experimental study on humidity sensing using a fbg sensor with polyimide coating, *Proceedings of SPIE in Optical Sensors and Biophotonics II*, Vol. 7990, SPIE, Shanghai, China, pp. 1–7.

[10] Geiger, H., Xu, M., Eaton, N. & Dakin, J. [1995]. Electronic tracking system for multiplexed fibre grating sensors, *Electronics Letters* Vol. 31(No. 12): 1006–1007.

[11] Gerlach, G. & Sage, K. [1994]. A piezoresistive humidity sensor, *Sensors and Actuators A: Physical* Vol. 53(No. 1-3): 181–184.

[12] Giles, C. [1997]. Lightwave applications of fiber bragg gratings, *Journal of Lightwave Technology* Vol. 15(No. 8): 1391–1404.

[13] He, J., Li, F., Xiao, H. & Liu, Y. [2008]. Fiber bragg grating sensor array system based on digital phase generated carrier demodulation and reference compensation method, *Proceedings of Optical Fiber Sensors Conference, APOS 2008*, Chengdu, China, pp. 1–4.

[14] Houhui, L., Yongxing, J. & Yu, Z. [2011]. High birefringence fiber loop mirror with polymer coating used as humidity sensor, *Proceedings of SPIE in Optical Sensors and Biophotonics III*, Vol. 8311, SPIE, Shanghai, China, pp. 1–6.

[15] Kersey, A. D., Davis, M. A., Patrick, H. J., LeBlanc, M., Koo, K. P., Askins, C. G., Putnam, M. A. & Friebele, E. J. [1997]. Fiber grating sensors, *Journal of Lightwave Technology* Vol. 15(No. 8): 1442–1463.

[16] Kronenberg, P., Rastogi, P., Giaccari, P. & Limberger, H. [2002]. Relative humidity sensor with optical fiber bragg gratings, *Optics Letters* Vol. 27(No. 16): 1385–1387.

[17] Kumar, A., Zhang, P., Vincent, A., McCormack, R., Kalyanaraman, R., Cho, H. & Seal, S. [2011]. Hydrogen selective gas sensor in humid environment based on polymer coated nanostructured-doped tin oxide, *Sensors and Actuators B: Chemical* Vol. 155(No. 2): 884–892.

[18] Kunzler, W., Calvert, S. & Laylor, M. [2003]. Measuring humidity and moisture with fiber optic sensors, *Sixth Pacific Northwest Fiber Optic Sensor Workshop*, Vol. 5278, SPIE, Troutdale, OR, USA, pp. 86–93.

[19] Liu, Y., Guo, Z., Z., L., Ge, C., Zhao, D. & Dong, X. [2000]. High-sensitivity fiber grating pressure sensor with polymer jacket, *Chinese Journal of Lasers* Vol. A27(No. 3): 211–214.

[20] Liyin, H., Lutang, W., Nian, F., Huishan, C. & Huang, Z. [2011]. Fbg moisture sensor system using soa-based fiber laser with temperature compensation, *Proceedings of SPIE in Optical Sensors and Biophotonics III*, Vol. 8311, SPIE, Shanghai, China, pp. 1–6.

[21] Lu, P., Men, L. & Chen, Q. [2008]. Tuning the sensing responses of polymer-coated fiber bragg gratings, *Journal of Applied Physics* Vol. 104(No. 116110): 1–3.

[22] Luo, Z., Zhong, W., Wen, Y., Cai, Z. & Ye, C. [2008]. Multiwavelength fiber lasers based on soa and double-pass mach-zehnder interferometer, *Proceedings of 2008 IEEE PhotonicsGlobal@Singapore (IPGC)*, IEEE, Singapore, Singapore, pp. 1–4.

[23] Lutang, W., Nian, F., Fuxin, D. & Zhaoming, H. [2010]. Simultaneous measurements of vibration, temperature, and humidity using a soa-based fiber bragg grating laser, *Proceedings of SPIE in Optical Sensors and Biophotonics II*, Vol. 7990, SPIE, Shanghai, China, pp. 1–9.

[24] Majumder, M., Gangopadhyay, T., Chakraborty, A., Dasgupta, K. & Bhattacharya, D. [2008]. Fibre bragg gratings in structural health monitoring–present status and applications, *Sensors and Actuators A: Physical* Vol. 147: 150–164.

[25] Meltz, G. & Morey, W. W. [1991]. Bragg grating formation and germanosilicate fiber photosensitivity, *Proceedings of SPIE, International Workshop on Photoinduced Self-Organization Effects in Optical Fiber*, Vol. 1516, SPIE, pp. 185–199.

[26] Men, L., Lu, P. & Che, Q. [2008]. A multiplexed fiber bragg grating sensor for simultaneous salinity and temperature measurement, *Journal of Applied Physics* Vol. 103(No. 053107): 1–6.

[27] Men, L., Lu, P. & Chen, Q. [2008]. Intelligent multiparameter sensing with fiber bragg gratings, *Applied Physics Letters* Vol. 93(No. 071110): 1–3.

[28] Philipp, L., Mario, W., Sascha, L. & Katerina, K. [2010]. Distributed humidity sensing based on rayleigh scattering in polymer optical fibers, *Proceedings of SPIE in Fourth European Workshop on Optical Fibre Sensors*, Vol. 7653, SPIE, Porto, Portugal, pp. 1–4.

[29] Ping, L., Liqiu, M. & Qiying, C. [2009]. Polymer-coated fiber bragg grating sensors for simultaneous monitoring of soluble analytes and temperature, *Sensors Journal, IEEE* Vol. 9(No. 4): 340–345.

[30] Qiu, Y. Y., Azeredo-Leme, C., Alcacer, L. & Franca, J. E. [2001]. Characterization of a cmos humidity sensor using different polyimides as sensing films, *Proceedings of SPIE in Electronics and Structures for MEMS II*, Vol. 4591, SPIE, Adelaide, Australia, pp. 310–315.

[31] Qiying, C. & Ping, L. [2008]. Fiber bragg gratings and their applications as temperature and humidity sensors, *in* L. Chen (ed.), *Atomic, Molecular and Optical Physics: New Research*, Nova Science Publishers, Inc., pp. 235–260.

[32] Reilly, S., James, S. & Tatam, R. [2002]. Dual wavelength fiber bragg grating external semiconductor laser sources for sensor applications, *Optical Fiber Sensors Conference Technical Digest, Ofs 2002*, Vol. 1, IEEE, pp. 281–284.

[33] Ryszard, H. & Henryk, J. W. [1996]. Fiber optic technique for relative humidity sensors, *Proceedings of SPIE in Optoelectronic and Electronic Sensors II*, Vol. 3054, SPIE, Szczyrk, Poland, pp. 145–150.

[34] Sager, K., Schroth, A., Nakaldal, A. & Gerlach, G. [1996]. Humidity-dependent mechanical properties of polyimide films and their use for ic-compatible humidity sensors, *Sensors and Actuators A: Physical* Vol. 53(No. 1-3): 330–334.

[35] Sun, A., Qiao, X., Jia, Z., Guo, T. & Chen, C. [2004]. The study of fiber bragg grating pressure sensor with high pressure resistance, *ACTA Photonica Sinica* Vol. 33(No. 7): 823–825.

[36] Tao, L., Xinyong, D., Chun-Liu, Z. & Yang, L. [2010]. Polymer-coated hybrid fiber grating for relative humidity sensing, *Proceedings of SPIE in Advanced Sensor Systems and Applications IV*, Vol. 7853, SPIE, Beijing, China, pp. 1–6.

[37] Udd, E. [2007]. Review of multi-parameter fiber grating sensors, *Proceedings of SPIE in Fiber Optic Sensors and Applications V*, Vol. 6770, SPIE, Boston, MA, USA, pp. 1–10.

[38] Venugopalan, T., Yeo, T. L., Basedau, F., Henke, A., Sun, T., Grattan, K. T. V. & Habel, W. [2009]. Evaluation and calibration of fbg-based relative humidity sensor designed for structural health monitoring, *Proceedings of SPIE, 20th International Conference on Optical Fibre Sensors*, Vol. 7503, SPIE, Edinburgh, United Kingdom, pp. 1–5.

[39] Wada, A., Tanaka, S. & Takahashi, N. [2009]. High-sensitivity vibration sensing using in-fiber fabry-perot interferometer with fiber-bragg-grating reflectors, *Proceedings of SPIE, 20th International Conference on Optical Fibre Sensors*, Vol. 7503, SPIE, Edinburgh, United Kingdom, pp. 1–4.

[40] Wang, J., Huo, D., Chang, J., Liu, X., Liu, J., Ma, L., Liu, T. & Wang, C. [2009]. A temperature self-adapting fbg vibrating sensor system, *Proceedings of SPIE, 20th International Conference on Optical Fibre Sensors*, Vol. 7503, SPIE, Edinburgh, United Kingdom, pp. 1–4.

[41] Wang, Y., Cui, Y. & Yun, B. [2006]. A fiber bragg grating sensor system for simultaneously static and dynamic measurements with a wavelength-swept fiber laser, *IEEE Photonics Technology Letters* Vol. 18(No. 14): 1539–1541.

[42] Xing, L., Gino, R., Muthukumaran, P. & Ion, S. [2004]. Polyimide-based fiber optic humidity sensor, *Proceedings of SPIE, Photonics North 2004: Photonic Applications in Telecommunications, Sensors, Software, and Lasers*, Vol. 5579, SPIE, Ottawa, Canada, pp. 205–212.

[43] Yeo, T., Sun, T., Grattan, K., Parry, D., Lade, R. & Powell, B. [2005a]. Characterisation of a polymer-coated fibre bragg grating sensor for relative humidity sensing, *Sensors and Actuators B: Chemical* Vol. 110(No. 1): 148–156.

[44] Yeo, T., Sun, T., Grattan, K., Parry, D., Lade, R. & Powell, B. [2005b]. Polymer-coated fiber bragg grating for relative humidity sensing, *Sensors Journal, IEEE* Vol. 5(No. 5): 1082–1089.

[45] Yeo, T., Tong, S., Grattan, K., Parry, D., Lade, R. & Powell, B. [2005]. Polymer-coated fiber bragg grating for relative humidity sensing, *Sensors Journal, IEEE* Vol. 5(No. 5): 1082–1089.

[46] Zhang, C., Chen, X., Webb, D. & Peng, G. [2009]. Water detection in jet fuel using a polymer optical fibre bragg grating, *Proceedings of SPIE, 20th International Conference on Optical Fibre Sensors*, Vol. 7503, SPIE, Edinburgh, United Kingdom, pp. 1–4.

Chapter 9

Semi-Alicyclic Polyimides: Insights into Optical Properties and Morphology Patterning Approaches for Advanced Technologies

Andreea Irina Barzic, Iuliana Stoica and Camelia Hulubei

Additional information is available at the end of the chapter

http://dx.doi.org/10.5772/53607

1. Introduction

Polyimides (PIs) represent a very important class of high performance synthetic polymers, characterized by stability at elevated temperatures, low thermal expansion, chemical resistance, low dielectric constant, high breakdown voltage, good optical transparency, high adhesion and dimensional stability [1-3]. Therefore, they have found widespread applicability: in microelectronics - as interlayer dielectrics, in aircraft industry - as thermo-resistant components and in optoelectronics - as optical waveguides. In recent years, some types of PIs were proven to be biocompatible [4], and they have been used in biomedical fields as flexible and implantable intracortical electrode arrays [5] and as microstructured substrates for contact guidance of cell growth [6], on condition of being patterned at micro or nanoscale. However, the strong intermolecular forces between the aromatic PI chains make them difficult to process because they decompose prior to melting and are insoluble in organic solvents [7-9]. One the other hand, the formation of intra- and intermolecular charge transfer complex (CTC) between the electron-donating diamine and the electron-accepting dianhydride moieties explains the coloration ranging from pale yellow to deep brown, causing strong absorption in the visible region [10,11], and enhance the dielectric constant. All these problems greatly limit the widespread use of aromatic PIs in areas where colorlessness and transparency are required, for example, as covers for solar cells, orientation films in liquid crystal display (LCD) devices, optical waveguides for communication interconnectors and other high-tech fields. An intensive research effort has been undertaken to counteract these shortcomings by designing their structure, including: (1) increasing flexibility along the polymer chain by employment of flexible (*e.g.* –O–,–SO_2–, –CH_2–, NHCO–) or less symmetric links, such as meta- or ortho-catenated aromatic units, (2) introducing bulky pendant groups and biphenyl units onto the backbone, and (3) utilizing fluorinated monomers [12]. One of the most successful approaches to enhance transparency, to decrease the dielectric constant and simultaneously to improve solubility consisted in incorporation of aliphatic monomers in the PI backbone, which lead to development of a new type of PIs, namely aliphatic ones [13-17], but in this case the thermal stability is compromised [14]. Partially aliphatic PIs are occupying an intermediate position between aromatic and aliphatic derivatives; they combine the advantages of both. Cycloaliphatic monomers impart the thermal properties similar to that of aromatic ones since they foster less probability of main chain scission because of the presence of multibonds, while they offer improvements in transparency and dielectric constant since they are free of CTC [13]. Moreover, it is well known that utilization of biologically inert and nontoxic monomers leads to a greater mechanical stability during exploitation, allowing the polymer existence in a living organism without negative consequences. Comparatively to other aliphatic tetracarboxylic dianhydrides used as raw materials for PIs, Epiclon (5-(2,5-dioxotetrahydrofuryl)-3-methyl-3-cyclohexene-1,2-dicarboxylic acid anhydride) is a unique bulky, flexible and asymmetrical chemical structure, rendering new properties to the resulting compounds [18-25]. Refractive index, optical transparency, energies describing the absorption edge and surface morphology patterning approaches of some polyimides based on Epiclon and aromatic diamines containing flexible sequences are presented in this chapter, in correlation with their potential applications.

2. Synthesis approach and thermal stability of some new semi-alyciclic PIs based on Epiclon

The adopted strategy used to obtain novel semi-alicyclic PIs is focused on obtaining a proper balance between thermal and physical properties and solubility. The synthesis approach involves the control of chain flexibility and segmental mobility through the incorporation of cycloaliphatic groups and flexible linkages, so that to weaken the strong intra- and intermolecular interactions by disrupting coplanarity and conjugation, reducing symmetry, and separating electronic segments [26]. In this context, in order to obtain soluble PIs, with enhanced transparency and insulation properties, the backbone was chemically modified/designed by:

- incorporation of alicyclic dianhydride: Epiclon (98% purity, Merk) to increase the solubility (due to the less polymer–polymer interactions) and thermal stability (due to the multibond and rigidity of the alicyclic structure) [27];
- incorporation of flexible linkages (–O–, –CH_2–, – $(CH_2)_2$–) to introduce "kinks" to the polymer chain which decrease the rigidity, inhibit the packing, reduce the intermolecular interactions, weaken the intensity of the yellow color, and increase the solubility [28].

The new polyimides **PI1-PI5** were synthesized by solution polycondensation of equimolar amounts of Epiclon with five diamines containing different flexible sequences, namely: *p*-phenylenediamine (PPD), 4,4'-methylenedianiline (DDM), 4,4'-ethylenedianiline (DDE), 4,4'-oxydianiline (ODA) and 1,4-bis(*p*-aminophenoxy)benzene (PDA). The reactions were carried out in N-methylpyrrolidinone (NMP), as solvent, under inert atmosphere (Scheme 1) [22-25]. Concentration of the reaction mixture was adjusted to 20% total solids. The first step of this reaction performed under completely anhydrous conditions at 15-20°C, led to the poly(amic acid)s (**PAA1-PAA5**). In the second step, the polymer solution was heated to 180-190°C to perform the cyclodehydration of the poly(amic acid) to the corresponding polyimide structure (**PI1-PI5**).

Polyimide films were prepared through imidization of PAA films cast on a glass substrate, which were placed overnight in an 80 °C oven to remove most of the solvent. The semidried PAA films were further dried in an oven and transformed into PIs, by the following heating program: 120 °C, 160 °C, 180 °C, 210 °C and 250 °C for 1h at each temperature. After stripping the films in hot water, the resulting samples were dried at 65 °C in a vacuum oven for 24 h. The thermal imidization leads to cross-linked, partially aliphatic PI films, since the special structure of the Epiclon dianhydride (Scheme 1) [22-25,29] is different from that of conventional PIs.

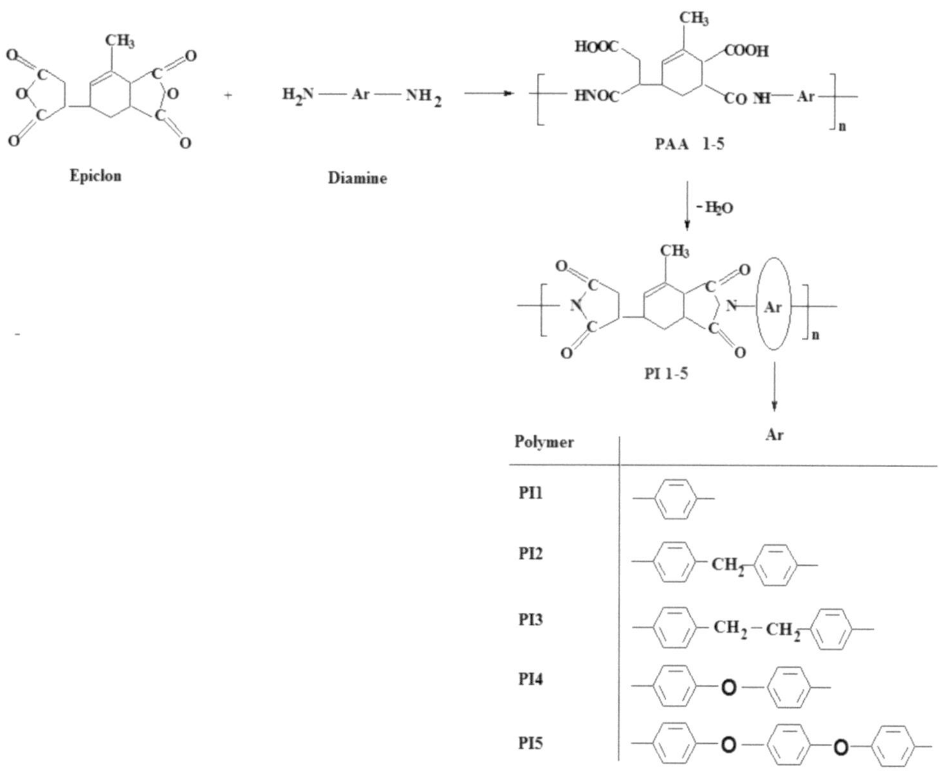

Scheme 1. Synthesis of the Epiclon-based polyimides

The number average molar mass, M_n, and the polydispersity index, M_w/M_n, for the PIs (Table 1) were determined by gel permeation chromatography (GPC) measurements in N,N-dimethylformamide (DMF), on a PL EMD-950 evaporative light scattering detector apparatus. Considering that the polymerization degrees of the PAAs and of the corresponding PIs are identical, the number-average molar mass, M_n, of the PAAs was determined from the polymerization degree of the polyimides (M_n/m_{PI}) and the molar mass of the structural units of poly(amic acid)s, m_{PAA}, according to Table 1 [30].

Conversion of PAAs to fully cyclized PIs was determined by FTIR vibrational spectroscopy. The absorption bands most frequently used for determining imide formation are 1780 cm^{-1} (C=O asymmetrical stretching), 1720 cm^{-1} (C=O symmetrical stretching), 1380 cm^{-1} (C-N stretching), and 725 cm^{-1} (C=O bending) [31,32]. The amide bands at 1660 cm^{-1} (C=O) and 1550 cm^{-1} (CONH), which may also appear as broad peaks, are useful for qualitative assessment of the imidization degree [33,34]. The polymers containing Epiclon moieties showed characteristic peaks at 2930–2920 cm^{-1}, associated with the aliphatic sequences. In Figure 1 is exemplified the FTIR spectra of the PI1 and PI4. It can be observed that the broad absorption band at 3350–3450 cm^{-1}, characteristic to amidic NH, and the narrow absorption peak at 1650–1660 cm^{-1}, due to the C=O groups in amide linkage, entirely disappeared, indicating completion of thermal imidization of the intermediate polyamidic acid into a final PI structure, confirming the successful synthesis of polymers.

Sample	m_{PAA}	M_n, g/mol	Sample	m_{PI}	M_n, g/mol	M_w/M_n	IDT,°C	$T_{g, PI}$,°C
PAA1	372	63 000	PI1	336	57 000	1.28	270	258
PAA2	462	53 000	PI2	426	50 000	1.59	273	255
PAA3	476	75 000	PI3	440	70 000	1.27	274	246
PAA4	462	76 000	PI4	426	48 000	1.31	281	235
PAA5	556	77 000	PI5	520	75 000	3.36	280	230

Table 1. Molar mass of chain units, m, number average molar mass, M_n and polydispersity indices, M_w/M_n, of poly(amic acid)s and polyimides [21,23,25,29,30]

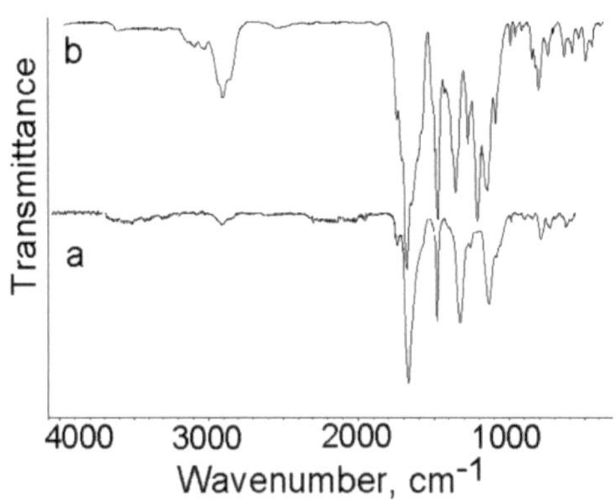

Figure 1. FTIR spectra for (a) **PI1** and (b) **PI4** samples

The thermal stability of the Epiclon-based PIs (Table 1) was evaluated by thermogravimetric analysis (TGA) and differential scanning calorimetry (DSC). The PIs exhibit an initial decomposition temperature (IDT) in the range of 270-281°C. The glass transition temperature (T_g) of PIs was probed the by monitoring the heat capacity as a function of temperature. The T_g could be considered as the temperature at which a polymer undergoes extensive cooperative segmental motion along the backbone. Table 1 presents the T_g's of resulting PIs that differ in diamine structure. In particular, the chain stiffness has been found to have an important influence on the T_g values. The presence of flexible linkage increased the polymer chain flexibility by decreasing the energy of internal rotation [35], thereby lowering the T_g as is evident from Table 1, showing that the flexible diamines containing ether-linkages exhibit a little lower T_g values. The cycloaliphatic structure of Epiclon renders to the resulting PIs good thermal resistance and also improves other properties, especially the optical ones [36].

3. Refractive index, optical gap and absorption edges

Since in applying PIs to optical and electrical use the refractive index is one of great importance, this parameter must be evaluated in correlation with the chemical structure. Ellipsometry is a modern technique that uses polarized light to probe the refractive index, absorption coefficient and optical gap. The method allows evaluation of these parameters from the correlation between the incidence angle, the amplitude of reflected and incident electric field ratio, and their phase difference, which are related to the ratio of Fresnel reflection coefficients. Variable angle ellipsometry performs measurements as a function of wavelength and angle of incidence. Because ellipsometers measure ratios of two values they are accurate and reproducible, even at low light levels, and no reference material is necessary. The dependence of the ellipsometric parameters on the incidence angle is important in electing the correct mathematical model for determination of the optical constants. The basic ellipsometry equation is [37]:

$$\tan\Psi e^{i\Delta} = \rho = \frac{r_p}{r_s} \tag{1}$$

where r_p and r_s are the complex Fresnel reflection coefficients for p and s polarized light (polarized in the incidence plane or perpendicular to it, respectively).

The refractive index, n, and extinction coefficient, k, values were calculated from the measured values of ellipsometric angles Ψ and Δ, using equation (2):

$$(n+jk)^2 = \sin^2\theta\left\{1+\tan^2\theta\left[\frac{1-\tan\Psi e^{j\Delta}}{1+\tan\Psi e^{j\Delta}}\right]^2\right\} \tag{2}$$

where θ is the incidence angle.

Figure 2 presents the dependence of the ellipsometric parameters Ψ and Δ on the incidence angle for the Epiclon-derived PIs. The shape of this dependence is important in electing the correct mathematical model for determination of the optical constants. Using the principal angle ellipsometry method ($\Delta = -\pi/2$) [37], the equation (2) becomes as shown in relation (3):

$$(n+jk)^2 = \sin^2\theta_p\left[1+\tan^2\theta_p\left(\cos 4\Psi_p + j\sin 4\Psi_p\right)\right] \tag{3}$$

where: θ_p is the principal incidence angle.

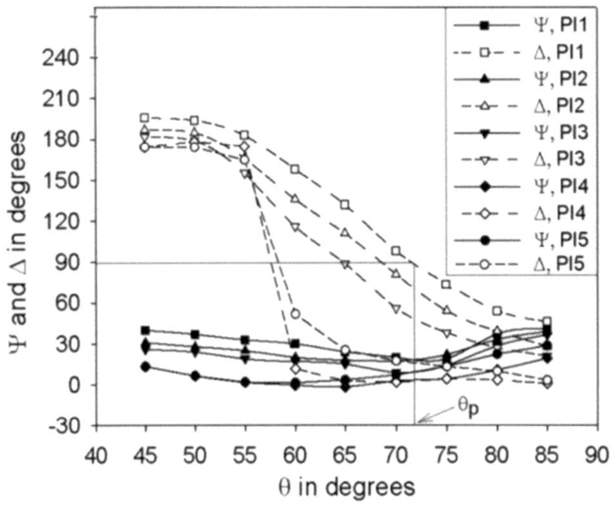

Figure 2. Dependence of the ellipsometric parameters Ψ and Δ on the incidence angle for Epiclon-derived PIs

The low polarizability provided by the alicyclic Epiclon moieties leads to PIs with low refractive index. The ellipsometricaly measured values listed in Table 2 show that refractive indices of this type of PIs measured at 632.8 nm correspond to transparent materials, *i.e.* 1.6-1.7, and are in good agreement with theoretical estimations based on molar refraction and molar volume derived from group contributions theory [29,36]. This approach is based on the assumption that the molar volume, V_u, and the molar refraction, R_u, of the chain repeating unit are additive functions of composition:

$$V_u = \sum_i a_i \cdot V_i \tag{4}$$

$$R_u = \sum_i a_i \cdot R_i \tag{5}$$

where V_i and R_i are the contributions of group, and a_i is the number of groups i in the repeating unit.

The increments of various substructures R_i and V_i were taken from the literature [38,39]. The refractive index of Epiclon based polyimides were determined by equation proposed by Lorenz-Lorentz [40,41]:

$$R_{u,LL} = V_u\left(n_{LL}^2 - 1\right)/\left(n_{LL}^2 + 2\right) \tag{6}$$

The experimental values of the refractive index for these semi-alicyclic PIs slightly differ from those calculated with the Lorenz-Lorentz equation, on considering the corresponding group contributions to the molar refraction and to the molar volume.

Both refractive index and extinction coefficient, *k*, are influenced by the aromatic character of the diamines used in synthesis. The decrease of the charge transfer interactions is reflected by a small probability of optical transitions and implicitly by low values of n and k.

Samples	R_i (cm^3/mol)	V_i (cm^3/g)	n_{LL}	$n_{ellipsom}$	k	$\varepsilon' = (n_{ellipsom})^2$
PI1	82.076	209.163	1.713	1.725	2.883	3.273
PI2	111.815	290.603	1.691	1.683	2.209	3.115
PI3	116.319	306.131	1.685	1.646	1.808	2.980
PI4	108.936	284.127	1.693	1.615	0.097	2.869
PI5	135.796	359.091	1.681	1.611	0.092	2.855

Table 2. Theoretical and experimental refractive index, extinction coeficient and dielectric constant values of Epiclon-derived PIs

The control of refractive index, n, of polyimides also means the control of dielectric constant (ε') according to Maxwell's relation $\varepsilon' = n^2$. Therefore, ε' around 1 MHz is evaluated to be equal about 1.1 times n square including an additional contribution of approximatively 10% from the infrared absorption, that is $\varepsilon' = 1.1n^2$. Materials with relatively low dielectric constants provide effective electrical barrier properties. A lower dielectric constant indicates that the respective material is suited for protection and insulation in microelectronics. In multilevel electronics packaging, an important function of a dielectric material is to block the electromagnetic interactions between parallel metal conducting lines, which operate independently. For assuming an effective blocking the material should have low ε'. The dielectric constant is used to describe a material's ability to store the charge, when used as a capacitor device. Materials with dielectric constants below 4.0, which correspond to the value of the standard SiO_2 insulator, have been recognized by the electronics industry as being superior in electrical performance to ceramics. For this reason, when polyimide films are used as interlayer dielectrics in microelectronic packaging, the pulse propagation delay is proportional to ε' of the polymer and accordingly, the reduction of ε' allows faster machine time concurrently with lower crosstalk noise.

Generally, the literature shows that for polyimides the dielectric constant decreases gradually with increasing frequency. This variation is attributed to the frequency dependence of the polarization mechanisms which include the dielectric constant. The magnitude of the dielectric constant is dependent on the ability of the polarizable units in a polymer to orient fast enough to keep up with the oscillations of an alternating electric field. At optical frequencies (approx. 10^{14} Hz), only the lowest mass species, the electrons, are efficiently polarized. At lower frequencies, the atomic polarization of nuclei, which move more slowly, also, contribute to the dielectric constant. Atomic polarization of induced dipoles such as a carbonyl group may occur in infrared (10^{12} Hz) or lower frequency regimes. Dipole polarization represents the redistribution of charge when a group of atoms with a permanent dipole align in response to the electric field. In solid state, the alignment of permanent dipoles requires considerably more time than electronic or atomic polarization, occurring at microwave (10^9 Hz) or lower frequencies. The contribution of each polarization mode to the dielectric constant is expressed in equation (7):

$$\varepsilon = \varepsilon_{electronic} + \varepsilon_{atomic} + \varepsilon_{dipolar} \tag{7}$$

At optical frequencies, when only electronic polarization occurs, the dielectric constant is related to the refractive index by Maxwell's identity. A comparison between ε' values at different frequencies may assure a basic understanding of the influence of the molecular structure on the dielectric properties of Epiclon-based PIs.

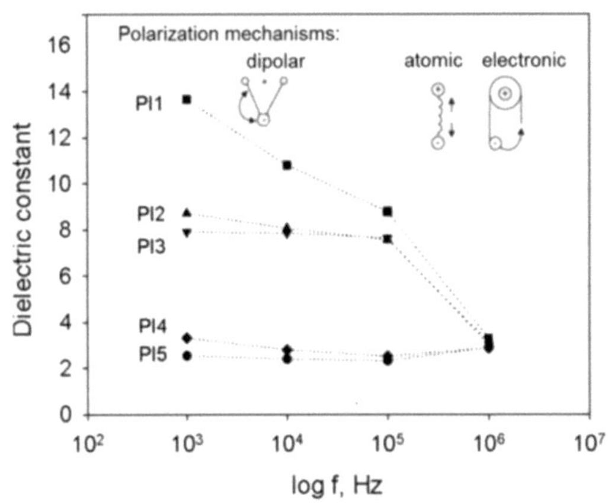

Figure 3. Dielectric constant dependence on frequency for Epiclon-derived PIs. The ε' values at 1 kHz, 10 kHz and 100 kHz were experimentally obtained on a LCR METER, while those at 1 MHz were estimated by Maxwell's identity using ellipsometrical measurements

Introduction of different Ar structures in PI backbone determines variation of the free volume by changing of the polarizable group number per unit volume, which leads to the modification of polymer densities and, implicitly, of the dielectric constants. Thus, according to literature data [42,43], a higher free volume determines a lower dielectric constant. In Figure 3 it can be observed that dielectric constants of Epiclon-derived PIs containing ether bridges are around 2.8 at 1 MHz, recommending this type of PIs as interlevel dielectric layers in the fabrication of semiconductor chips and multichip packaging structures [44].

On the other hand, PIs containing cycloaliphatic structures are expected to have better transparency than aromatic ones, due to the prohibition of electron conjugation by the introduction of alicyclic moieties [45]. This property is a key factor in visualization of the cell culture morphology development on the patterned polymer substrate, but it is also required to obtain liquid crystal devices with good quality. Optical transparency of Epiclon-derived PI films was evaluated in the range of 290-1100 nm (Figure 4). The combination in the PI backbone of cycloaliphatic Epiclon moieties with aromatic diamine residues containing different flexible sequences leads to a transmittance of appreciatively 90% over visible domain. The transparency band can be extended to UV region depending on the diamine chemical structure. The cutoff wavelength (at which transmittance is 1%) is lower as the diamine chain length is shorter, namely 316 nm – value comparable to that of a fully alicyclic PI [14].

To obtain the energy gap and other energies describing the absorption edge, the method proposed by Tauc for amorphous semiconducting materials is applied. Then, on the basis of these values, the influence of polymer chain structure and structural disorder on the optical properties and probable electronic transitions can be considered. In order to determine the absorption coefficient a from the transmission data, the expression (8) was used:

$$\alpha = (1/d)\ln(1/T) \tag{8}$$

where d is the film thickness (approx. 40 μm) and T is the transmittance.

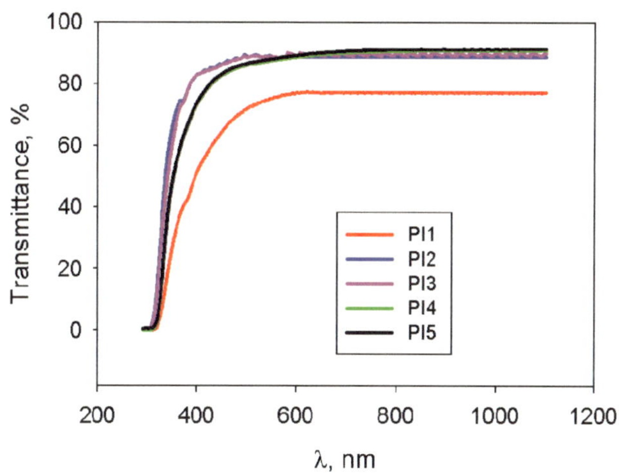

Figure 4. Typical overall transmission spectra of Epiclon-based polyimide films

Generally, for a typical amorphous semiconductor, three domains are evident in the variation of the absorption coefficient *versus* photon energy:

- in the first region, the absorption coefficient due to interband transition near the band-gap describes the optical gap energy E_G in amorphous semiconductors;
- in the second region, absorption at photon energy below the optical gap depends exponentially on photon energy, which defines the Urbach edge energy, E_U. The theories of the Urbach edge are based on the idea that the sharp absorption edge is broadened by some kind of mechanisms. For example, in semiconductors, the exponential edges are due to the electric fields produced by charged impurities. Besides the charged impurities, there are also other possible sources of internal electric fields. One of them is represented by density fluctuations via the piezoelectric effect in

semiconductors, with a piezoelectric constant different from zero. In amorphous materials, these density fluctuations do not change with time, so that the exponential edge can be thought of as due to frozen-in longitudinal optical phonons;

- the third region describes the optical absorption generated by defects appearing at an energy lower than the optical gap energy. This energy, E_T, the so-called Urbach tail, refers to the weak absorption tail and describes the defect states; this energy is rather sensitive to the structural properties of the materials. The absorption tail lies below the exponential part of the absorption edge (the second region), and its strength and shape were found to depend on the preparation, purity and thermal history of the material, varying only slightly with its thickness.

From the transparency data, the absorption coefficient was plotted in Figure 5 for Epiclon-based polyimide films, as a function of photon energy, according to equation (9):

$$\alpha = \alpha_0 \exp(E / A) \tag{9}$$

where α_0 is a constant and E is the photon energy.

The shape of all curves is very similar to the behavior proposed by Tauc for a typical amorphous semiconductor [46-48], the level of absorption (10^1–10^3 cm^{-1}) is lower than for amorphous, inorganic thin films. These results agree with other literature data [49,50], which assume that a lower absorption in polymer materials is due to a lower degree of bonding delocalization. An absorption edge is a sharp discontinuity in the absorption spectrum by an element appearing when the photon energy corresponds to the energy of a shell of the atom. Each of the absorption edges in Figure 5 exhibits two different slopes, and a saturation region for higher energies. Parameter A becomes either E_U, in the high-energy region, or E_T, in the low-energy region of absorption, according to the slopes exemplified in Figure 5 for Epiclon-derived PIs.

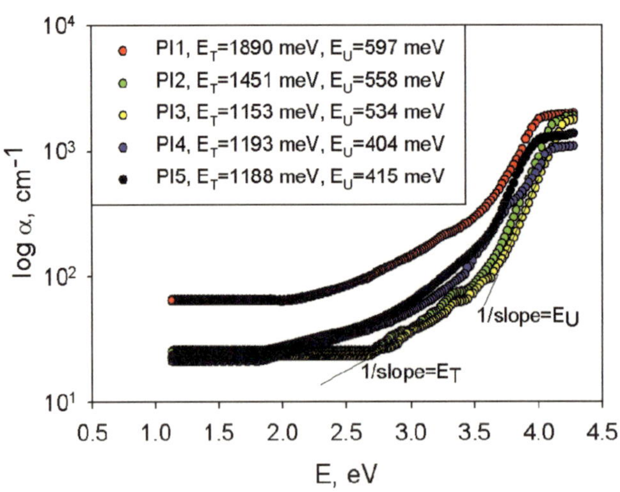

Figure 5. Absorption coefficient dependence on photon energy for Epiclon-based polyimide films

In the case of the Epiclon-based polyimide films, the absorption edges were found to follow the Tauc power law expressed in equation (10), in the range over which photon energy was higher than optical gap energy, E_G:

$$(\alpha \cdot E)^{1/2} = B(E - E_G) \tag{10}$$

where B is a constant.

The dependencies plotted in Figure 6 were used to obtain the Tauc optical gap energy, E_G. This approach, typical for amorphous semiconductors, has been also applied by us earlier for amorphous polymer films [51]. The values of optical gap energy higher than 3.26 eV are a good indicative of transparent polyimide films [50]. Also, Urbach and Tauc energies are distinctly lower as the conjugation in the aromatic structure and/or the intermolecular and intramolecular charge-transfer interactions are diminished [52]. These two optical energies are describing the absorption edges in rapport with the localized states induced by the polymer atomic structure. Such possible structure defects like the break, the abbreviation or the torsion of polymer chains, seem to be responsible for low-energy absorption, described by the Tauc parameter. Therefore, optical properties can be correlated with surface morphological aspects, offering useful insights for obtaining polyimide films for advanced microelectronic applications and bio-technologies.

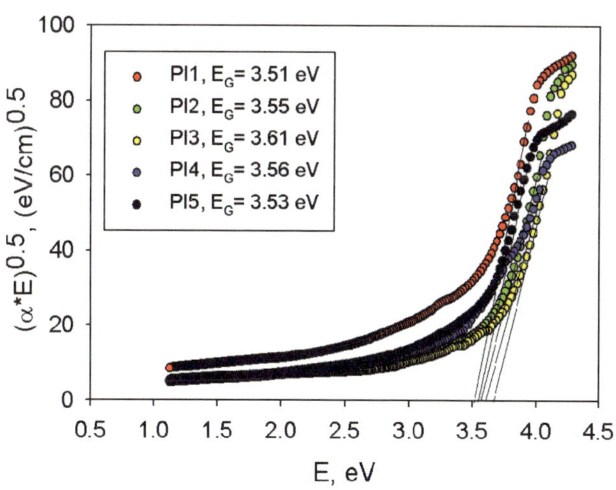

Figure 6. The Tauc dependence of Epiclon-based polyimide films

4. New insights on polyimide surface morphology and on patterning approaches

The characterization of PIs morphology involves a variety of different microscopic techniques such as optical microscopy (OM), scanning electron microscopy (SEM) and atomic force microscopy (AFM). Among them, AFM has proved to be a useful and attractive tool to examine the surface topography and the possible existing defects. Moreover, AFM in tapping mode is suitable for soft PI films and can provide high resolution surface topographical and surface distribution data on the nanoscale. Besides the preparation history, the chemical structure determines the developed PIs morphology, which can be more carefully examined through the surface texture parameters. They can be accurately evaluated only using this method. In analyzing PI surface characteristics there are no studies (to our knowledge), which take into consideration these parameters.

The morphological features of PIs play an important role in various applications, such as tissue engineering - where the cell morphology can be controlled through polymer topography - or in liquid crystal displays (LCDs) - where the tilt angle of the liquid crystal (LC) molecules can be optimized by PI substrate. However, for these purposes the PI surface must be properly adapted [53]. In this sub-chapter are presented several surface processing techniques particularized for the above discussed semi-alicyclic PIs. Discussion of the results will try to highlight new aspects in explaining the surface ordering mechanisms and in patterning approaches.

4.1. Surface texture parameters

Three-dimensional surface topography techniques and parameters can be successfully applied in characterizing polyimide surfaces, even after they have been subjected to certain surface modification processes. Recently, specific 3D parameters, also called S-parameters, have been implemented. They are calculated from three-dimensional topographic data measurements [54]. In the last ten years efforts to standardize these 3D parameters have been made. Therefore, the European consortium divided roughness parameters into four general categories: amplitude parameters (based on overall height), spatial parameters (based on frequencies of occurrence of characteristics), hybrid parameters (based on the combination of height of the formations and frequencies of occurrence) and functional parameters (based on suitability for certain applications). Some of them are presented in Table 3 and they are used to emphasize the character of the surface formations, the wear ability of the surface, the fluid retention and the orientation of structuring.

Therefore, atomic force microscopy (AFM) was used to examine the Epiclon-based polyimide films and to measure their surface topography. Figure 7 plots the 3x3 μm^2 bidimensional representation of the surface morphology (small image from the right corner), material ratio curves and height histograms obtained for polyimides **PI1-PI5** and used for surface texture parameters determinations. As it can be observed, all the samples showed uniform and smooth surface morphology, derived mainly from the characteristics of the polymer chains that govern aggregation and molecular ordering, which occur during drying and thermal imidization processes. Bearing curve illustrating the three areas of interest used for calculation of the functional parameters: Sbi, Sci and Svi is presented in Figure 8.

Roughness parameters	Description	Symbol	Unit
Amplitude parameters	Average height	Ha	nm
	Average roughness	Sa	nm
	Root mean square roughness	Sq	nm
	Peak Height	Sp	nm
	Valley Depth	Sv	nm
	Peak to Valley Height	St	nm
Spatial parameters	Texture Direction of a surface	Std	deg
	Texture Direction Index	Stdi	
	Texture Aspect Ratio of a surface	Str	
Hybrid parameters	Surfaces Area Ratio	Sdr	%
	Projected Area	S2A	nm^2
	Surface Area	S3A	nm^2
Functional parameters	Surface Bearing Index	Sbi	
	Core Fluid Retention Index	Sci	
	Valley Fluid Retention Index	Svi	
	Peak Material Volume	Vmp	ml/m^2
	Core Material Volume	Vmc	ml/m^2
	Core Void Volume	Vvc	ml/m^2
	Valley Void Volume	Vvv	ml/m^2

Table 3. Tridimensional surface roughness parameters listed by their name, symbol and unit

Height distribution analysis was used to estimate the peak height (defined as the largest peak height value from the mean plane within the sampling area), the valley depth (defined as the largest valley depth value from the mean plane within the sampling area) and the peak to valley height (defined as the sum of the largest peak height value and the largest valley depth from the mean plane within the sampling area). The higher values of these parameters obtained for **PI1** and **PI5** films (Figure 7) were reflected also by a higher rms roughness (the root mean square of the surface departures from the mean plane within the sampling area) of the 2.35 nm and 1.60 nm and average roughness (the arithmetic mean of the absolute distances of the surface points from the mean plane) of 1.82 nm and 1.27 nm (Table 4). For the other PIs the roughness values were below 1 nm. Material ratio curve, also known as Abbot-Firestone curve (Figure 7, gray plot), is the integral of the amplitude distribution function (ADF/Surface Histogram). It is a cumulative probability distribution and a measure of the material to air ratio expressed as a percentage at a particular depth below the highest peak in the surface. At the highest peak, the material to air ratio is 0%, where the material to air ratio at the deepest valley is 100%.

The functional parameters [55], such as surface bearing index (the ratio of the RMS deviation over the surface height at 5% bearing area) core fluid retention index (the ratio of the void volume of the sampling area at the core zone divided by the RMS deviation) and valley fluid retention index (the ratio of the void volume of the sampling area of the valley zone divided by the RMS deviation) are computed directly from this curve. The evaluation of Sbi, Sci and

Svi indicates that the bulky and asymmetric molecule of Epiclon induces a good fluid retention in the core and valley zone and good bearing properties of the corresponding polyimides – aspect useful in membrane industry and in cell culture substrates.

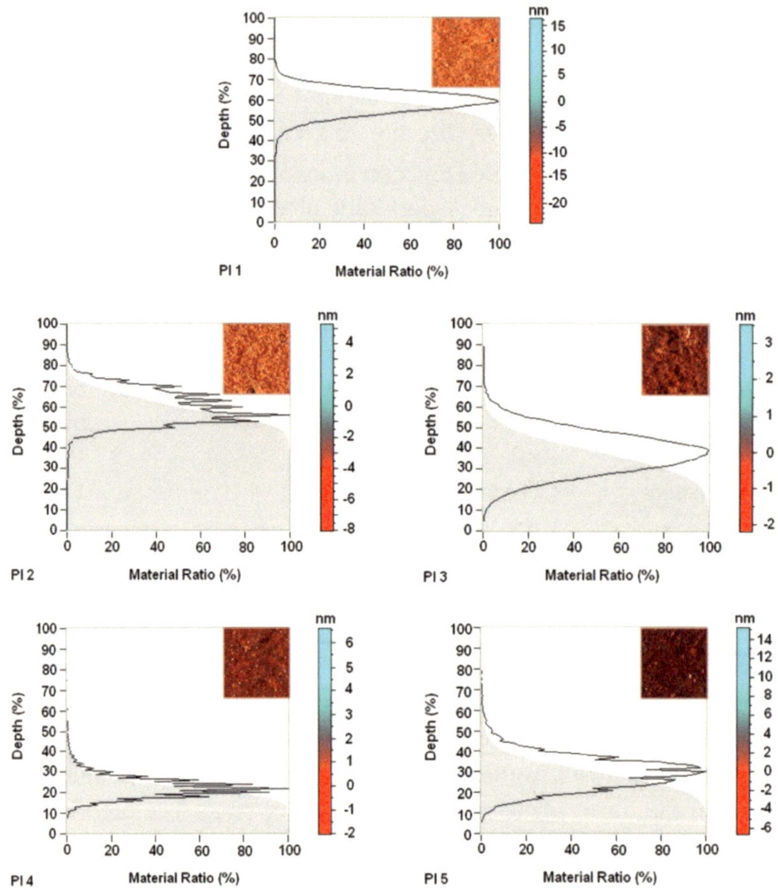

Figure 7. Amplitude distribution curves and height histograms used for surface texture parameters determinations for Epiclon-derived PIs. Each insert shows corresponding 3x3 μm² bidimensional topographical image

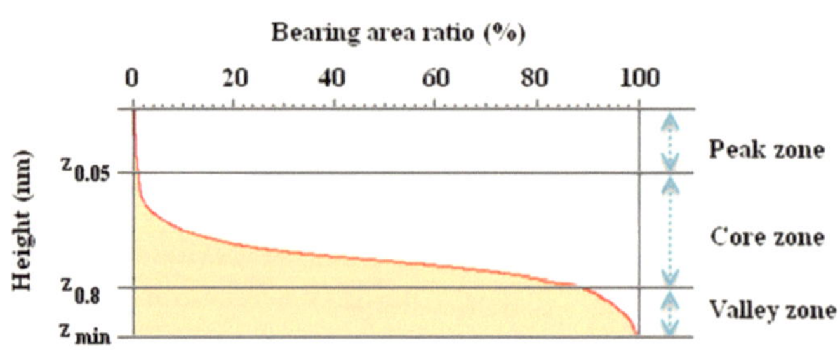

Figure 8. Bearing curve illustrating the three areas of interest used for calculation of the functional parameters: Sbi, Sci and Svi

Surface texture parameters	Samples				
	PI 1	PI 2	PI 3	PI 4	PI 5
Sa (nm)	1.82	0.81	0.42	0.33	1.27
Sq (nm)	2.35	0.97	0.53	0.48	1.60
Sp (nm)	15.31	4.77	3.15	5.28	8.75
Sv (nm)	10.95	4.37	1.77	1.95	5.01
St (nm)	26.25	9.14	4.92	7.23	13.76
Stdi	0.83	0.82	0.70	0.75	0.72
Str	0.70	0.64	0.38	0.54	0.65
Sbi	0.20	0.31	0.24	0.11	0.26
Sci	1.38	1.64	1.58	1.53	1.54
Svi	0.13	0.07	0.12	0.08	0.11

Table 4. Surface texture parameters for polyimides **PI1-PI5** calculated form height histograms, amplitude distribution curves and angular spectra

Tridimensional AFM images and corresponding angular spectra depicted in Figure 9 were used to estimate the spatial properties, related to the frequencies of occurrence of some features.

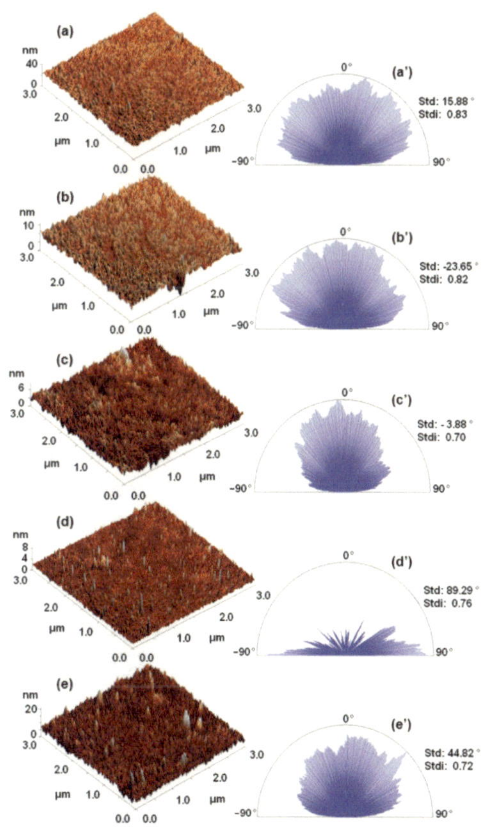

Figure 9. Tridimensional AFM images and corresponding angular spectra obtained for **PI1** (a, a'), **PI2** (b, b'), **PI3** (c, c'), **PI4** (d, d') and **PI5** (e, e') polyimide films

These can be described by parameters such as texture direction [55] (defined as the angle of the dominating texture in the image), texture direction index [56] (defined as the average amplitude sum divided by the amplitude sum of the dominating direction) and texture aspect ratio (used to identify the uniformity of the texture of the surface). For all the films, the texture direction index values (Stdi) were close to 1 (Table 4), indicating that the surfaces do not have a predominant texture direction (there is no significant lay to the surface) and the amplitude sum of all direction is almost similar. Moreover, the Epiclon-based PIs have the Str parameter higher than 0.3, revealing the isotropic character of the morphology (the surface has the same characteristics in every direction). AFM studies reveal that the enhancement of roughness is generated by the structural peculiarity of these polyimides which contain bulky, alicyclic dianhydride moieties and aromatic diamine residues with different flexibility [57]. The functional and spatial properties obtained for the discussed PI films indicate that these materials present good adhesion with circuit inorganic components, being suited as interlevel dielectric layers.

4.2. Novel aspects in patterning polyimide surface morphology

Depending on the nature of their precursors – the poly(amic acid)s – the polyimide microstructures can be typically patterned, either by direct photolithography of photosensitive type, or by lithography plus dry or wet etching of non-photosensitive type. From the variety of PI surface modification techniques the discussions were focused on main efficient ones, namely rubbing with textile materials and dynamic plowing lithography (DPL). Also, a novel approach is proposed, which consists in imprinting PI surface topography with the texture of a sheared LC matrix. The processing techniques are discussed in correlation with the remaining unsolved problems in the field, attempting to bring some contributions from this point of view.

4.2.1. Influence of textile fiber characteristics on pi morphology patterned by rubbing

Modification of polyimide morphology is one of key factors in designing LCD devices with wide viewing angle and high resolution [58]. Generally, they are unidirectional rubbed with a cloth to uniformly align the LC molecules and to generate an appropriate LC pretilt angle defined as the angle between the LC director and the AL surface plane, preventing the creation of disclinations in the LC cells [59]. The PI chemical structures greatly affect the pretilt angles. Although at industrial scale rubbing is the most used technique for topography patterning, there are still some drawbacks that could be overcome so that the LC alignment mechanism would be fully elucidated. For instance, the detailed mechanism controlling the LC alignment and the pretilt angles are not yet fully understood. Many studies have been reported on this problem by taking into consideration the following aspects: rubbing parameters (force, density and strength) [59,60], annealing [61], soaking solvent [62], surface wetting characteristics [63,64]. One possibility is that alignment is induced by grooves formed mechanically on the polymer surface by the rubbing process [65]. An alternative concept is that alignment acts through the van der Waals interactions between LC molecules and the polymer molecules, since rubbing causes the anisotropic orientation of the polymer molecules [66-68]. In this context, there are fewer studied aspects, such as the nature and the texture of the fabric fiber and the polymer surface hardness, which are factors that have to be considered in explaining the LC alignment. It is presumed that the texture and toughness of the fabric fibers and surface mechanical properties of the polymer are strongly related with the stability and dimensions of the features created on the PI surface, thus leading to a better control to LC orientation.

In order to elucidate the influence of these factors, a fundamental research has been employed on the Epiclon-based PIs containing ether linkages in the structural unit [69], since they exhibit good transparency and the lowest dielectric constants from the above discussed PIs. The **PI4** and **PI5** films were rubbed with two types of velvet: a natural one from cotton (V_C) and a artificial one made from cellulose diacetate (V_{CD}). The texture of these fabrics is different, as shown in SEM images from Figure 10. The cotton velvet is constituted from fibers of 950 μm length and 15 μm thickness, while the cellulose one has 1.5 mm length and 25 μm thickness. Figures 12 and 13 present the AFM images of the semi-alicyclic samples **PI4** and **PI5** before rubbing treatment. The rubbing process was performed

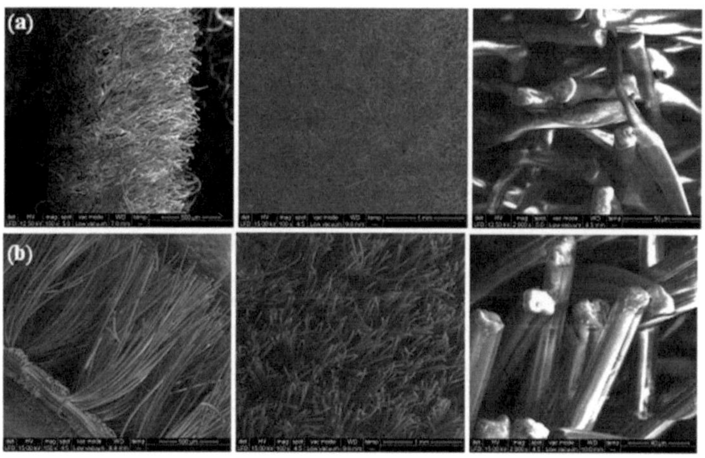

Figure 10. SEM images obtained for (a) cotton velvet and (b) velvet of cellulose diacetate

with a device, consisting in a cylinder on which was attached the velvet and the same experimental conditions were kept (100 rot/min). The surface ordering induced by rubbing is maintained over large areas (> 20x20 μm^2), but for a detailed analysis of the textile fiber effect on PI surface topography, the dimensions of the created features were examined on smaller scan areas (Figures 11 and 12). For both **PI4** and **PI5** there were obtained two dimensional and three-dimensional topographic images before and after surface ordering by rubbing. On the basis of these images, the average roughness (Sa) and percentage of surface excess (Sdr) were calculated. On each 3D image was made a diagram of the corresponding height profile of random lines, for a better visualization of the surface formations in section. Cotton velvet, although has shorter fibers compared with the artificial one, produced deeper grooves. This effect can be explained by taking into account the flexibility of the textile fibers. Therefore, natural velvet fibers are more rigid and better penetrate the PI surface, resulting in more intense tracks. On the other hand, although artificial velvet fibers are longer, have greater flexibility leading to a greater flexion during rubbing, reducing the size of the surface grooves created on polyimides. For both samples, a better regularity of surface topography is achieved when rubbing with cotton velvet. Average roughness and surface excess percentage varied for unmodified and rubbed **PI4** and **PI5** films, according to the data presented in Table 5.

Sample	Sa (nm)	Sdr (%)	Stdi	Str	Isotropy (%)	Svi	Vmp (ml/m²)	Vmc (ml/m²)	Vvc (ml/m²)	Vvv (ml/m²)
PI4	0.3	0.12	0.79	0.35	65.80	0.07	$5.0 \cdot 10^{-5}$	$3.3 \cdot 10^{-4}$	$5.4 \cdot 10^{-4}$	$4.1 \cdot 10^{-5}$
PI4V_C	4.0	3.59	0.20	0.03	5.91	0.13	$2.1 \cdot 10^{-4}$	$4.5 \cdot 10^{-3}$	$5.6 \cdot 10^{-3}$	$6.7 \cdot 10^{-4}$
PI4V_{CD}	2.2	1.06	0.28	0.04	6.49	0.12	$1.3 \cdot 10^{-4}$	$2.4 \cdot 10^{-3}$	$3.2 \cdot 10^{-3}$	$3.5 \cdot 10^{-4}$
PI5	1.3	1.61	0.68	0.54	63.2	0.09	$1.2 \cdot 10^{-4}$	$1.5 \cdot 10^{-3}$	$2.0 \cdot 10^{-3}$	$1.6 \cdot 10^{-4}$
PI5V_C	4.7	3.21	0.19	0.05	5.01	0.11	$3.0 \cdot 10^{-4}$	$5.2 \cdot 10^{-3}$	$7.2 \cdot 10^{-3}$	$6.7 \cdot 10^{-4}$
PI5V_{CD}	3.0	1.36	0.23	0.06	9.10	0.13	$3.1 \cdot 10^{-4}$	$3.0 \cdot 10^{-3}$	$4.6 \cdot 10^{-3}$	$5.6 \cdot 10^{-4}$

Table 5. Roughness parameters calculated from AFM data for unmodified and rubbed **PI4** and **PI5**

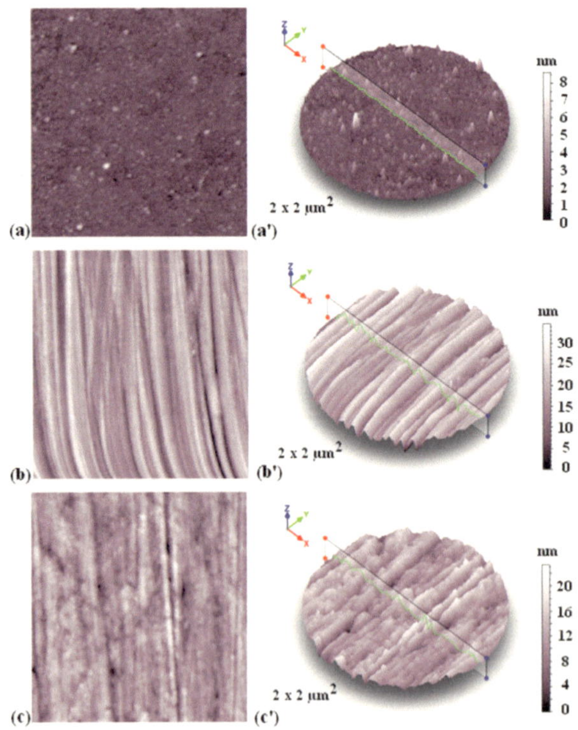

Figure 11. Bi-dimensional and three-dimensional tapping mode AFM images obtained for **PI4** polyimide film before (a, a') and after nanostructuring process induced by rubbing with a cotton velvet (b, b') and a velvet of cellulose diacetate (c, c')

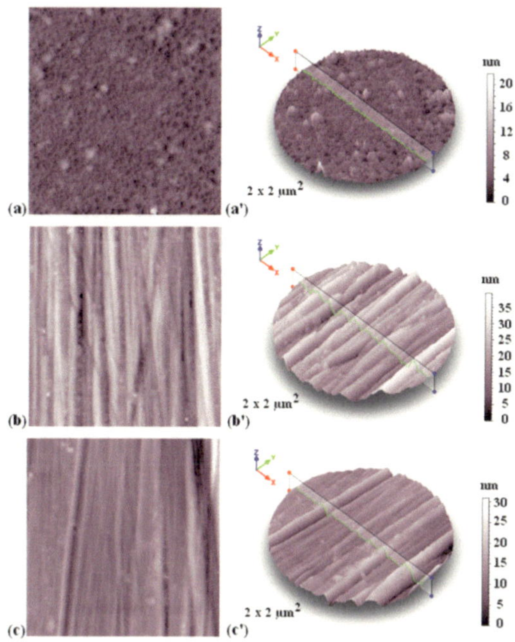

Figure 12. Bi-dimensional and three-dimensional tapping mode AFM images obtained for **PI5** polyimide film before (a, a′) and after nanostructuring process induced by rubbing with a cotton velvet (b, b′) and a velvet of cellulose diacetate (c, c′)

Regarding this, in both PI films, the utilization of cotton velvet led to nanostructuring of the surface which changed significantly roughness parameters.

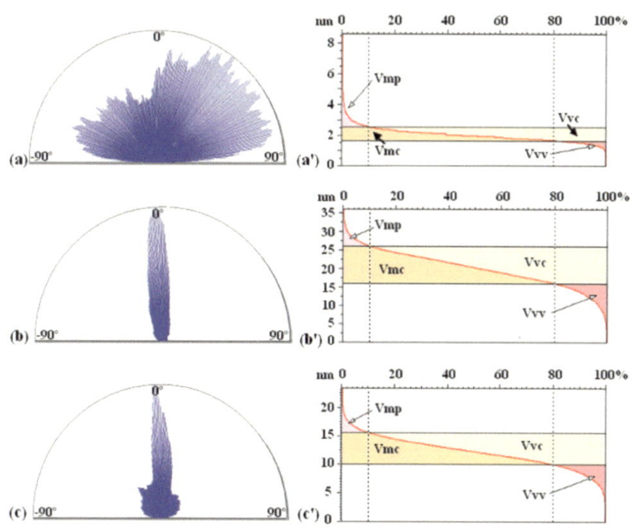

Figure 13. Angular spectra used for determination of the spatial parameters and Abbott-Firestone curves used for calculation of the volume functional parameters, obtained for **PI4** film before (a, a′) and after nanostructuring process induced by rubbing with a cotton velvet (b, b′) and a velvet of cellulose diacetate (c, c′)

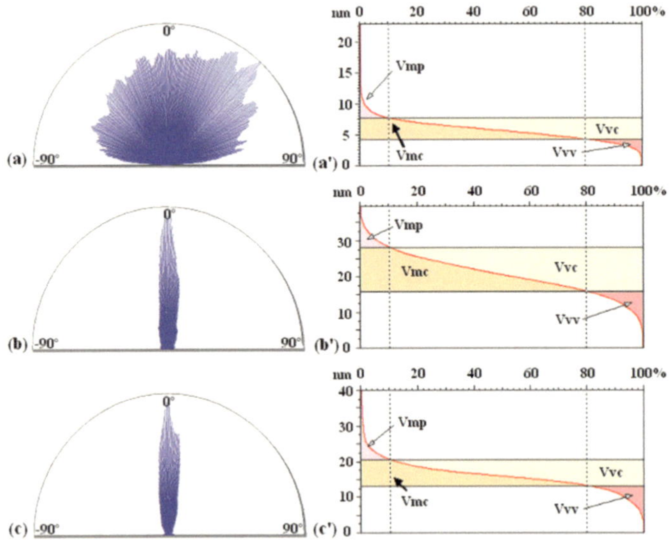

Figure 14. Angular spectra used for determination of the spatial parameters and Abbott-Firestone curves used for calculation of the volume functional parameters, obtained for **PI4** film before (a, a') and after nanostructuring process induced by rubbing with a cotton velvet (b, b') and a velvet of cellulose diacetate (c, c')

A possible mechanism of surface ordering by rubbing is that during this process local heating is produced. Thus, the mechanical stress, and thereby increasing temperature generally results in a increase in malleability of the polyimide film surface. Moreover the alignment between LC molecules and the polymer molecules acts not only through the van der Waals forces, but also through the electrostatic forces induced by rubbing.

The angular spectra from Figures 13 and 14 for **PI4** and **PI5** reveal the generation of a surface anisotropy after rubbing process. It can be observed that for unmodified samples, the formations at the film surface are not oriented by a preferred direction, but randomly. The surface isotropy of **PI4** is of 65.80%, while for **PI5** is 63.20%. Surface isotropy has decreased considerably after the rubbing process of PI films, thus the surfaces become anisotropic due to the occurrence of ordered nanostructures in the direction of rubbing. Patterning the topography with cotton velvet leads to a more regular structure, this being confirmed by lower values obtained for isotropy of 5.91% (for **PI4**) and 5.01% (for **PI5**), compared with those obtained using cellulose diacetate velvet namely: 6.49% and 9.10% respectively for **PI4** and **PI5**. These values are supported by the Stdi index, which shows the degree of surface orientation, and by Str index which accounts for the ratio aspect of the texture. Orientation indices are smaller for unmodified PIs, according to Table 5. A similar tendency is observed also for the texture indices. These parameters indicate that a more pronounced anisotropy is achieved by rubbing with cotton velvet. Volume functional parameters (Vmp, Vmc, Vvc, Vvv), calculated from Abbott-Firestone curve, as plotted in Figures 13 and 14 for Epiclon-derived PIs, were used to evidence the improvements in surface adhesion after rubbing. For both **PI4** and **PI5**, these parameters increased during surface nanostructuring with cotton velvet, comparatively with unmodified samples.

Another influencing factor on the surface ordering is represented by the PI chemical structure. Grooves produced on the **PI5** film surface are slightly smaller and denser than on the film surface **PI4**, regardless of the material used in rubbing process. For both PIs, a great difference in the depth and number of grooves created with the two types of textiles it can be noticed: the distinct topologies could be correlated with the flexibility or different mechanical properties of the two structures. During the rubbing process, a tough polymer will be less distorted, producing a weak structure, whereas a polymer with high ductility will be slightly distorted, forming a better shaped structure. Therefore, in the case of **PI5**, which is slightly more flexible than **PI4** sample, one can distinguish more surface features. It can be concluded that the pattern created on PIs based on Epiclon is finer comparatively with other PIs processed by rubbing [70]. All these mentioned aspects recommend these semi-alicyclic PIs as orientation layers for LCDs.

4.2.2. Importance of mechanical surface properties on PI morphology patterned by DPL

Another method of patterning polyimide morphology implies a modern concept based on atomic force interactions between the AFM tip and the material. Dynamic Plowing Lithography (DPL) leads to the creation of nanometric channels on film surface. The dimension of the resulted pattern depends not only on the applied force, but also on the structural organization of the polymer. As observed in Figure 15 from height profiles the nanochannels created in **PI4** film surface are of 45 nm depth and 50 nm wide, while for the **PI5** film they have smaller dimensions, namely 10 nm depth and 20 nm wide. The surface ordering induced by DPL in case of Epiclon-derived PIs is more pronounced comparatively with other commercial PIs [71], making them more suitable for LCDs applications. The differences regarding the size of the nanochannels for **PI4** and **PI5** can be correlated with their surface mechanical properties. These characteristics were evaluated from withdraw force after AFM tip penetration into the PI surface. AFM lithography was performed in contact mode using a Si cantilever, characterized by a spring constant K_e=0.03 N/m (provided by the manufacturer). Topography images were used to obtain force-distance curves over an entire image frame. Figure 16 shows two representative experimental force curves, describing the approach and retract of the AFM tip from the PI surface. Thus, after the tip is approaching the sample surface and applies a constant and default force upon the surface that leads to sample indentation and cantilever deflection, the tip tries to retract and to break loose from the surface.

Various adhesive forces between the sample and the AFM tip hamper its retraction. These adhesive forces can be calculated directly from the force-distance curves, as seen in Figure 16. Finally, the tip withdraws and looses contact to the surface upon overcoming of the adhesive forces (at level 0). Therefore, the surface hardness can be investigated through the force-distance curves, offering information on flexibility and chain packing. The withdraw force is equal with the adhesion force and can be determined from Hooke's law:

$$F = K_e \cdot \Delta x \tag{11}$$

where Δx is the cantilever shift in rapport with PI film surface.

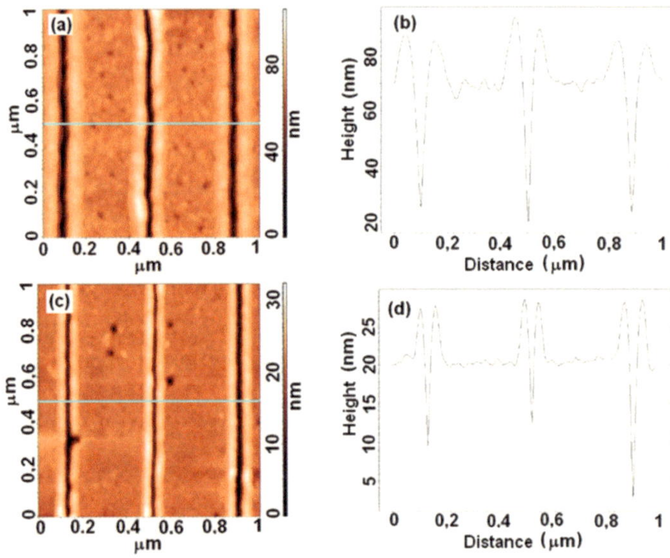

Figure 15. AFM images of (a,b) **PI4** and (c,d) **PI5** films patterned by DPL

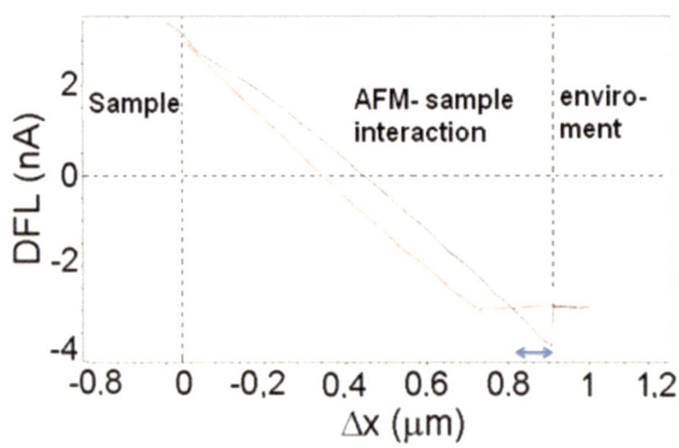

Figure 16. Typical force-distance curve showing the adhesion force for **PI4** film

The adhesion forces for **PI4** and **PI5** films, determined from expression (11), were 2.94 nN and 1.51 nN, respectively. It can be observed that adhesion force for **PI4** is slightly higher than that of **PI5**, which is in agreement with the data concerning the surface features obtained by rubbing. In this way, it can be appreciated that **PI5** exhibits, due to its higher flexibility, a better structural organization, thus explaining the smaller depths of the nanochannels produced by DPL. This type of polyimide has a patterned morphology suitable for LCDs or cell culture substrates.

4.2.3. Patterning PI surface with sheared LC matrix

A new approach for patterning polyimide, starting from its precursor, consists in using a lyotropic liquid crystal template, , namely hydroxypropyl cellulose (HPC), which under shear conditions exhibits a banded texture [72,73] (Figure 17). Besides the specific interactions, such as the hydrogen bonds between Epiclon-based PAA corresponding to **PI4** and the LC component, a slight cross-linking produced in the precursor structure by UV irradiation can be applied to stabilize the resulting morphology. The effect of UV exposure and the influence of **HPC** weight ratio on the developed morphology are reflected in the different intensities and dimensions of the induced bands. The generated texture in the PI precursor film can be distinguished even at low lyotropic content and is also maintained after its removal with a selective solvent, namely acetone.

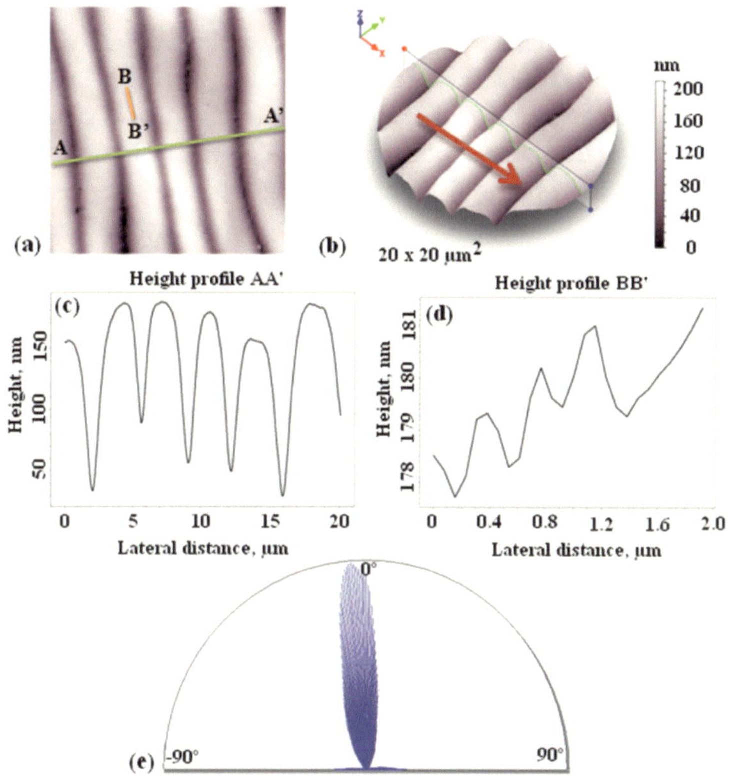

Figure 17. AFM images for **HPC** in lyotropic phase: (a) 2D-image, (b) 3D-image, (c) cross-section profiles parallel (AA') and (d) perpendicular (BB') to shear direction and (e) angular spectrum

At an equal ratio of **50 PAA/50 HPC**, the possibility of interaction between the two components increases, leading to a higher degree of ordering reflected through the formation of the third set of bands, observed only for this composition (the profile taken along the CC' direction from the 2D AFM image in Figure 18. The primary bands of the 50/50 PAA/HPC film are characterized by 6.1 μm width and 43 nm height.

The resulted surface structure can be explained using the schematic representation from Figure 19. It is well known that the free surface of a sheared HPC consists in a fibrillar morphology, with the fibrils (which are considered to be made of oriented HPC molecules) running sinusoidally along the shearing direction [74].

The primary bands are formed as a direct consequence of the sinusoidal variation in the fibrillar trajectory. When a polarizer is parallel to the shear direction, bright areas are observed where the fibril is at an angle to the shear direction, and dark areas are observed at the peaks and valleys of the sinusoidal trajectory of the fibril where the fibril is parallel to the shear direction. Thus, when large numbers of fibrils are stacked together, a banded structure is observed, with the bands running normal to the direction of shear. Bright bands are separated by dark lines which are remarked where the peaks and valleys of adjacent fibrils are registered. The secondary bands which are called "torsads" are believed to be produced as a consequence of the hydrodynamics of solvent evaporation. This situation corresponds to the structure from Figure 17. By addition of **PAA** in the system the dimensions of the large bands increase because the hydrogen bonds might lead to the stretching of the **HPC** serpentine fibers. The effect is accentuated as the content of PAA is increased – situation sustained by AFM images [72,73]. The imidization process does not affect the aspect of the final surface texture, as presented in Figure 20. The average width of the "large" bands is approximately the same before taking out **HPC**, namely 6.6 μm, while the height is larger than 120 nm. The secondary pattern is easily modified, while the third set of bands is no longer visible. The distance between the primary bands is of 4.5 μm, being a little larger after **HPC** removal. In case of directly patterning the polyimide (by this method), the band texture is larger comparatively with the poly(amic acid), and also increases with decreasing the amount of liquid crystal matrix. However, the shape of the bands is less defined probably due to the lack of hydrogen bonds. Given the surface molecular oriented Epiclon-based poly(amic acid) film, the induced alignment mechanism might be included in the category of anisotropic LC orientational elasticity in connection with an induced surface pattern. The alignment behavior of N-(4-methoxybenzylidene)-4-butylaniline (MBBA) nematic liquid crystal on patterned PI precursor was investigated (Figure 21).

Polarized Light Microscopy (PLM) revealed successively disposed dark and bright states, which change their luminosity by rotating the sample between crossed polarizers. The dark region in Figure 21a was obtained at 0° and 90° rotation of the sample director with respect to the crossed polarizers, indicating that the liquid crystal director is aligned parallel to the polarizer transmission direction. By rotating the sample from this position at 45° and 135° with respect to the crossed polarizers, bright states are observed because electric field components passing through the easy direction of MBBA give the highest resultant on the analyzer transmission direction, as revealed in Figure 21b. This behaviour is characteristic for homogeneous alignment of a liquid crystal. The high contrast between the dark and bright images indicates that the liquid crystal alignment is quite uniform.

Pure and applied research related to the shear induced morphology and structural relaxation after cessation of shear in polymer/LCP blends will become more and more important for developing high performance alignment layers used in display devices and cell culture substrates with tuned surface morphology.

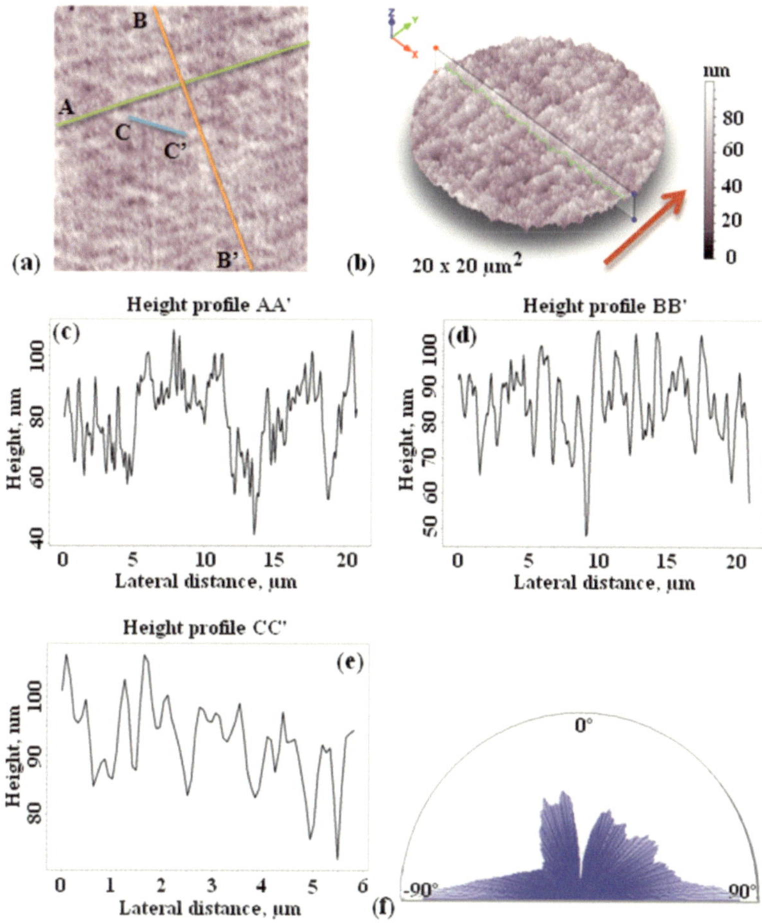

Figure 18. AFM images for **50 PAA/50 HPC**: (a) 2D-image, (b) 3D-image, (c) cross-section profiles parallel (AA′) and (d) perpendicular (BB′) to shear direction, (e) profile after CC′ and (f) angular spectrum

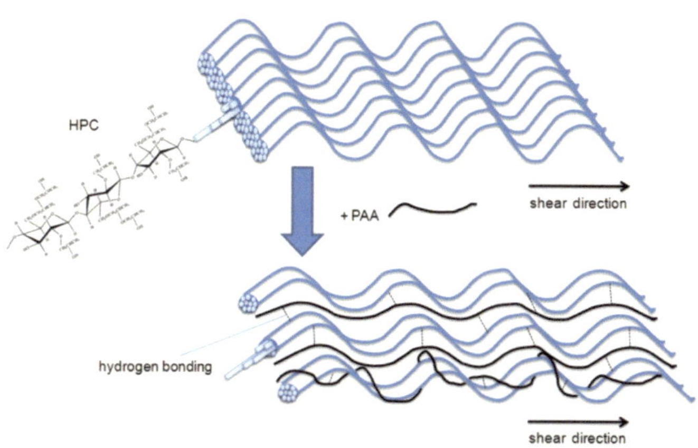

Figure 19. Schematic representation of the formation of surface banded structure in PAA/HPC

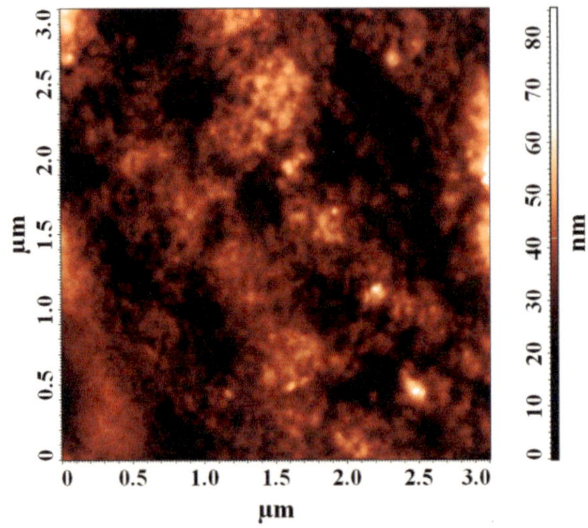

Figure 20. AFM image of imidized **PAA** after patterning with **HPC** and its removal with acetone

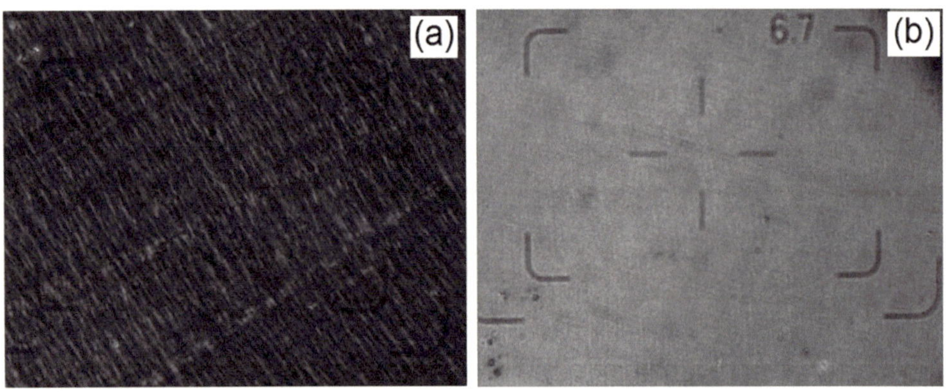

Figure 21. PLM images of MBBA on the molecular orientated **PAA** revealing (a) dark states at 0° rotation of the sample director with respect to the crossed polarizers and (b) bright states at 45° rotation of the sample director with respect to the crossed polarizers

5. Conclusion

The introduction of flexible, alicyclic Epiclon structure in the PI backbone in combination with some flexible aromatic diamines reduces the inter- and intramolecular CTC interactions, and consequently the polarizability. This approach leads to transparent materials as revealed from the refractive index values (1.6-1.7) and optical gap energies (> 3.26 eV). Also, the extinction coefficient evaluated by ellipsometry indicates that these semi-alicyclic PIs do not absorb visible light. The dielectric constant values for the samples containing ether bridges in the structural unit are lower than the classic dielectrics, namely 2.8 at 1 MHz. Moreover, their good transparency can be exploited in many applications such as LCDs or cell culture substrates. For these goals, the morphology of PIs was patterned by rubbing, DPL and by imprinting with a sheared lyotropic LC matrix. The dimensions of the created pattern depend on nature and the texture of the fabric fiber, but also on the polymer surface mechanical properties. The greater structural organization of PI5 allows finer pattern features at the surface, both by rubbing and DPL. Orientation of PI4 precursor morphology through a LC matrix is a low cost and simple approach to obtain a banded topography, with dimensions tunable by the LC content. Also, hydrogen bonds play an important role in stabilization of the desired morphology. The dimensions of the bands are not affected after the removal of HPC with acetone and after the imidization of PAA. Future studies will be focused on the influence of diamine structure on the developed morphology with this method. Also, cell culture studies will be performed and the efficiency of employed patterning techniques will be investigated.

Author details

Andreea Irina Barzic, Iuliana Stoica and Camelia Hulubei
"Petru Poni" Institute of Macromoleculari Chemistry,
Iasi, Romania

Acknowledgement

Part of this material represents the subject of the PhD thesis of Irina Barzic, elaborated under the guidance of Dr. Silvia Ioan – Senior Researcher from "Petru Poni" Institute of Macromolecular Chemistry. This author whishes to express the most sincere and deepest gratitude to her supervisor, which by her academic achievement and contribution had remarkable influence on her student entire career, being a model as scientist, mentor, and teacher. Also, the Project PN-II-ID-PCE-2011-3-0937, No. 302/5.10.2011 is acknowledged for the financial support of this work.

6. References

[1] Ghosh MK, Mittal KL. Polyimides: fundamentals and applications. New York: Marcel Dekker Inc; 1996.

[2] Kim SI, Ree M, Shin TJ, Jung JC. Synthesis of new aromatic polyimides with various side chains containing a biphenyl mesogen unit and their abilities to control liquid-crystal alignments on the rubbed surface. J. Polym. Sci. Part A 1999;37 2909-2921.

[3] Mathews AS, Kim I, Ha C-S.Synthesis, Characterization, and Properties of Fully Aliphatic Polyimides and Their Derivatives for Microelectronics and Optoelectronics Applications. Macromol. Res. 2007;15 114-128.

[4] Richardson RR Jr, Miller JA, Reichert WM. Polyimides as biomaterials: preliminary biocompatibility testing. Biomaterials 1993;14 627-635.

[5] Rousche PJ, Pellinen DS, Pivin DP Jr, Williams JC, Vetter RJ, Kipke DR. Flexible polyimide-based intracortical electrode arrays with bioactive capability. IEEE Trans Biomed Eng 2001;48 361-371.

[6] Charest JL, Bryant LE, Garcia AJ, King W (2004) Hot embossing for micropatterned cell substrates. Biomaterials 25:4767-4775.

[7] Kreuz JA, Hsiao BS, Renner CA, Goff DL. Crystalline Homopolyimides and Copolyimides Derived from 3,3',4,4'-Biphenyltetracarboxylic Dianhydride/1,3-Bis(4-aminophenoxy)benzene/ 1,12-Dodecanediamine. 1. Materials, Preparation, and Characterization. Macromolecules 1995; 28 6926-6930.

[8] Koning C, Delmotte A, Larno P, Van Mele B. Influence of polymerization conditions on melt crystallization of partially aliphatic polyimides. Polymer 1998;39 3697-3702.

[9] Eichstadt AE, Ward TC, Bagwell MD, Farr IV, Dunson McGrath JE. Structure-property relationships for a series of amorphous partially aliphatic polyimides. J. Polym. Sci. Polym. Phys. Ed. 2002;40 1503-1512.

[10] Yang CP, Chen RS, Hunk KS. Synthesis and properties of soluble colorless poly(amide–imide)s based on N,N′-bis(3-carboxyphenyl)-4,4′-oxydiphthalimide and various aromatic diamines. Polymer 2001;42 4569-4577.

[11] Jin Q, Yamashita T, Horie K. Polyimides with alicyclic diamines. II. Hydrogen abstraction and photocrosslinking reactions of benzophenone-type polyimides. J. Polym. Sci. A 1994;32 503-511.

[12] Hsiao S-H, Chen W-T. Syntheses and Properties of Aromatic Polyimides Based on 1,1-Bis[4-(4-aminophenoxy)phenyl]-1-phenyl-2,2,2-trifluoroethane and 1,1-Bis[4-(4-aminophenoxy) phenyl]-1-phenylethane. J. Polym. Res. 2003;10 95-103

[13] Schab-Balcerzak E, Grobelny L, Sobolewska A, Miniewicz A. Cycloaliphatic–aromatic polyimides based on diamines with azobenzene unit. Eur. Polym. J. 2006;42 2859-2871.

[14] Matsumoto T, Kawabata S, Takahashi R. Alicyclic Polyimides based on bicyclo[2.2.1]heptane- 2,3,5,6-tetracarboxylic 2,3-5,6-dianhydrides. High Perform. Polym. 2006; 18 719-726.

[15] Eichstadt AE, Ward TC, Bagwell MD, Farr IV, Dunson DL, McGrath JE. Synthesis and Characterization of Amorphous Partially Aliphatic Polyimide Copolymers Based on Bisphenol- A Dianhydride. Macromolecules 2002;35 7561-7568.

[16] Kim EH, Moon IK, Kim HK, Lee MH, Han SG, Yi MH. Synthesis and characterization of novel polyimide-based NLO materials from poly(hydroxy-imide)s containing alicyclic units (II). Polymer 1999;40 6157-6167.

[17] Chung CM, Cho SY, Kim SY, Moon SY. Photosensitive polyimides having N-sulfonyloxyimide and N-carbonyloxyimide groups in the main chain. Opt. Mater. 21:421-424.

[18] Mallakpour S, Zamanlou MR (2004) Synthesis of new optically active poly(amide-imide)s containing EPICLON and L-phenylalanine in the main chain by microwave irradiation and classical heating. J. Appl. Sci. 2002;91 3281-3291.

[19] Chung EY, Choi SM, Sim HB, Kim KK, Kim DS, Kim KJ, Yi MH. Synthesis and characterization of novel photosensitive polyimide based on 5-(2,5-

dioxotetrahydrofuryl)-3- methyl-3-cyclohexene-1,2-dicarboxylic anhydride. Polym. Adv. Technol. 2005;16 19-23.

[20] Hamciuc E, Lungu, R, Hulubei C, Bruma M. New Poly(Imide-Ether-Amide)s Based on Epiclon. J. Macromol. Sci. A 2006;43 247-258.

[21] Ioan S, Cosutchi AI, Hulubei C, Macocinschi D, Ioanid G. Surface and Interfacial Properties of Poly(amic acid)s and Polyimides. Polym. Eng. Sci. 2007;47 381-389.

[22] Hulubei C, Hamciuc E, Bruma M. New imide type polymers based on epiclon. Rev. Roum. Chim. 2007;52(8–9) 891-898.

[23] Hulubei C, Hamciuc E, Bruma M. New polyimides based on Epiclon. Rev. Roum. Chim. 2007;52(11) 1063-1069.

[24] Hulubei C, Popovici D. Novel polyimides containing alicyclic units. Synthesis and characterization. Rev. Roum. Chim. 2011;56(3) 209-215.

[25] Cosutchi AI, Nica S-L, Hulubei C, Homocianu M, Ioan S. Effects of the aliphatic/aromatic structure on the miscibility, thermal, optical and rheological properties of some polyimide blends. Polym. Eng. Sci. 2012; 52(7) 1429-1439.

[26] Matsumoto T. Nonaromatic Polyimides Derived from Cycloaliphatic Monomers. Macromolecules 1999;32 4933-4939.

[27] Matsumoto T, Mikami D, Hashimoto T, Kaise M, Takahashi R, Kawabata S. Alicyclic polyimides – a colorless and thermally stable polymer for opto-electronic devices. J. Phys. Conference Series 2009;187 012005, doi:10.1088/1742-6596/187/1/012005.

[28] Ogura T, Higashihara T, Ueda M. Low-CTE photosensitive polyimide based on semialicyclic poly(amic acid) and photobase generator. J. Polym. Sci. Part A 2010;481317-1323.

[29] Cosutchi AI, Hulubei C, Stoica I, Dobromir M, Ioan S. Structural and dielectric properties of some epiclon-based polyimide films. e-Polymers 2008;068:1-15. http://www.e-polymers.org/journal/abstract_shw.cfm?Abstract_id=2388

[30] Macocinschi D, Taranu A, Hulubei C, Ioan S. Solution properties of poly(amic acid)s and polyimides. Rev. Roum. Chim. 2006;51 1001-1009.

[31] Yang CP., Hsiao SH, Wu KL. Organosoluble and light-colored fluorinated polyimides derived from 2,3-bis(4-amino-2-trifluoromethylphenoxy)naphthalene and aromatic dianhydrides. Polymer 2003;44 7067-7078.

[32] Park SJ, Cho KS, Kim SH. A study on dielectric characteristics of fluorinated polyimide thin film. J. Colloid Interf. Sci. 2004;272 384-390.

[33] Huang JC, He CB, Xiao Y, Mya KY, Dai J, Siow YP. Polyimide/POSS nanocomposites: interfacial interaction, thermal properties and mechanical properties. Polymer 2003;44 4491-4499.

[34] Khalil M, Saeed S, Ahmad Z. Properties of Binary Polyimide Blends Containing Hexafluoroisopropylidene Group. J. Macromol. Sci. A 2006;44 55-63.

[35] Hariharan RS, Bhuvana M, Sarojadevi S. Structural Characterization and Properties of Organo-soluble Polyimides, Bismaleimide and Polyaspartimides Based on 4,4′-Dichloro-3,3′- Diamino Benzophenone. High Perform. Polym. 2006;18 163-184.

[36] Cosutchi AI, Hulubei C, Ioan S. Optical and dielectric properties of some polymers with imidic structure. J. Optoelectron. Adv. Mater. 2007;9 975-980.

[37] Azzam RMA, Bashara NM. Ellipsometry and Polarized Light. Amsterdam: North-Holland Physics Publishing; 1986.

[38] van Krevelen DW. Properties of Polymers. Amsterdam: Elsevier; 1972.

[39] Groh W, Zimmermann A.What is the lowest refractive index of an organic polymer?. Macromolecules 1991;24 6660-6663.

[40] Lorenz LV. Wied. Ann. Phys. 1880;11 70-103.

[41] Lorentz HA. Wied. Ann. Phys. 1880;9 641-65.

[42] Ye YS, Chen YI, Wang YZ. Synthesis and properties of low-dielectric-constant polyimides with introduced reactive fluorine polyhedral oligomeric silsesquioxanes. J. Polym. Sci. Part A 2006;44 5391-5402.

[43] Hougham G, Tesoro G, Viehbeck A. Influence of Free Volume Change on the Relative Permittivity and Refractive Index in Fluoropolyimides. Macromolecules 1996;29 3453-3456.

[44] Cosutchi AI, Hulubei C, Buda M, Botila T, Ioan S. Effects of chemical structure on the electrical properties of some polymers with imidic structure. e-Polymers 2007;067: 1-11. http://www.e-polymers.org/journal/abstract_shw.cfm?Abstract_id=1885

[45] Yi MH, Huang H, Choi K-Y. Soluble and Colorless Polyimides from Alicyclic Diamines. J. Macromol. Sci. A 1998;35 2009-2022.

[46] Thorpe MF, Tichy L. Properties and applications of amorphous materials. London: Kluwer Academic Publishers; 2000.

[47] Tauc J, Menth A, Wood DL. Optical and Magnetic Investigations of the Localized States in Semiconducting Glasses. Phys. Rev. Lett. 1970;25 749-752.

[48] Tauc J, Menth A. States in the gap. J. Non-Cryst. Solids 1972;8-10 569-585.

[49] Jarząbek B, Weszka J, Burian A, Pocztowski G. Optical properties of amorphous thin films of the Zn-P system. Thin Solid Films 1996;279 204-208.

[50] Jarząbek B, Schab-Balcerzak E, Chamenko T, Sek D, Cisowski J, Volozhin A. Optical properties of new aliphatic–aromatic co-polyimides. J. Non-Cryst. Solids 2002;299 1057-1061.

[51] Jarząbek B, Orlik T, Cisowski J, Schab-Balcerzak E, Sezk D. Optical properties of semiladder polymer foils. Acta Phys. Pol. A 2000;98 655-659.

[52] Barzic AI, Stoica I, Hulubei C. New insights correlating transparency, absorption edges with morphological features of some semi-alicyclic polyimides. send for publication in J. Phys. Chem. B 2012

[53] Chae B, Kim SB, Lee SW, Kim SI, Choi W, Lee B. Surface Morphology, Molecular Reorientation, and Liquid Crystal Alignment Properties of Rubbed Nanofilms of a Well-Defined Brush Polyimide with a Fully Rodlike Backbone. Macromolecules 2002;35 10119-10130.

[54] Gadelmawla ES, Koura MM, Maksoud TMA, Elewa IM, Soliman HH. Roughness parameters. J. Mater. Process. Technol. 2002;123 133-145.

[55] Stout KJ, Sullivan PJ, Dong WP, Mainsah E, Luo N, Mathia T, Zahouani H. The development of methods for the characterization of roughness on three dimensions. Publication no. EUR 15178 EN of the Commission of the European Communities, Luxembourg; 1994

[56] Barbato G, Carneiro K, Cuppini D, Garnaes J, Gori G, Hughes G, Jensen CP, Jørgensen JF, Jusko O, Livi S, McQuoid H, Nielsen L, Picotto GB, Wilkening G. Scanning tunneling microscopy methods for roughness and micro hardness measurements, Synthesis report for research contract with the European Union under its programme for applied metrology, 109 pages European Commission Catalogue number: CD-NA-16145 EN-C, Brussels Luxemburg; 1995.

[57] Stoica I, Barzic AI, Hulubei C, Timpu D. Statistical analysis on morphology development of some semi-alicyclic polyimides using atomic force microscopy. send for publication in Microsc. Res. Tech. 2012

[58] Yang F, Zoriniants G, Ruan L, Sambles JR. Optical anisotropy and liquid-crystal alignment properties of rubbed polyimide layers. Liq. Cryst. 2007;34 1433-1441.

[59] Paek SY. Comparative study of rubbing parameters on polyimide alignment layers and liquid crystal alignment. J. Ind. Eng. Chem. 2001;7 316-325.

[60] Takatoh K, Hasegawa M, Koden M, Itoh N, Hasegawa R, Sakamoto M. Alignment Technologies and Applications of Liquid Crystal Devices. London and New York: Taylor and Francis; 2005.

[61] Hirosawa I, Koganezawa T, Ishii H, Sakai T. Effects of Annealing on Rubbed Polyimide Surface Studiedby Grazing-Incidence X-Ray Diffraction. IEICE Transactions on Electronics 2009; E92.C 1376-1381.

[62] Hirosawa I, Matsushita T, Miyairi H, Saito S. Effect of Soaking in Solvents on Molecular Orientation of Rubbed Polyimide Film. Jpn. J. Appl. Phys. 1999;38 2851-2855.

[63] Zheng W, Wang CC, Wang SP, Ko CL. Surface Wettability of Rubbed Polyimide Thin Films. Mol. Cryst. Liq. Cryst. 2009;512 81-90.

[64] Son PK, Choi SW. Preferred Pretilt Direction of Liquid Crystal Molecules on Rubbed Polyimide Surfaces. J. Korean Phys. Soc. 2010;57 207-209.

[65] Castellano JA. Surface Anchoring of Liquid Crystal Molecules on Various Substrates. Mol. Cryst. Liq. Cryst. 1983;94 33-41.

[66] Fukuro H, Kobayashi S. Newly Synthesized Polyimide for Aligning Nematic Liquid Crystals Accompanying High Pretilt Angles. Mol. Cryst. Liq. Cryst. 1988;163 157-162.

[67] Oo TN, Iwata T, Kimura M, Aakahane T. Surface alignment of liquid crystal multilayers evaporated on photoaligned polyimide film observed by surface profiler. Sci. Technol. Adv. Mater. 2005;6 149-157.

[68] Xiao L, Jia L, Li H, Hong T, Xu SY. Electric force microscopy study of the surface electrostatic property of rubbed polyimide alignment layers. Thin Solid Films 2000;370 238-242.

[69] Stoica I, Barzic AI, Hulubei C. The impact of rubbing fabric type on surface roughness and tribological properties of some semi-alicyclic polyimides evaluated from atomic force measurements. send for publication in Appl. Surf. Sci. 2012

[70] Chae B, Lee SW, Lee B, Choi W, Kim SB, Jung YM, Jung JC, Lee KH, Ree M. Sequence of Rubbing-Induced Molecular Segmental Reorientations in the Nanoscale Film Surface of a Brush Polymer Rod. J. Phys. Chem. B 2003;107 11911-11916.

[71] Kim JH, Yoneya M, Yamamoto J, Yokoyama H. Nano-rubbing of a liquid crystal alignment layer by an atomic force microscope: a detailed characterization. Nanotechnology 2002;13 133-137.
[72] Cosutchi AI, Hulubei C, Stoica I, Ioan S. Morphological and structural-rheological relationship in epiclon-based polyimide/hydroxypropylcellulose blend systems. J. Polym. Res. 2010;17 541-550.
[73] Cosutchi AI, Hulubei C, Stoica I, Ioan S. A new approach for patterning epiclon-based polyimide precursor films using a lyotropic liquid crystal template. J. Polym. Res. 2011;18 2389-2402.
[74] Patnaik SS, Bunning TJ, Adams WW. Atomic force microscopy and high-resolution scanning electron microscopy study of the banded surface morphology of hydroxypropyl cellulose thin films. Macromolecules 1995;28 393-395.

Chapter 10

Sensor Applications of Polyimides

Aziz Paşahan

Additional information is available at the end of the chapter

http://dx.doi.org/10.5772/53456

1. Introduction

A sensor is a converter that measures a physical quantity and converts it into a signal which an be read by an observer or by an instrument. Nowadays common sensors convert measurement of physical phenomena into an electrical signal.

Polyimides have inherently high mechanical properties, good chemical resistance, a low dielectric constant and high thermal stability [1]. In recent years, polyimide-based sensor materials have received much attention due to their simple preparation method, chemical inertness, mechanical and thermal stability, and high biocompatibility[2–4]. Thus they have a wide range of diverse and potential applications in several major technologies. Among the polymers, polyimides (PI) possess reliable high-temperature stability, excellent chemical resistance, high hydrolytic stability, adhesive properties, and good mechanical strength [7–9]. Polysulfones and phosphine oxidecontaining polyimides have excellent adhesive properties without sacrificing thermal stability [10–12].

Polyimides have found applications as enzyme immobilization membrane, due to their good chemical stabilitiy and low reactivity[13–16]. The main criteria for selecting a specific polyimide as a membrane material to be used in a given commercial application are diverse and complex but, besides mechanical and thermal stability, other factors are important, such as manufacturing reproducibility, tolerance to contaminants and other economical issues, without omission of permeation rate and selectivity, which are the most important characteristics[17–19]. Flexible electronics has developed greatly in recent years. Many flexiblematerials have been used in flexible lower cost sensors, larger area manufacturing, and packages that are significantly thinner, lighter, and more compact. It is worth noting that the polyimide (PI) material is beginning to be applied in flexible fields due to its good mechanical strength, higher glass transition temperature, electrochemical stability, and flexibility in polymer materials[20].

Wessa et al. described two methods for the direct coupling of proteins to polyimide. With the cyano-transfer technique, anti-glucose oxidase was coupled to the polymer surface for the detection of glucose oxidase. In a second approach, glucose oxidase was derivatized with photoreactive groups via 3-trifluoromethyl 3-(m-isothiocyanophenyl)-diazirine (TRIMID) and immobilized on the polyimide layer via photochemistry. The subsequent binding of the corresponding antibody was observed. Sakong et al. suggest a SAW biosensor that is supplied with a fluidic system consisting of a microfluidic polyimide tubing system. The system is designed for the detection of DNA[21]. Ekinci and coworkers, the modified polyimide electrodes are widely used in the field of permselective membranes for the electroactive species (dopamine and H_2O_2) and biosensor [22-26]. All sensor applications with polyimide as a shielding layer were done with surface acoustic wave devices(SAW's). This sensor type closely resembles the 'classic' SAW technique for chemical analysis in the gas phase [27].

2. Amperometric Polyimide Sensors

In recent years, polyimide-modified electrodes are widely used in the field of sensor and biosensor. Among the electroactive species, dopamine (DA) has been of interest to neuroscientists and chemists. A loss of DA containing neurons may result in some serious diseases such as Parkinsonism. Therefore, the determination of the concentration of this neurochemical is important. Dopamine in central nervous system coexists with ascorbic acid, whose oxidation peak potential is close to that of dopamine. Therefore, a significant problem faced in electrochemically determination of dopamine is the presence of electroactive ascorbic acid, which reduces the selectivity and sensitivity [28-30].

Ekinci and coworkers fabricated two kinds of PI membranes composed of various diamines and dianhydrides for determination of dopamine. Phosphine oxide-containing polyimides and piperazine - containing polyimides utilized to synthesize polyimides to be used as a selective membrane for dopamine in the presence of ascorbic acid. Polyimide solutions were prepared by dissolving 70 mg polyimide in 2 mL NMP. Then, polyimide films were prepared onto the Pt surface by dropping of polyimide solution and allowed to dry at room temperature for 2 days. Prior to permselectivity tests, the polyimide electrode had reached to steady state and, then the injection of electroactive substances into the PBS solution was made. The amperometric responses of the polymeric electrodes toward the electroactive substances (dopamine and ascorbic acid) were recorded as a function of time at a potential of 0,7 V vs. Ag/AgCl. The polyimides films showed excellent permselectivity behavior toward dopamine in the presence of ascorbic acid. In other words, it means that the chemically-modified polyimides can be used as dopamine-selective sensor in the presence of ascorbic acid.[21-22]

Ekinci and coworkers pyrimidine-based hyperbranched and aromatic polyimides were tested as hydrogen peroxide-selective membrane in the presence of interferents. It was found that the polyimide films can be applicable to amperometric sensing of hydrogen peroxide in the presence of the mentioned electroactive and non-electroactive interferents[23-24].

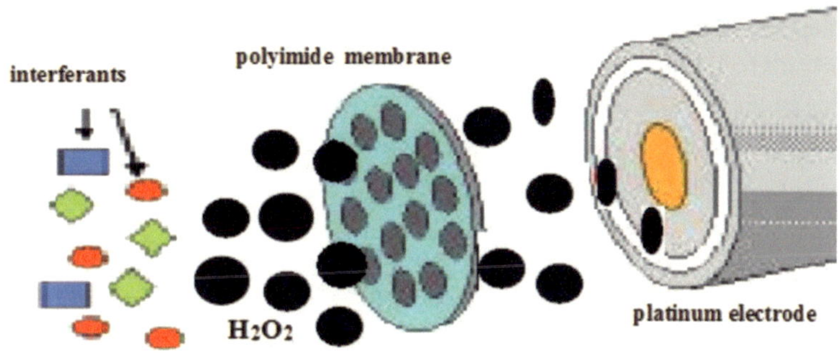

Figure 1. The preparation of the polyimide electrodes and its H_2O_2 sensing.

Since the determination of glucose in biological fluids is very important for the diagnosis and management of diabetes mellitus, the search for the ideal glucose sensor continues to be a focus of biosensor research. During the preparation of an enzyme electrode, immobilization of the enzyme at the electrode surface is a key step in the fabrication of a high-performance biosensor. However, there are several disadvantages of the enzyme-modified electrodes, such as instability, high cost of enzymes, and complicated immobilization procedure. Polymeric films have been widely used to immobilize enzymes. One of the most important problems faced in hydrogen peroxide detecting enzymatic biosensor applications is the presence of interferents. Electrochemical biosensors have found a wide range of application areas in recent decades because of their ability to combine the high specificity of biologic material with the high accuracy of electrochemical systems. The most frequently employed glucose biosensors are based on glucose oxidase (GOD). Fast, accurate, easy to-use, and specific determination of glucose level are qualities that are especially important for diabetic patients. It catalyzes the electron transfer from glucose to oxygen accompanied by the production of gluconolactone and hydrogen peroxide as represented by:

$$GOx\left(ox\right) + \beta - D - glucose \text{ ¾}{\rightarrow} GOx\left(red\right) + D - gluconicacid \tag{1}$$

$$GOx\left(red\right) + O_2 \text{ ¾}{\rightarrow} GOx\left(ox\right) + H_2O_2 \tag{2}$$

The formed hydrogen peroxide produced in the above reaction then diffuses towards the electrode surface where it is amperometrically detected by electrochemical oxidation around 0,7 V (Ag/AgCl). The quantification of glucose can be achieved via electrochemical detection of the enzymatically liberated H_2O_2. The voltage for the oxidation/reduction of H_2O_2 at solid electrodes can be considerably reduced with the incorporation of Au and Pt nanoparticles that also allow proteins to retain their biological activity.

One of the major problems affecting most biosensors is the immobilization of the enzyme in a membrane well adherent to the electrode surface to retain its activity and to enhance the stability as longer as possible. To prevent the enzyme release a membrane was then applied on the electrode surface. The immobilization of enzyme was performed by two different methods. In case one-step process, the polyimide solution was mixed with the GOD solution in the proportions1:1). Later, final solution was spread onto the bare Pt electrodes and was allowed to dry in air for at least 18 h. On the other hand, for the two-step process, the GOD solution was spread onto Pt electrode and allowed to dry in air for 10–15 min. Then, the polyimide solution spread onto it and was again allowed to dry for at least 18 h. The schematic diagram of enzyme electrodes prepared by one-step process and two-step process and its glucose sensing is displayed in Scheme 1. Using the two-step process, it was suitable for immobilization of GOD. The results of study reveal that the polyimide films is very promising substrates for the immobilization and stabilization of enzymes and the development of highly stable biosensors. Polyimide films could be used as a polymeric membrane in the detection of glucose because of its selectivity and strong adherence to electrode surface and chemical stability[31-32].

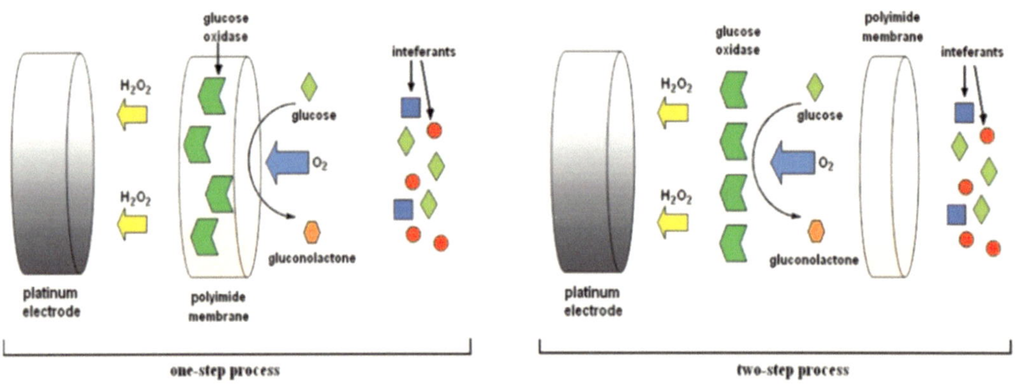

Figure 2. The preparation methods of the enzyme electrodes and their glucose sensing.

3. Polyimide gas sensors

Over the last 20 years polymeric gas separation membranes have grown from a research laboratory subject into a significant industry. In particular, polyimide membranes have been commercialized for the separation of a range of gases and considerable quantities of research has been published on the gas transport properties of these materials. Potential future uses of polyimide membranes includes the separation of carbon dioxide from flue gases for geosequestration; however, for this to become economical, significant advances in the efficiency and lifetimes of polymeric membranes will be required. In particular, polyimides have excellent intrinsic CO_2/CH_4 separation properties and have robust mechanical properties to withstand high-pressure natural gas feeds. Despite their attractive intrinsic properties, susceptibility to plasticization also causes performance declines for polyimides which negatively impacts process economics (increased methane loss) and reliability[33-34].

A commercially available polyimide, Matrimid® 5218, exhibits a combination of selectivity and permeability for industrially significant gas pairs superior to that of most other readily available polymers. Its permeation properties, combined with its processability (i.e., solubility in common solvents) makes it an ideal candidate for gas separation applications.

Furthermore, its mechanical strength and high glass transition temperature, *Tg*, suit it better for more rigorous working environments than polysulfone, especially in high temperature applications[35].

Past reports have also pointed out that the addition of bulky CF_3 groups in polyimide (PI) results in a good gas permeability and selectivity. Further, Hofmann et al.14 fabricated perfluorinated polymeric membranes containing different percentages of tetrafluoroethylene (TFE). In their report, they adopted positron annihilation lifetime spectroscopy (PALS) to analyze the membrane free volume and compared these data with the gas permeability. It was found that the membranes containing more TFE structures provided much greater voids and gas permeability. Hirayama et al. synthesized 32 kinds of PI membranes composed of different diamines and dianhydrides to evaluate their gas permeability. They analyzed the cohesive energy density (CED) of PI membranes and correlated these CED values with the diffusion coefficients. From their analyses, they suggested that polymer chains containing polar substituents would change the segment mobility and then control the gas diffusion mechanisms. Wang et al. also fabricated seven kinds of aromatic PI membranes composed of various diamines and dianhydrides. They noted that the bulky groups inhibited polymer chain packing and then formed a larger fractional free volume (FFV) in the membranes. The increased FFV of the membrane prompted higher gas permeability. Liu et al.18 prepared mono-PI, 6FDA-durene, 6FDA-2,6-diaminotoluene (2,6-DAT), and co-PI, 6FDA-durene/2,6-DAT membranes with various diamine ratios. They found that the increase in 6FDA-2,6-DAT in co-PI decreased the gas diffusivity and solubility, resulting in improved intrasegment packing, which lowered the gas permeation. Today, molecular simulation techniques supply a new potential method for obtaining an in-depth understanding on a molecular scale. Smit et al. analyzed the motion types of gas molecules in PI membranes using a molecular dynamics (MD) simulation. Two different types of diffusion behavior, residual time and flying time, in the membrane were observed. In addition, an increase in the temperature raised the small-molecule thermal motion and thus contributed to a higher diffusivity. Tung et al. investigated the effect of PMMA tacticity and casting solvent types on free volume morphology and the gas transport mechanisms using MD and Monte Carlo (MC) techniques. Their simulated values coincided well with the experimental data. Heuchel et al. simulated the gas permeability, solubility, and diffusivity of O_2 and N_2 in PI membranes[36].

There were two specific designs for the micro-heating elements realised on polyimide sheets, one for a resistive gas sensor and the other one for the thermal actuator. Figure 3a illustrates the design for a resistive gas sensor with the platinum electrodes and the heater patterned on the top side and on the bottom side of the 50 μm-thick polyimide sheet, respectively. This design involves simplified processing steps to realise fully flexible micro-hotplates for resistive gas sensors. However, some changes have to be brought to the standard packaging procedure of the chips on TO headers to ensure thermal insulation (by suspending the chip in air) and to be able to contact the heater on the backside of the chip. This designwas fabricated but not tested due to the constraints that were just mentioned before. Another design proposed and presented in Figure 3b consists of an Upilex sheet on

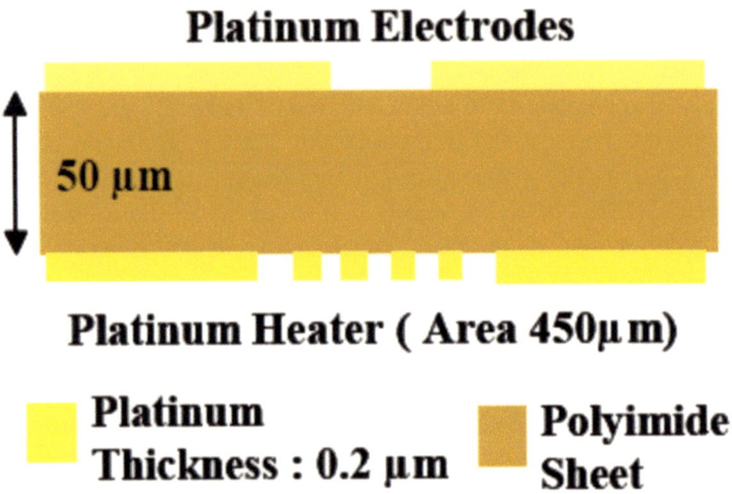

(a)

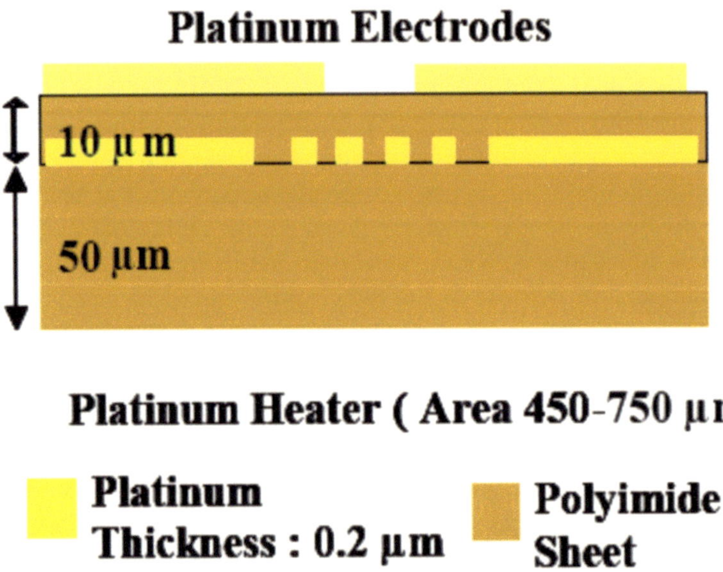

(b)

Figure 3. (a) Design of the gas sensing structure realised using the both sides of a polyimide sheet. (b) Design of the gas sensing structure realised only on the top side of the polyimide sheet[37].

which is patterned a heating element (500×500μm^2) covered by a 10 μm-thick spin coating layer of photosensitive polyimide (PI 2731 from HD MicroSystems, Tg > 350°C) with on top platinum electrodes. This design allows having the electrical contacts, both for the heater and the electrodes, on the same side, and therefore this design was chosen to realise polyimide hotplates for resistive type gas sensor. The micro-hotplate design for the thermal actuator is presented in Figure 4. An aluminium film is patterned to define the heating element and the rim for the anodic bonding of the Pyrex cavity, used to store the paraffin, on the top side of a 50 μm-thick polyimide sheet. Metal to glass anodic bonding was the technique chosen to fix the Pyrex chip on the polyimide hotplate. To achieve at the same time the anodic bonding of the Pyrex chip on the Al rim and on the two interconnections of the Al heater, an Al line linking both structures has been added to the design to electrically connect them during the anodic bonding. This line was cut afterwards to allow the independent electrical operation of the heater afterwards[37].

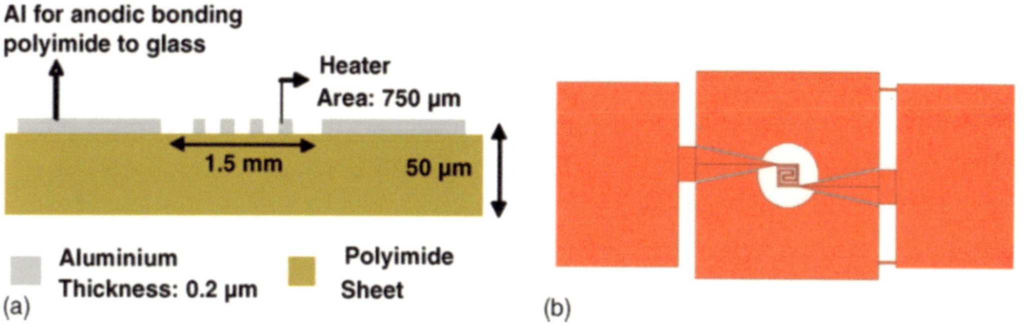

Figure 4. Design of the micro-heating element made of aluminium realised on a polyimide sheet for a thermal actuator (a) cross-section view (b) top view[37].

The calorimetric gas sensor consists of a chip-based differential set-up of a catalytically activated and a passivated temperature-sensitive thin-film resistance, introduced in [38]. If the calorimetric gas sensor is exposed in H_2O_2 atmosphere, a temperature difference between the activated and passivated resistance can be detected that is caused by the exothermal decomposition of hydrogen peroxide on the catalytic surface. In place of an active heating structure, the sensor utilizes the elevated temperature of the sterilization process. As sensor substrate, a polyimide foil (Kapton® HN from DuPontTM) with a thickness of 25 μm instead of conventional silicon was envisaged due to its expedient thermal and chemical properties (low thermal conductivity and thermal endurance up to 400 °C, ample resistance to hydrogen peroxide and to elevated humidity). Platinum with a thickness of 200 nm was deposited and photolithographically patterned as meander-shaped thin-film resistances on the polyimide foil (s. Figure 5(a)). The width of the meander-shaped paths is 40 μm and the area of one thin-film resistance amounts to be 1,0 mm^2 that results in a theoretical resistance value of 200 Ω at room temperature. The sensor was afterwards passivated with SU-8 photo resist, which is stable up to 350 °C as well as in highly concentrated H_2O_2 atmosphere, and catalytically activated with a dispersion of manganese(IV) oxide (Figure 5(b)). The reaction mechanisms of the H_2O_2 decomposition on the catalyst involves two pathways: i) a redox reaction with electron exchange with the catalytic surface creating free hydroxyl radicals, and ii) a chain reaction between free radicals and hydrogen peroxide in which the final products, namely water and oxygen, are formed and reaction heat is released [39-40].

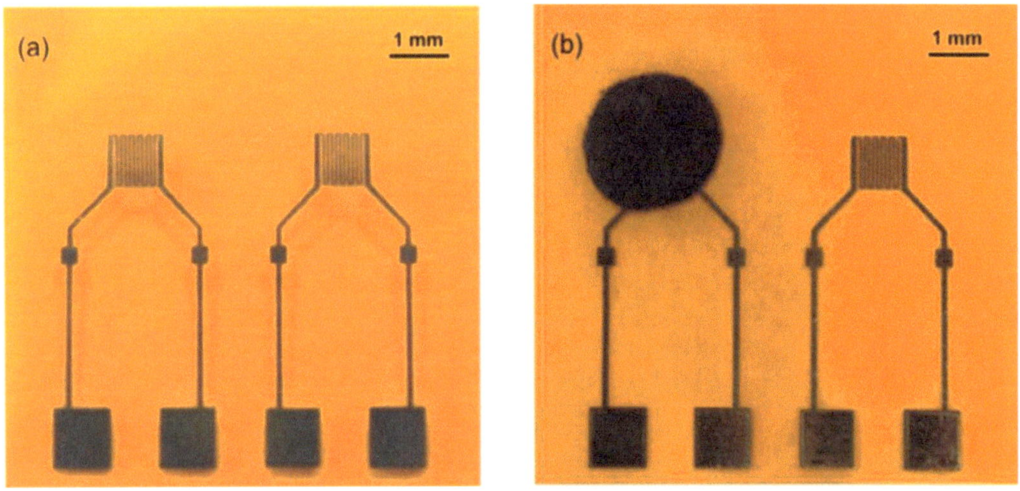

Figure 5. Calorimetric gas sensor on polyimide foil; (a) fabricated thin-film resistances with contact pads on polyimide; (b) thin-film resistances with SU-8 photo resist as passivation film and manganese(IV) oxide as catalytically active layer (chip size: 10 x 10 mm^2)[40].

4. Polyimide humidity sensors

Humidity sensors have been widely used in order to make living and working environment more comfortable in our daily life including process control, meteorology, agriculture and medical equipment. Conventional humidity sensors have some drawbacks such as large volume, low sensitivity, and slow response. To overcome these drawbacks, there has been growing interest in miniaturization of humidity sensors with high performance and low fabrication cost. In this respect, micro electro mechanical system (MEMS) technology is now very popular to miniaturize humidity sensors[41].

A number of parameters have an influence on the response time of polymer-based humidity sensors, such as the dimensions of the sensor, moisture diffusivity in the film, the film thickness, and the ambient temperature. Once the sensing material is chosen, however, the response speed can only be enhanced by modifying the geometry of the device. Two different humidity sensors, Design A and Design B in Figure 6 were designed and fabricated. A high-speed capacitive humidity sensor (Design A) is achieved by introducing a micro-bridge structure with several holes created by using the front-side anisotropic and isotropic dry etching and by allowing moisture to diffuse into both top and side surfaces of polyimide film. In order to compare the sensitivity and speed, the conventional parallel-plate structure (Design B), consisting of a polyimide film and two electrodes, is designed (Figure 6b). [42]

Polyimides are compatible with IC processes and have both chemical stability and long-term stability in a presence of moisture and heat, in addition to the good hygroscopic and

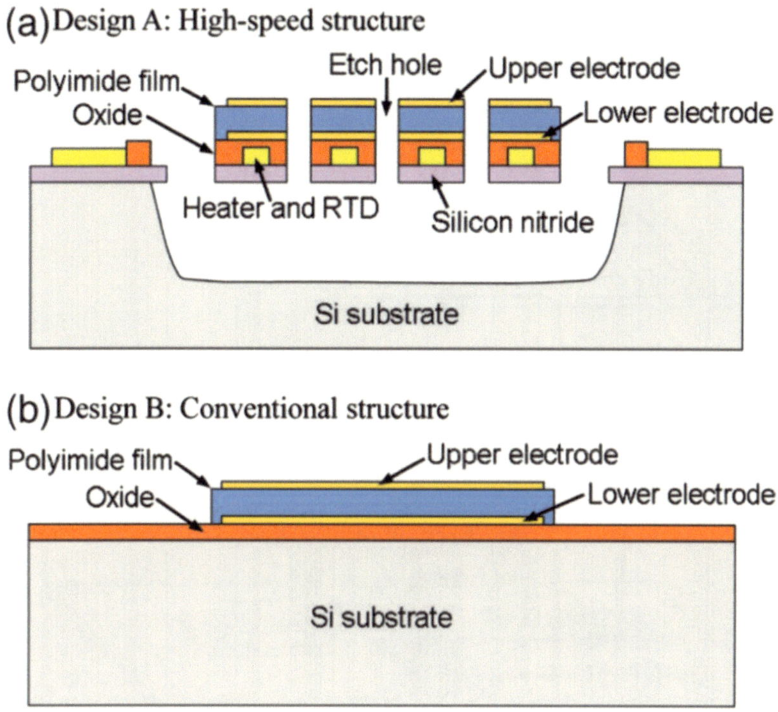

Figure 6. Schematic cross-sectional views of two capacitive humidity sensors (a) Design A: High-speed structure (b) Design B: Conventional structure[42].

dielectric properties. Therefore the capacitive type polyimide sensors have been extensively investigated. However, the typical polyimide humidity sensors suffered from slow response and substantial long-term drift and they were improved by introducing the cross-linked structure and fluorine atoms in the polyimide unit

5. Polyimide tactile sensors

Piezoelectric materials are one of the key materials utilized in tactile sensing applications. Typically, a tactile sensing array is fabricated to provide sensory feedback regarding contact with other objects or surfaces, as in Figure 7. It is desirable for many applications to be able to sense tactile information over a non-planar geometry, which requires a flexible, conformal sensor technology. One key element of a flexible sensing technology would be a polymer piezoelectric material. Previously, the only piezoelectric polymer to be applied to this technology has been PVDF [43]. PVDF has the advantages of low cost, ease of fabrication and flexibility. It's primary limitation is the its low temperature range of operation (< 80 ºC). It is desirable to be able to fabricate flexible polymer tactile sensor elements that can be operated over a larger range of temperatures. PVDF is limited to less than 80 ºC simply by the thermal stability of the film, causing film degradation at higher temperatures. Recent work has shown higher temperature piezoelectric response in newly developed polyimides[44].

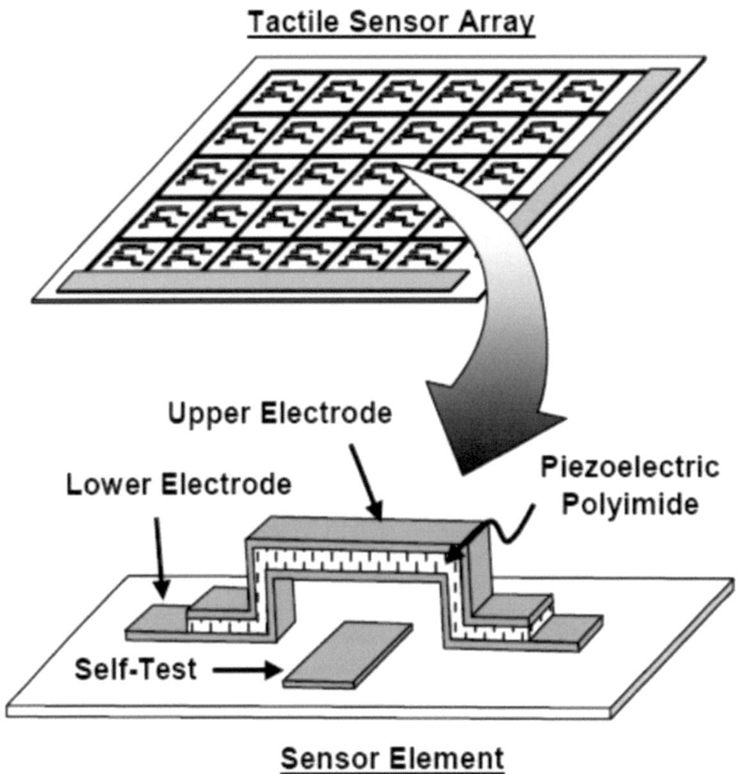

Figure 7. Tactile sensor array and an individual polyimide sensor element.[44]

Multimodal tactile sensor on flexible polyimide substrate capable of sensing the hardness, roughness, temperature and thermal conductivity of the object in contact has been developed. The sensor is constructed using a polyimide substrate (Figure 8) and consists of multiple sensor nodes arranged in an array format[45].

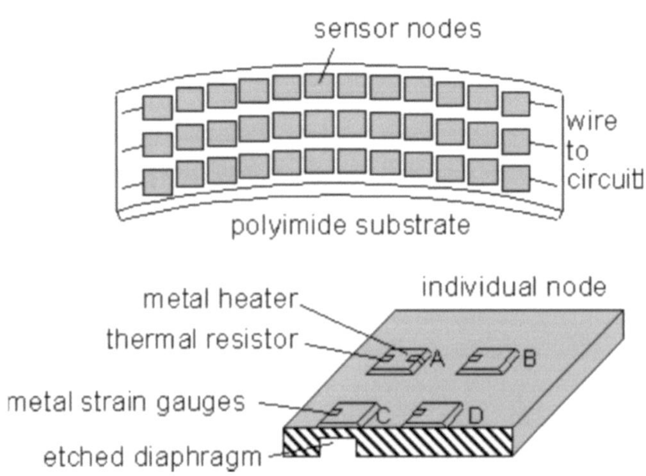

Figure 8. Schematic diagram of the polyimide multimodal tactile sensor[45].

6. Pressure sensors based polyimide

Kapton is an example of an insulating polymer that can be laser-carbonized to form conducting filaments. This substrate is flexible, inexpensive, durable and easy to manufacture. While the Kapton polyimide is a good insulator, research has shown that the laser-carbonized filaments are fair conductors. Additional opportunities to increase the effectiveness of some sensor designs lie in the fact that the carbonized filaments are porous and their resistivity can be manipulated easily during processing[46].

The fabrication sequence of the capacitive pressure sensor array is illusturates in Figure 9. The process starts on square, stainless steel substrates that are each 5.7 cm on a side, 0.5 mm thick, and have surface roughness of approximately 6–8 μm. An array of 8×8 pressure inlet holes with a diameter of 2 mm, with 5mm center-to-center distances, is milled through the stainless steel substrate. Kapton polyimide film (Dupont, Kapton HN200, 50 μm thick) is laminated onto the milled stainless steel substrate using a hot press with a pressure of 8.65 MPa and a temperature of 175 °C for 30 min. The pressure-sensitive diaphragms will be the Kapton polyimide film in the regions suspended over the milled pressure inlet holes (Figure 9a). A triple metallic layer of Ti/Cu/Ti with a thickness of 100/2000/500Å is deposited by electron-beam evaporation and then patterned using a lift-off process to create bottom electrodes, electroplating seed layers, and bonding pads on the surface of the Kapton polyimide film (Figure 9a). Multiple layers of PI2611 polyimide (Dupont) are spun onto the patterned layer with a spin speed of 1200 rpm for 60 s, and hard-cured in a N_2 ambient at 200 °C for 120 min yielding a final thickness of polyimide of approximately 44–48 μm. The polyimide layer is anisotropically etched using reactive ion etching to create electroplating molds for the support posts of the fixed backplates, and to remove the uppermost titanium layer of the seed layer (Figure 9b). Nickel supports are then electroplated through the polyimide molds. A Ti/Cu/Ti metallic triple layer with a thickness of 300/2000/300Å is deposited using DC sputtering to act as a seed layer for the deposition of the backplate. Thick photoresist (Shipley SJR 5740) is spun on the seed layer with a spin speed of 1100 rpm for 30 s (yielding a final thickness of approximately 15 μm) and patterned to act as electroplating molds for the backplates. After removal of the uppermost Ti layer, nickel is electroplated through the thick photoresist electroplating molds to create the backplates (Figure 9). The thick photoresist electroplating molds and the remaining seed layer are removed. Finally, the polyimide molds for the backplate posts as well as polyimide sacrificial layers are isotropically etched to create air gaps between the fixed backplates and the pressure sensitive Kapton polyimide flexible diaphragms (Figure 9d and 9c). The isotropic dry etch is carried out in a barrel plasma etcher using CF_4/O_2 plasma with a RF power of 120W. Figure 9 illusturates photographs of a fabricated pressure sensor array, where (b) shows a side-view and (c) shows a close-up view of the gap defined between the fixed backplate and the diaphragm. Note that these sensors are operating in differential mode, with the side containing the backplate held at a pressure of 1 atm[47].

The readout are connected to a hydraulic restriction. The key processes of the device are a thermal bonding of a polyimide sheet to an already micromachined silicon wafer and the deposition of transducers onto the polyimide membrane. A schematic cross section is

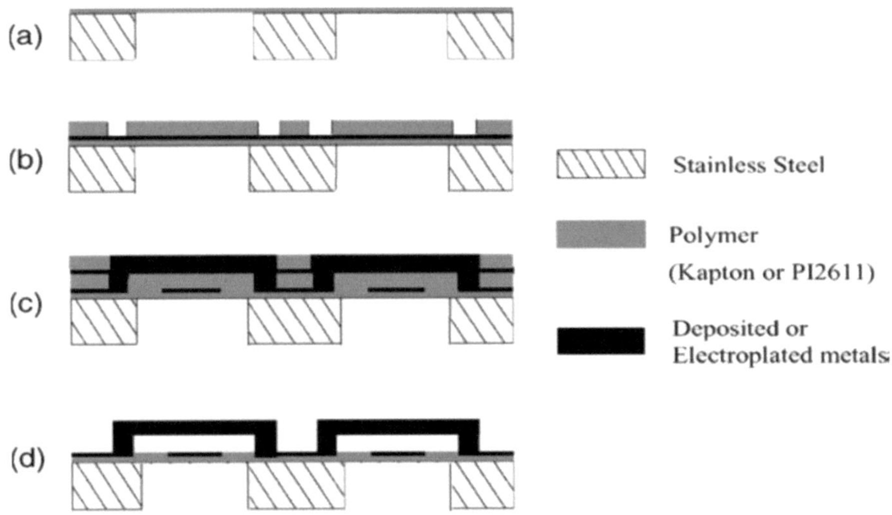

Figure 9. Fabrication sequence of pressure sensor based on Kapton polyimide diaphragm.

depicted in figure 10. The ZnO sensing layout is a ring on the top of a circular pressure membrane. The size of the ring is designed in such a manner that the piezoelectric ZnO layer is only under compressive strain. Highest sensitivity is reached by the ring arrangement because the compressive strain at the edge of the membrane is high and the electrode surface is large compared to other configurations. The top view of a fabricated ZnO piezoelectric transducer is shown in figure 11 The diameter of the pressure sensing membrane is 1 mm with a thickness of 25 μm.devices have been fabricated by using standard MEMS technologies. Two pressure sensors with piezoelectric[48].

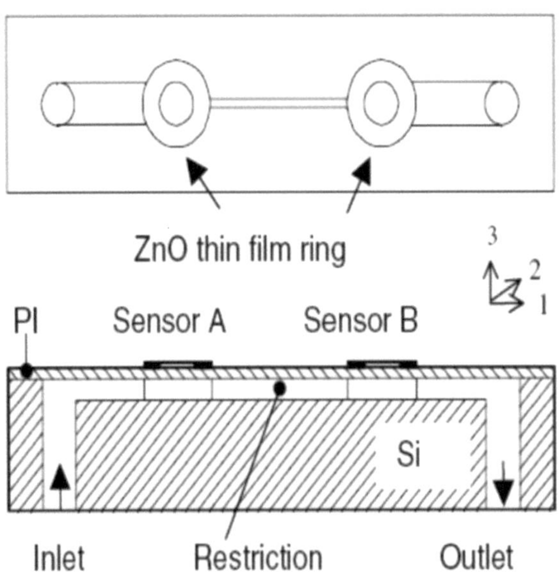

Figure 10. Schematic view of the device[48].

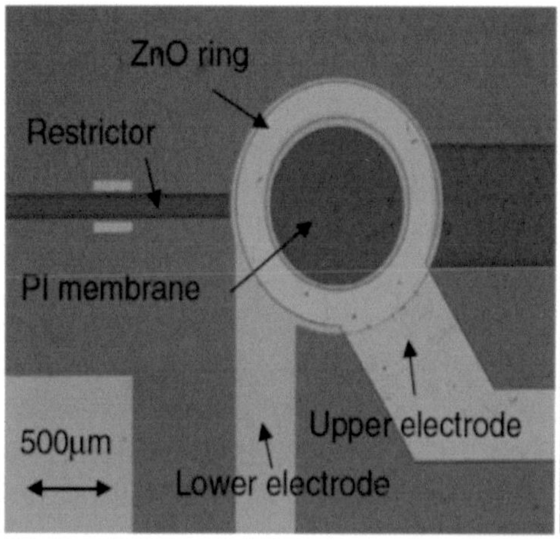

Figure 11. Top view of a pressure sensor[48].

7. Conclusions

Polyimides have excellent thermal stability, solvent resistance, radiation resistance, wear resistance, hydrolytic stability, low dielectric constant, high mechanical properties, good chemical resistance and a low dielectric constant. Due to these superior properties, the application field of polyimide has generally been enlarged from printed circuit boards and electrical insulation layers in microelectronics to functional layers of humidity sensors, shielding layers for sensor surfaces and novel platforms for thermal sensor devices, temperature sensor arrays, micro-hotplates integrated into gas sensors and biosensors.

Author details

Aziz Paşahan
İnönü University, Faculty of Sciences and Literature, Department of Chemistry, Malatya, Turkey

8. References

[1] Koytepe S, Vural S, Seckin T (2009) Molecular design of nanometric zinc borate-containing polyimide as a route to flame retardant materials. Mater. Res. Bull. 44: 369–376.

[2] Ghosh MK, Mittal KL (1996) Polyimides: fundamentals and applications. New York: Marcel Dekker. 126–130 pp..

[3] Mittal KL (1984) Polyimides: synthesis, characterization and applications. New York: Plenum Press. 1–10 pp.

[4] Wilson D, Stenzenberger HD, Hergenrother PM (1990) Polyimides. New York: Blackie. 1–21 pp.

[5] Xu JJ, Zhang XQ, Yu ZH, Fang HQ, Chen HY (2001) A stable glucose biosensor prepared by co-immobilizing glucose oxidase into poly(p-chlorophenol) at a platinum electrode Fresenius J Anal Chem. 369::486-490.

[6] Eldın MSM, De Maio, A, Di Martıno, S, Bencivenga, U, Rossi, S, Duva, A, Gaeta, FS, Mıta DG (1999) Immobilization of b-Galactosidase on Nylon Membranes Grafted with Diethylenglycol Dimethacrylate (DGDA) by g-Radiation: Effect of Membrane Pore Size. Advances in Polymer Technology. 18:109-123.

[7] kinci E, Köytepe, S, Pasahan, A, Seckin T (2006) Preparation and characterization of an aromatic polyimide and its use as a selective membrane for H_2O_2. Turk. J. Chem. 30:277–285.

[8] Abasıyanık MF, Mehmet S (2010) Immobilization of glucose oxidase on reagentless ferrocene-containing polythiophene derivative and its glucose sensing application. J. Electroanal. Chem. 639:21–26.

[9] aby TT, Aravind SSJ, Arockiadoss T, Rakhi RB, Ramaprabhu S (2010) Metal decorated graphene nanosheets as immobilization matrix for amperometric glucose biosensor. Sensor. Actuat. B-Chem. 145:71–77.

[10] Chi Q, Dong S (1993) Flow-injection analysis of glucose at an amperometric glucose sensor based on electrochemical codeposition of palladium and glucose oxidase on a glassy carbon electrode. Anal. Chim. Acta. 278:17–23.

[11] Yu JH, Liu SQ, Ju HX (2003) Glucose sensor for flow injection analysis of serum glucose based on immobilization of glucose oxidase in titania sol–gel membrane. Biosens. Bioelectron. 19, 401–409.

[12] Cosnier, S, Fombon, JJ, Labbe P, Limosin D (1999) Development of a PPO-poly(amphiphilic pyrrole) electrode for on site monitoring of phenol in aqueous effluents. Sens. Actuator B-Chem. 59:134–139.

[13] Ghaemy M, Nasab SMA (2010) Synthesis and identification of organosoluble polyimides: Thermal, photophysical and chemiluminescence properties. Polym. J. 42:648–656.

[14] Chou WY, Kuo CW, Chang CW, Yeh BL, Chang MH (2010) Tuning surface properties in photosensitive polyimide. Material design for high performance organic thin-film transistors. J. Mater. Chem. 20: 5474–5480.

[15] Jiang LZ, Liu JG, Wu DZ, Li HQ, Jin RG (2006) A methodology for the preparation of nanoporous polyimide films with low dielectric constants. Thin Solid Films 510, 241–246.

[16] Jin XZ, Ishii H (2005) Novel positive-type photosensitive polyimide with low dielectric constant. J. Appl. Polym. Sci. 98:15–21.

[17] Ba CY, Economy J (2010) Preparation of PMDA=ODA polyimide membrane for use as substrate in a thermally stable composite reverse osmosis membrane. J. Membr. Sci. 363: 140–148.

[18] Yoon SJ, Choi JH, Hong YT, Lee SY (2010) Synthesis and characterization of sulfonated Poly(arylene ether sulfone) ionomers incorporating perfluorohexylene units for DMFC membranes. Macromol. Res. 18:352–357.

[19] Darvishmanesh S, Degreve J, Van der Bruggen B (2010) Performance of solvent-pretreated polyimide nanofiltration membranes for separation of dissolved dyes from toluene. Ind. Eng. Chem. Res. 49:9330–9338.
[20] Chang WY, Fang TH, Lin YC (2008) Physical characteristics of polyimide films for flexible sensors. Appl Phys. A 92: 693–701.
[21] Paşahan A, Köytepe S, Ekinci E, Seçkin T (2004) Synthesis, Characterization anc Dopamine Selectivity of 1,4-bis(3-aminopropyl)piperazine-Containing Polyimide. Polymer Bulletin. 51:351-358.
[22] Köytepe S, Paşahan A, Ekinci E, Alıcı B, Seçkin T (2008) Synthesis, characterization of phosphine oxide-containing polyimides and their use as selective membrane for dopamine. Journal Polymer Research. 15:249-257.
[23] Köytepe S, Paşahan A, Ekinci E, Seçkin T (2005) Synthesis, characterization and H_2O_2-sensing properties of pyrimidine-based hyperbranched polyimides. European Polymer Journal. 41:121-127.
[24] Ekinci E, Köytepe S Paşahan A, Seçkin T (2006) Preparation and Characterization of an Aromatic Polyimide and Its Use as a Selective Membrane for H_2O_2. Turkish Journal Of Chemistry. 30:277-285.
[25] Ekinci E, Emre FB, Köytepe S, Seçkin T (2005) Preparation, characterization and H_2O_2 selectivity of hyperbranched polyimides, containing triazine. J.Polym. Res. 12:205-210.
[26] Paşahan A, Köytepe S, Ekinci E (2011) Synthesis, characterization of poly(4, 4'-diaminophenylethanepyromellitimide) and its use as immobilized enzyme polyimidemembrane. High Perform Polym. 23:59-65.
[27] Rapp M, Boss B, Voight A, Gemmeke HJ, Ache HJ (1995) Development of an analytical microsystem for organic gas detection based on surface acoustic wave resonators, Fresenius J. Anal. Chem. 352: 699–704.
[28] Ciszewski A, Milczarek G (1999) Polyeugenol-modified platinum electrode for selective detection of dopamine in the presence of ascorbic acid Anal Chem 71:1055–1061
[29] Kaneko H, Yamada M, Aoki K (1990) Determination of Dopamine in Solution of Ascorbic Acid at Graphite-Reinforcement Carbon Electrodes by Differential Pulse Voltammetry) Anal Sci 6:439–442
[30] Niwa O, Morita M, Tabei H (1991) Highly sensitive and selective voltammetric detection of dopamine with vertically separated interdigitated array electrodes Electroanalysis 3:163–168.
[31] Pasahan A, Köytepe S, Ekinci E (2011) Synthesis, Characterization of a New Organosoluble Polyimide and Its Application in Development of Glucose Biosensor. Polymer-Plastics Technology and Engineering 50: 1239–1246.
[32] Pasahan A, Köytepe S, Ekinci E (2011) Synthesis, characterization of naphthalene based polyimides and their use as immobilized enzyme membrane. Polym. Advan. Technol. 22:1940–1947.
[33] Wind JD, Paul DR, Koros WJ (2004) Natural gas permeation in polyimide membranes. Journal of Membrane Science. 228: 227–236.

[34] Powell CE, Xavier J, Duthie XJ, Kentish SE, Qiao GG, Stevens GW (2007) Reversible diamine cross-linking of polyimide membranes Journal of Membrane Science 291: 199–209

[35] Ekiner OM, Hayes RA, Manos P (1992) Novel multicomponent fluid separation membranes, US Patent 5,085,676, E.I. du Pont de Nemours.

[36] Chang KS, Tung CC, Wang KS, Tung KL (2009) Free Volume Analysis and Gas Transport Mechanisms of Aromatic Polyimide Membranes: A Molecular Simulation Study J. Phys. Chem. B. 113:9821–9830.

[37] Briand D, Colin S, Gangadharaiah A, Velab E, Dubois P, Thiery L, de Rooij NF (2006) Micro-hotplates on polyimide for sensors and actuators. Sensors and Actuators A 132:317–324.

[38] Nather N, Henkel H, Schneider A, Schöning MJ (2009) Investigation of different catalytic active and passive materials for the realisation of a hydrogen peroxide gas sensor. Phys.Status Solidi A 206:449-54.

[39] Hiroki A, LaVerne JA (2005) Decomposition of hydrogen peroxide at water-ceramic oxide interfaces. J. Phys. Chem. B.109:3364-70.

[40] Kirchner P, Oberländer J, Friedrich P, Berger J, Rysstad G, Keusgen M, Schöning MJ (2010) Realization of a calorimetric gas sensor on polyimide foil for applications in aseptic food industry, Procedia Engineering. 5:264–267.

[41] Kim JH, Hong SM, Moon BM, Kim K (2010) High-performance capacitive humidity sensor with novel electrode and polyimide layer based on MEMS technology,Microsyst Technol 16:2017–2021

[42] Kim JS, Lee MY, Kang MS, Yoo KP, Kwon KH, Singh VR, Min NK (2010) Fabrication of high-speed polyimide-based humidity sensor using anisotropic and isotropic etching with ICP. Thin Solid Films, Procedia Engineering 5:264–267

[43] Ounaies Z, Harrison JS (2002) An Investigation of Piezoelectricity of a Nitrile-Substituted Polyimide. Journal of Polymer Science. Polymer Physics.

[44] Atkinson GM, Pearson RE, Ounaies, Z Park C, Harrison JS, Midkiff JA (2003) Piezoelectric Polyimide Tactile Sensors.

[45] By Chang Liu, (2007) Recent Developments in Polymer MEMS, Adv. Mater. 19:3783–3790

[46] Ingram JM, Grep M, Nicholson JA, Fountain AW (2003) Polymeric humidity sensor based on laser carbonized polyimide substrate Sensors and Actuators B 96:283–289

[47] Chang SP, Allen MG (2004) Demonstration for integrating capacitive pressure sensors with read-out circuitry on stainless steel substrate. Sensors and Actuators A. 116:195–204.

[48] Kuoni A, Holzherr R, Boillat M, de Rooij NF (2003) Polyimide membrane with ZnO piezoelectric thin film pressure transducers as a differential pressure liquid flow sensor. J. Micromech. Microeng. 13:103–107

Chapter 11

Auto-Reparation of Polyimide Film Coatings for Aerospace Applications Challenges & Perspectives

A.A. Périchaud, R.M. Iskakov, Andrey Kurbatov, T. Z. Akhmetov, O.Y. Prokohdko, Irina V. Razumovskaya, Sergey L. Bazhenov, P.Y. Apel, V. Yu. Voytekunas and M.J.M. Abadie

Additional information is available at the end of the chapter

http://dx.doi.org/10.5772/46005

1. Introduction

In the aerospace industry metal parts are substantially replaced by polymer matrix composites presenting many advantages compared with the metal parts they are replacing. However composite damage is very difficult to detect because it often forms a sub-laminate locations, invisible to the naked eye. To overcome these problems, research in the last two decades has led to the development of self healing polymeric materials that mimic some features found in biological systems. The development of self-healing polymeric materials, those that practically imitate self-healing process of wounds is a new and challenging problem for space applications to protect different types of devices made with composite materials and represents a new paradigm in materials design [1]. The pioneer works in auto-reparation of composites were developed by White S. R et al [2-6]. Their originality was to suggest the encapsulation of monomer – dicyclopentadiene and Grubbs catalysor disseminated uniformly into the composites. Under loading cracks are formed and microcapsules open up releasing monomer and initiator which fulfill the cracks by polymerization and crosslinking reactions and allow the composite to retrieve his original structure without losing its initial mechanical properties - Figure 1.

The concept is strongly dependant on the reactivity of the monomers used as well as the initiator systems [7].

A sub-orbital spaceflight (above 100 km) or a low Earth orbit (LEO) impose on the composites the following conditions : absence of oxygen, temperature variable from -120°C (dark side) to +300°C (solar side), UV and heavy ions radiation, low pressure 10^{-4} Pa. For example, the International Space Station is in a LEO that varies from 320 km (199 miles) to 400 km (249 miles) above the Earth's surface.

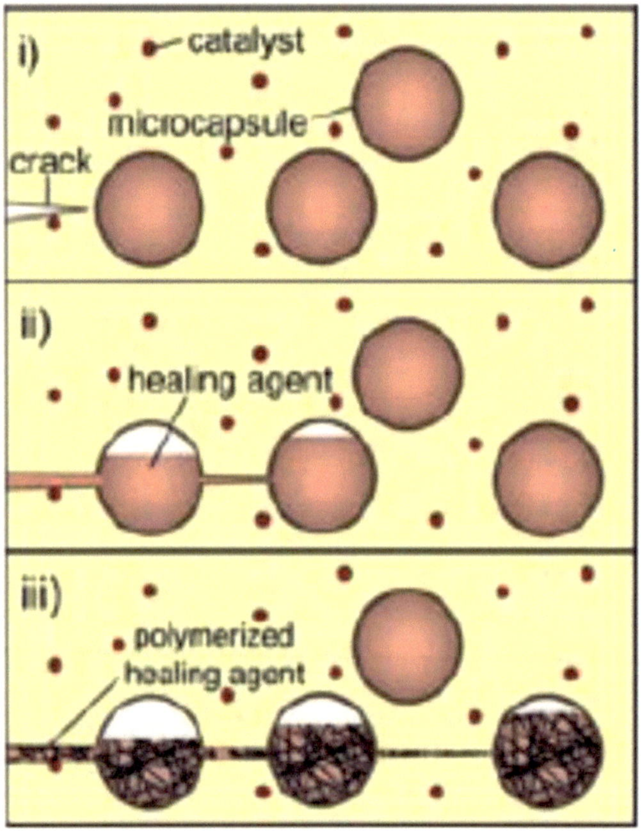

Figure 1. Concept of reservoirs developed by White S. R et al

The methodology developed was to test the feasibility of self-healing by using an encapsulated multifunctional acrylates, i.e. trimethylol propane triacrylate (TMPTA) in presence of a radical photoinitiator (Darocur® 1173).

The novelty of the work was to develop new polymeric film composites, based on Polyimide containing in the same reservoir microencapsulated active monomer and generator of active species, to prevent destruction of the damaged sample.

The new concept tested was to crosslink the monomer in presence of a radical photoinitiator which, upon exposure to UV, produces active centres, permitting the self-healing of the damaged composites.[8]. A schematic of our system is briefly described in Figure 2.

Three major problems have been studied:

1. Develop a suitable microcapsule which could, under loading, liberate the active principle in presence of initiator.

2. Distribute homogeneously the microcapsules into the polymeric film used as coatings
3. Test the new super-composite film and its auto-reparation once micro cracks have been formed, to prevent crack growth and macro damage of the composite.

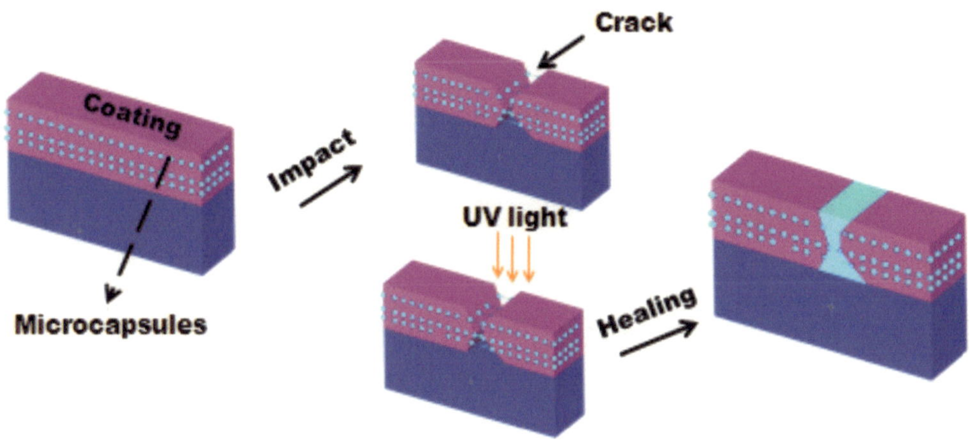

Figure 2. Self-healing of film coatings

2. Microencapsulation

Microcapsules shell should be resistant to high temperatures (> 300 °C) and have average diameters ranging 1-20 μm in order to put them in the host material.

Different models of microcapsules have been tested: Poly(urea/formaldehyde), Polyurethane and Silica gel

2.1. Synthesis of poly(urea-formaldehyde) microcapsules

The healing agent (trimethylol propane triacrylate TMPTA monomer) and the photoinitiator (Darocur® 1173) presented in – Figure 3, are encapsulated in poly(urea-formaldehyde) by *in situ* polymerization – Figure 4.

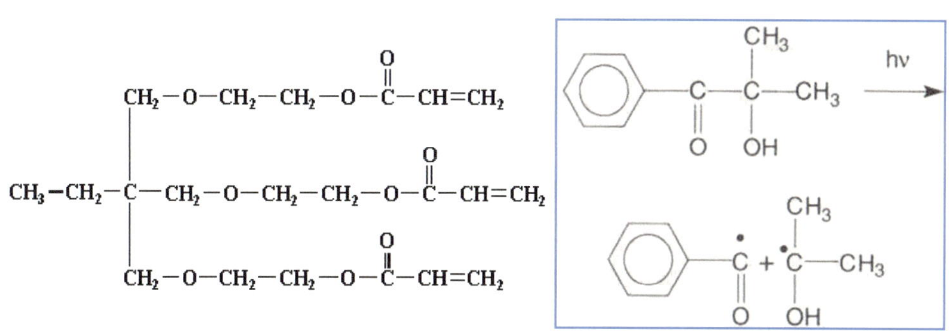

Figure 3. TMPTA monomer and Darocur® 1173 photoinitiator

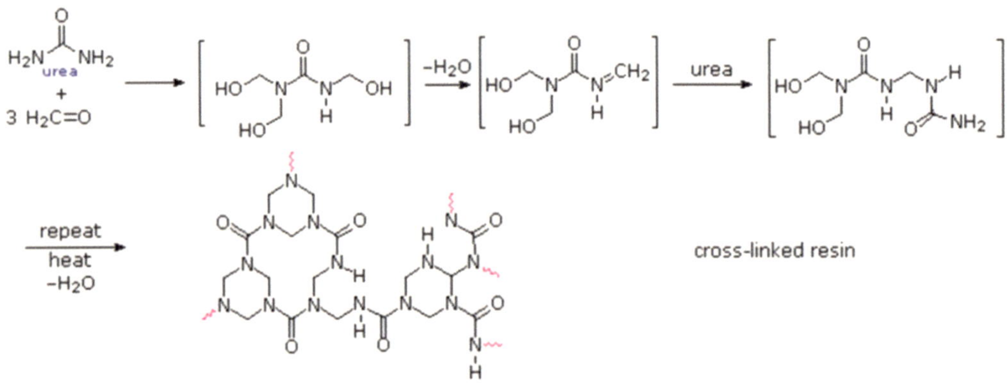

Figure 4. Poly(urea-formaldehyde) resin

These microcapsules are carried out by *in situ* polymerization in direct emulsion (Oil/Water). At room temperature, deionized water and aqueous solution of ethylene maleic anhydride (EMA) are mixed in a reactor. The reactor is placed in a temperature-controlled oil bath on a programmable hot plate with external temperature probe and the solution is agitated with a mechanical stirrer. Under agitation, urea, ammonium chloride and resorcinol were dissolved in the solution. The pH is raised from ~2.6 to 3.5 by drop-wise addition of sodium hydroxide (NaOH) and hydrochloric acid (HCl). One or two drops of 1-octanol are added to eliminate surface bubbles. Then, the healing agent (TMPTA monomer) and the photoinitiator (Darocur® 1173) are added slowly in the solution to lead to a stable emulsion. After 10 minutes of stabilization, an aqueous solution of formaldehyde (37 wt. %) is incorporated in the emulsion to obtain 1:1:9 molar ratio of formaldehyde to urea. The emulsion is covered and heated at 55 °C for 5 hours. After 5 hours of continuous agitation, the mixer and hot plate are switched off. Once cooled to ambient temperature, particles are filtered on Büchner and are washed with a water/ethanol solution (50 wt. %/50 wt. %). Then, particles are dried in air at room temperature for 24-48 h – Figure 5.

Scanning Electron Microscopy - The morphology of particles (surface and form) is investigated by Environmental Scanning Electron Microscopy (ESEM). The particles are spherical with average diameter in the range of 2-30 μm – Figures 6 (a) & 6 (b). The shell of microparticles presents two type of surface: one has a smooth surface – 6 (a), or a rough surface – Figure 6 (b). The Figure 6 (c) illustrates a poly(urea-formaldehyde) microcapsule which has been ruptured under physical constraint (pressure).

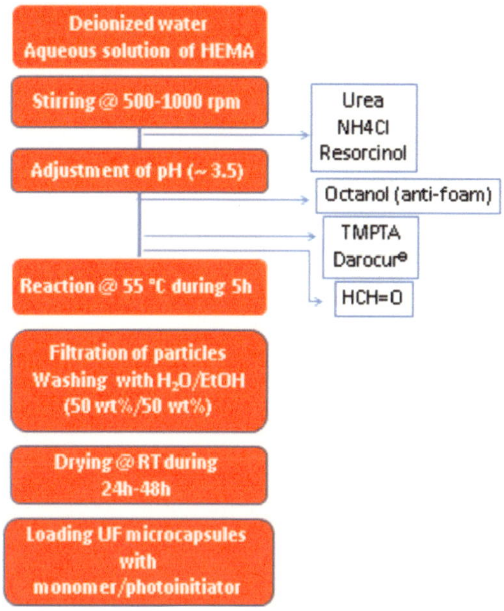

Figure 5. Schematic of the encapsulation process with poly(urea-formaldehyde)

Figure 6. ESEM image of (a) UF microcapsules loading in TMPTA monomer and Darocur® 1173 photoinitiator with a smooth surface, (b) UF microcapsules loading in TMPTA monomer and Darocur® 1173 photoinitiator with a rough surface and (c) ESEM image of ruptured Urea-Formaldehyde microcapsule

Infrared Analysis - The comparison between infrared spectra of UF empty microparticles – Figure 7 (a) and UF microparticles loaded in TMPTA – Figure 3 (b), shows the presence of new bands, in particular at 1722 cm^{-1} (C=O stretching vibration), 983 cm^{-1} (wagging of the $=CH_2$ group), and 808 cm^{-1} (twisting of the $=CH_2$ group), characteristic bands of TMPTA which indicates that this compound has been encapsulated in UF microcapsules.

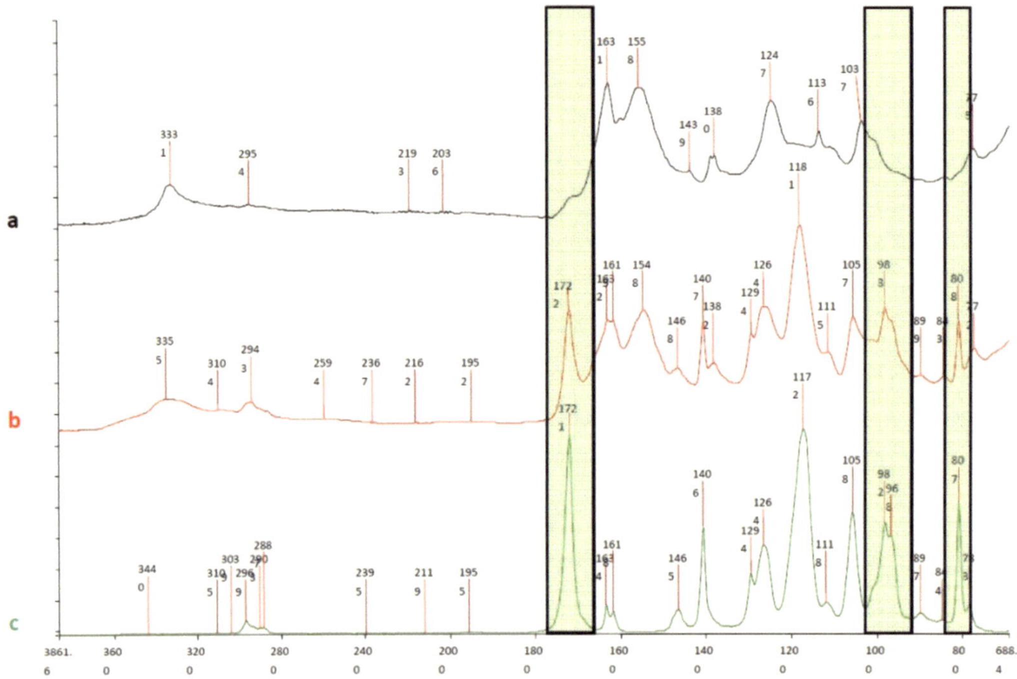

Figure 7. Infrared spectrum of (a) Empty UF microcapsules, (b) broken UF microcapsules loaded with TMPTA and (c) TMPTA monomer

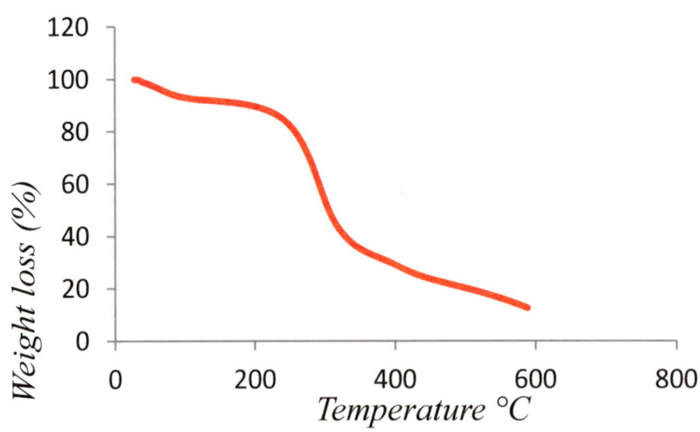

Figure 8. TGA of empty UF microcapsules

Thermogravimetric Analysis - Empty UF microcapsule shows – Figure 8, a weight loss (around 10 %) until 110 °C corresponding to the evaporation of free water in microcapsules, and an important decreasing in weight (around 80 %) starting from 180 °C due to the decomposition of urea-formaldehyde microparticles. Therefore the urea formaldehyde could not be used to self-heal spatial device composites because of polymer decomposition below 300 °C (maximum temperature in space).

2.2. Synthesis of poly(urethane) microcapsules by interfacial polycondensation

The TMPTA monomer and the Darocur® 1173 photoinitiator are encapsulated in polyurethane microcapsules by *in situ* polymerization. Microcapsules are obtained by *in situ* polymerization in direct emulsion (Oil/Water). At room temperature, deionized water and polyvinyl alcohol PVOH (3 wt. %) are mixed in a reactor equipped with a mechanical stirrer. Then, an organic solution of monomer (Hexamethylene diisocyanate HMDI/chloroform) containing the monomer (TMPTA) and photoinitiator (Darocur® 1173) to be encapsulated is added to the aqueous solution leading to a stable Oil/Water emulsion. After 5 minutes of stabilization, a hydrophilic monomer (Ethylene Diamine EDA in excess) is added to the emulsion – Figure 9. The reaction is continued until stabilization of the pH (around 5 hours).

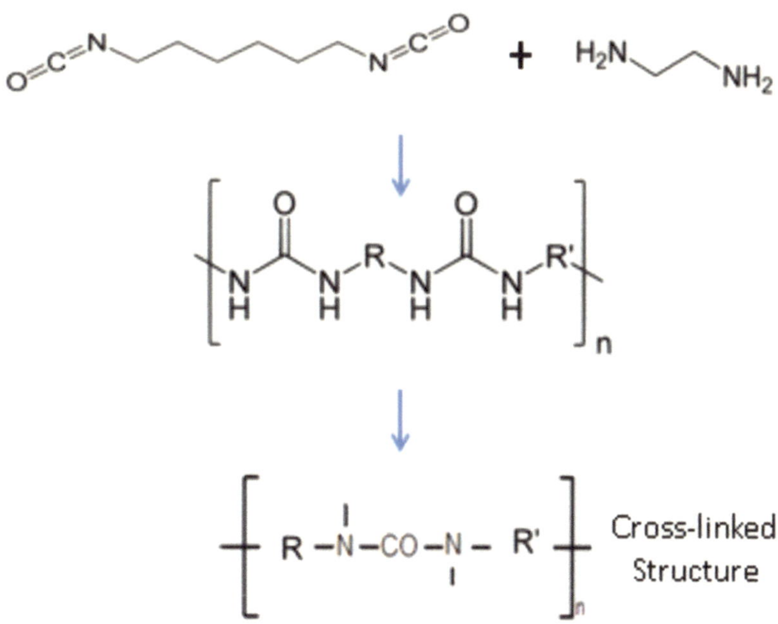

Figure 9. Cross-linked structure of polyurethane with R = $-(CH_2)_6-$ and R′ = $-(CH_2)_2-$

Then, particles are separated under vacuum with a Büchner system and are rinsed with an ethanol/deionized water mixture (50 wt. %/50 wt. %). Microparticles are collected and are air dried during 24-48 h – Figure 10.

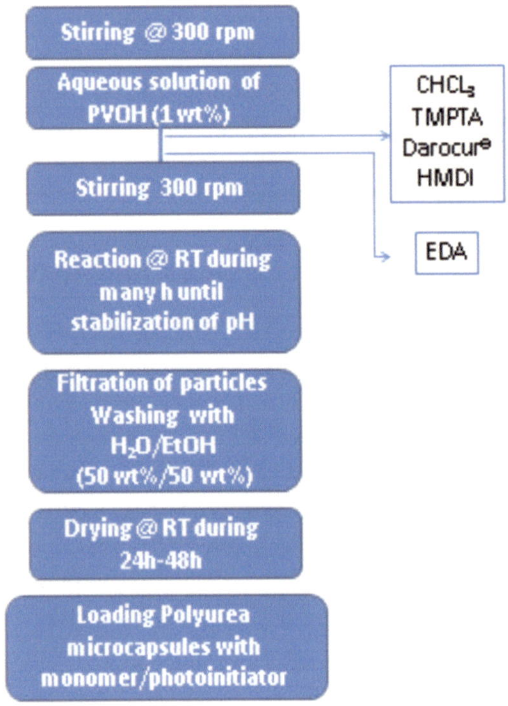

Figure 10. Schematic of the encapsulation process with polyurethane

Scanning electron microscopy - Particles were examined by ESEM. The particles are spherical with average diameter in the range of 200-300 μm and the membrane of microparticles is smooth, as shown in – Figure 11.

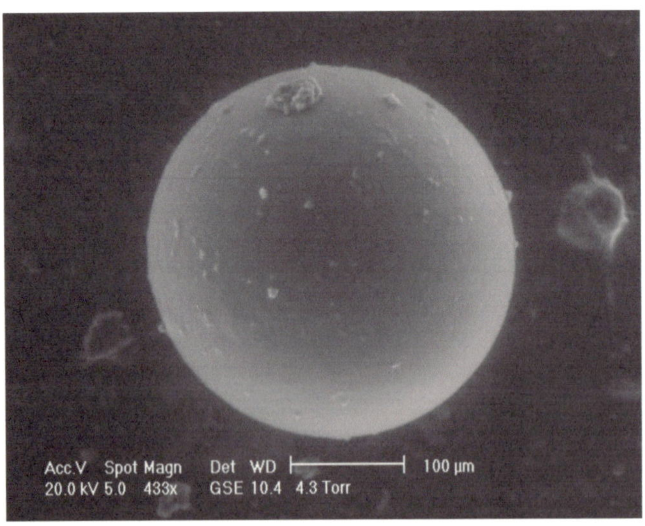

Figure 11. ESEM image of Poly(urethane) microcapsules loaded with TMPTA monomer and Darocur® 1173 photoinitiator

Infrared Analysis - The comparison between Infrared spectrums of the empty polyurethane microparticles - Figure 12 (b), and polyurethane microparticles loaded in TMPTA – Figure 12(c) shows the presence of new bands, in particular at 1721 and 983 cm^{-1} and confirm that TMPTA monomer has been encapsulated in polyurethane microcapsules.

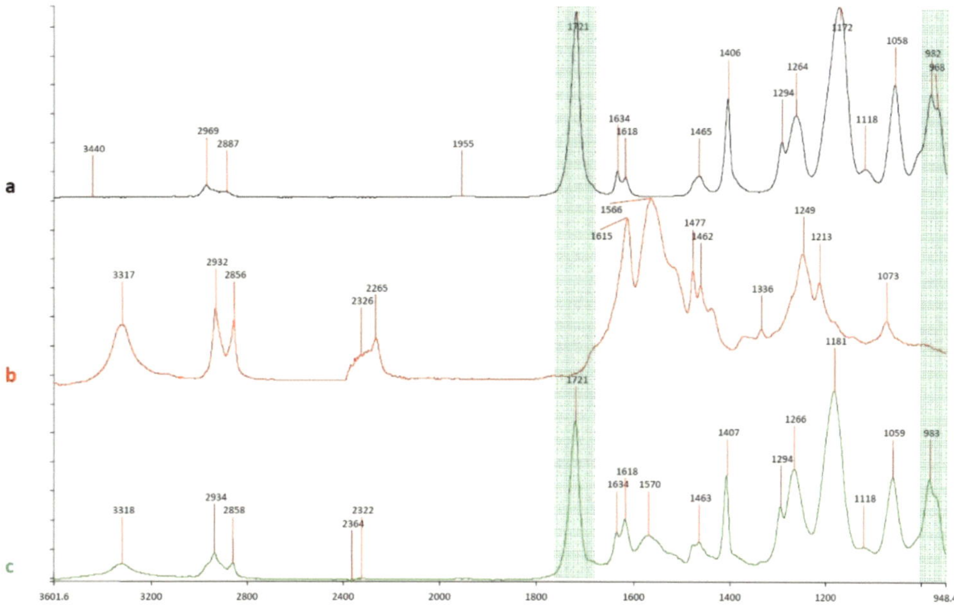

Figure 12. Infrared spectrum of (a) TMPTA, (b) empty polyurethane microparticles and (c) polyurethane microcapsules loaded with TMPTA monomer

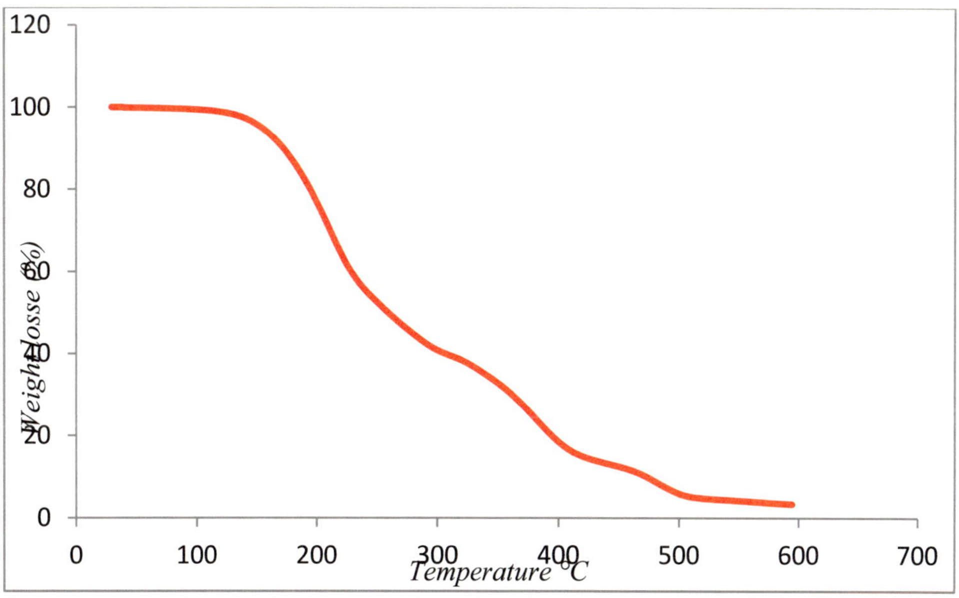

Figure 13. TGA of empty polyurethane microcapsules

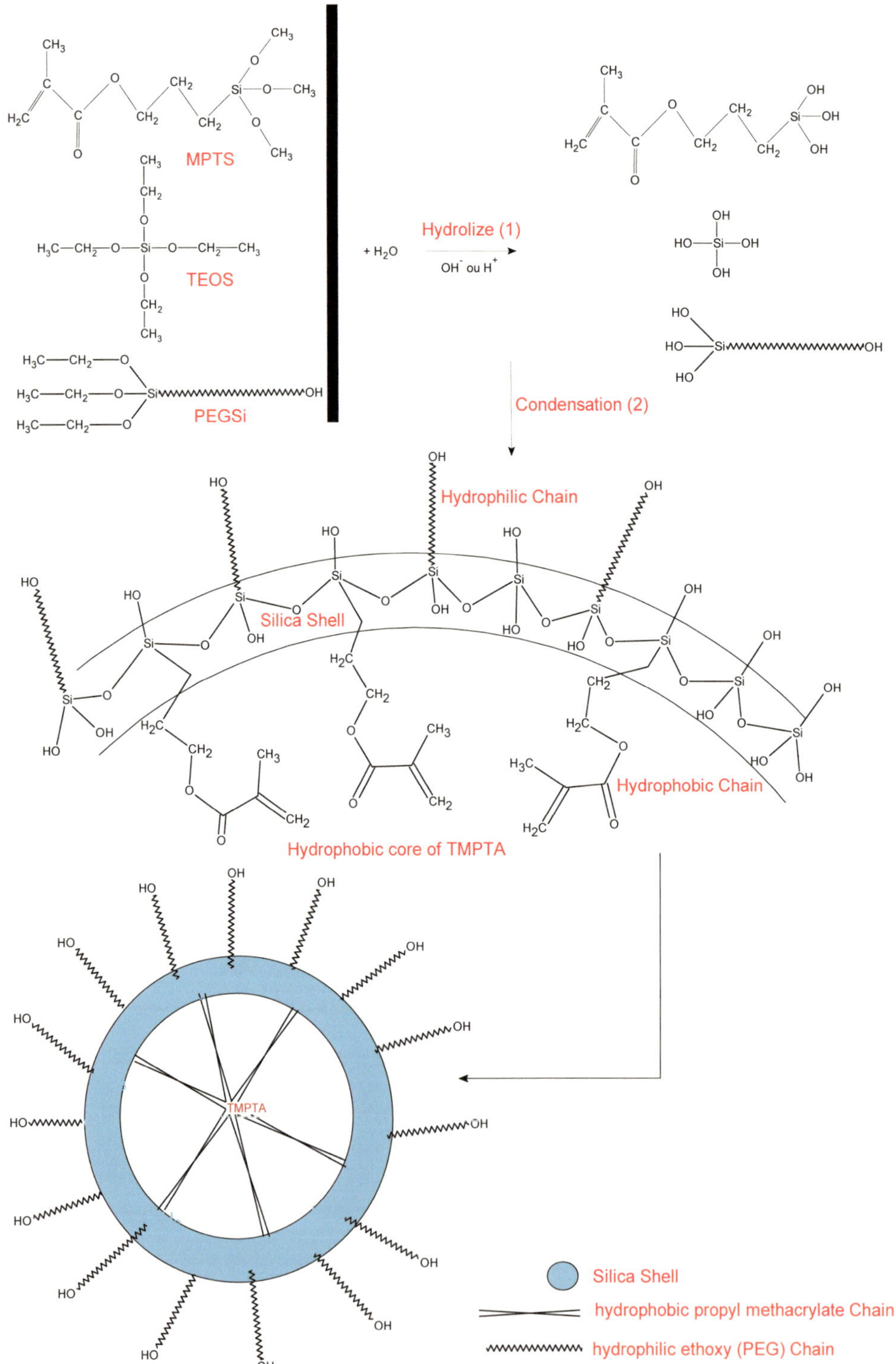

Figure 14. Schematic Road for Synthesis of Silica-Gel Microcapsules Loaded with TMPTA

Thermogravimetric Analysis - The thermogravimetric analysis of polyurethane microcapsules (without monomer) – Figure 13, presents a decreasing in weight (above 200°C) due to the deterioration of polymer. Therefore, this polymer is not a good candidate as well to encapsulate healed agent for aerospace applications.

2.3. Synthesis of silica-gel microcapsules by sol-gel polymerization process

The TMPTA monomer and the Darocur® 1173 photoinitiator are encapsulated in silica gels microcapsules by sol-gel polymerization. 3-(trimethoxysilyl)propyl methacrylate (MPTS), tetraethoxysilane (TEOS), trimethylol propane triacrylate (TMPTA), Darocur® 1173, silica surfactant PEGSi (compatibility agent to increase the stability between the shell and the aqueous solution) and ethanol are mixed with a magnetic stirrer in a bottle. Then, an aqueous solution of ammoniac (16 wt. %) is added drop by drop to the mixture. After 10 minutes of continuous agitation, this solution is added to an aqueous solution of polyoxyethylene 1,2- nonylphenyl ether (Igepal CO-720 (NP12) at 1 wt. %. After 2 hours of reaction, the suspension is filtered under vacuum. Microparticles are collected and are dried in air at room temperature for 24-48 h. Figure 14 illustrates the schematic road to synthesis of silica-gel microcapsules loaded with TMPTA monomer [9].

Scanning Electron Microscopy - Silica microcapsules are studied by ESEM. The spherical particles have average diameters ranging from 1-30 μm, shown in - Figure 15. The shells of microparticles present a smooth surface.

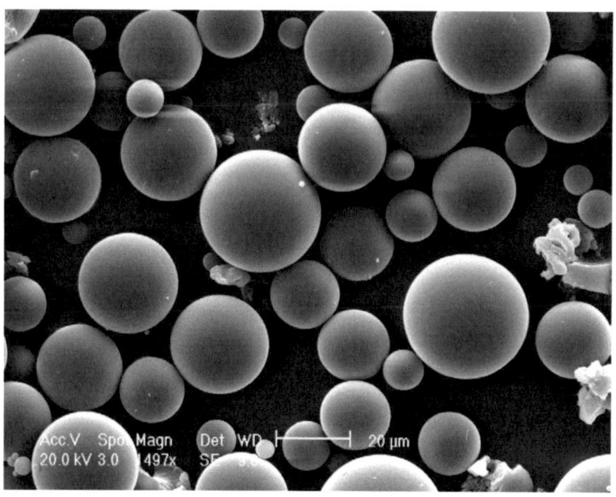

Figure 15. ESEM image of Silica-gel microcapsules loading with TMPTA monomer and Darocur® 1173 photoinitiator

Infrared Analysis - The comparison of IR spectra of empty silica-gel microparticules and silica-gel microparticles loaded with TMPTA do not enable us to prove the encapsulation of the TMPTA. In effect, the two spectra present the same absorption bands because silica-gel shell contains a methacrylate compound – Figure 16.

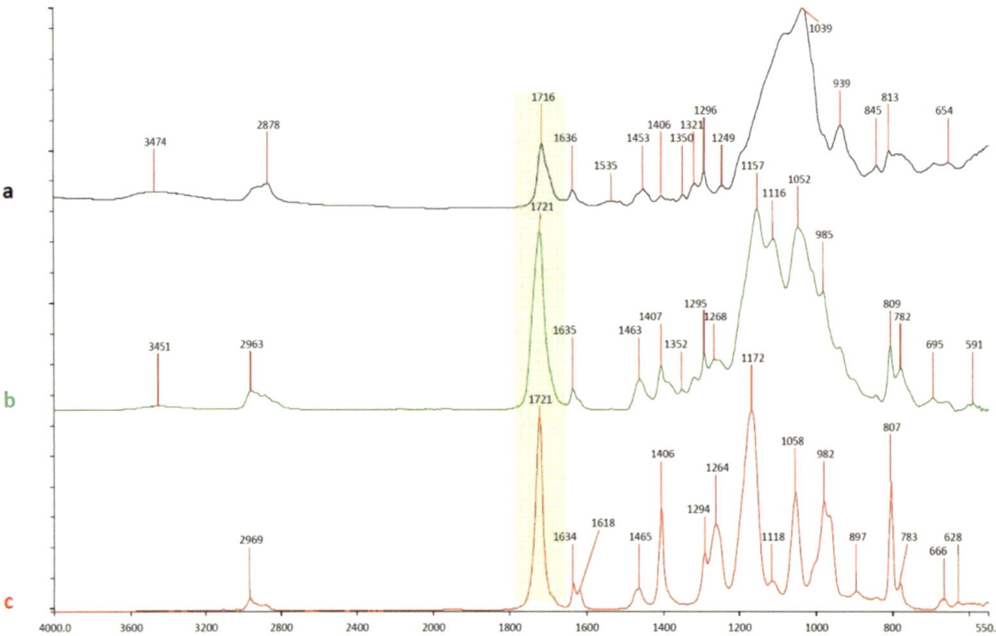

Figure 16. Infrared spectrum of (a) Empty silica-gel microcapsules (b) silica- gel microcapsules loaded with TMPTA monomer and (c) pure TMPTA

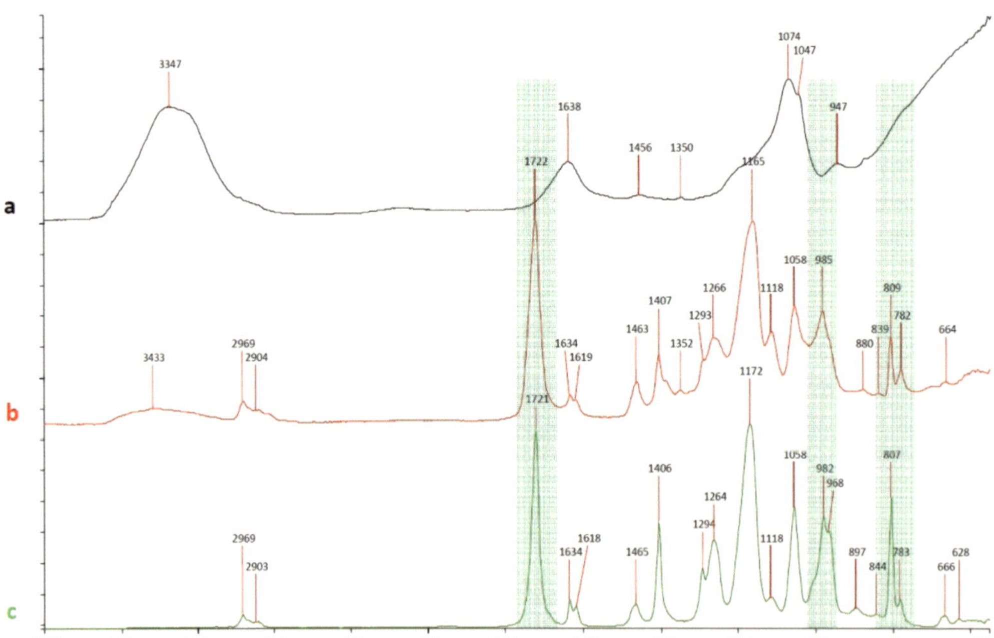

Figure 17. Infrared Spectrum of (a) empty silica-gel microcapsules with TEOS precursor alone, (b) silica-gel microcapsules with TEOS precursor alone loaded with TMPTA monomer and (c) pure TMPTA

For that we have used silica-gel microcapsules (empty microcapsules and microcapsules loaded with TMPTA) with TEOS precursor alone (not methacrylate chain in this compound). The comparison between Infrared spectra of the empty silica-gel microparticles with TEOS precursor alone – Figure 17 (a) and silica-gel microparticles loaded with TMPTA – Figure 17 (c), shows the presence of new bands, in particular at 1721, 983 and 807 cm^{-1}. These bands, characteristic of the TMPTA monomer, confirm the presence of TMPTA encapsulated in silica-gel microcapsules.

Thermogravimetric Analysis - The thermogravimetric analysis of silica-gel particles (without TMPTA) indicates a good thermal resistance (above 320°C) of silica-gel membranes – Figure 18.

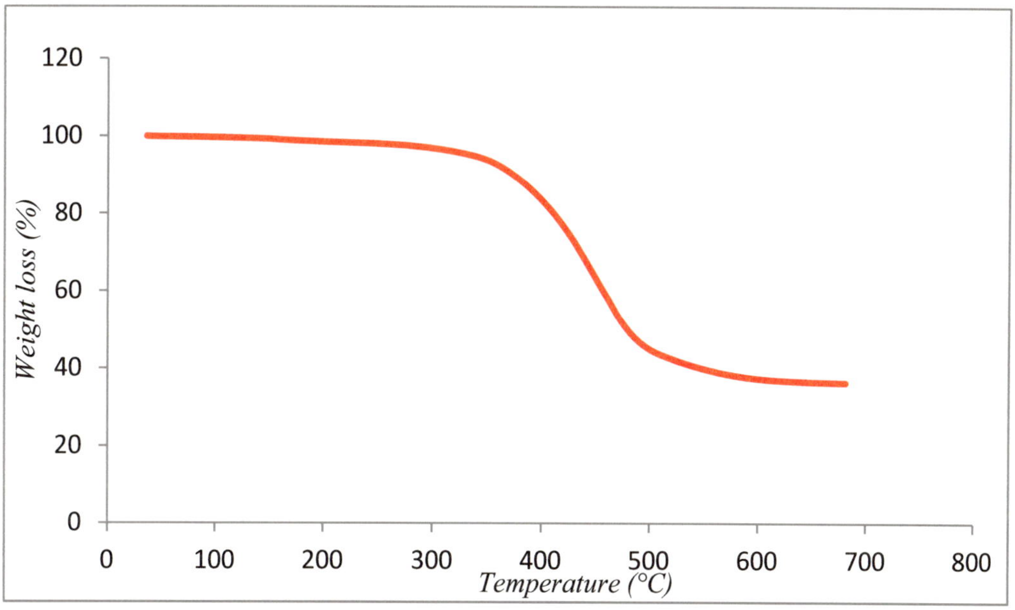

Figure 18. Thermogravimetric Analysis of Empty Silica-gel Microcapsules

We have encapsulated with success TMPTA monomer by different processes of microencapsulation. Two of three techniques used cannot be applicable for aerospace application as the microcapsules are not thermal resistant above 300 °C, presenting a temperature of degradation below 200°C. However, silica membrane is much more resistant due to their degradation temperatures exceeding 300°C.

3. Dispersion into the polyimide film

Polyimide films are based on dianhydride of dicyclodecene tetracarbonic acid and oxydianiline ODA – Figure 19. The dianhydride was synthesized by Solar-irradiation technology of a mixture of benzene/toluene fraction with maleic anhydride to produce highly reactive monomer – benzene adduct (AB) – Figure 20.

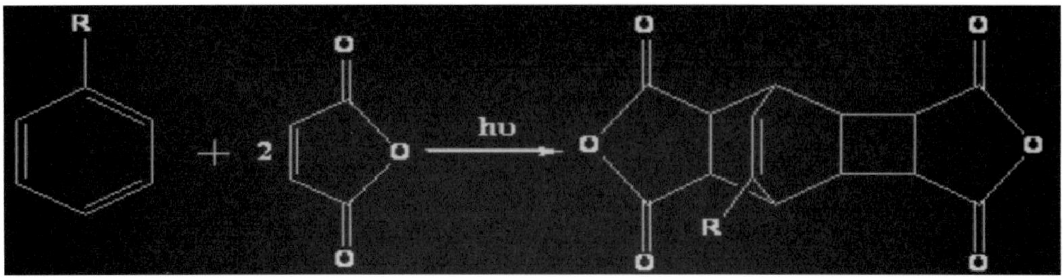

Figure 19. Photosynthesis of the dianhydride of dicyclodecene tetracarboxylic acis

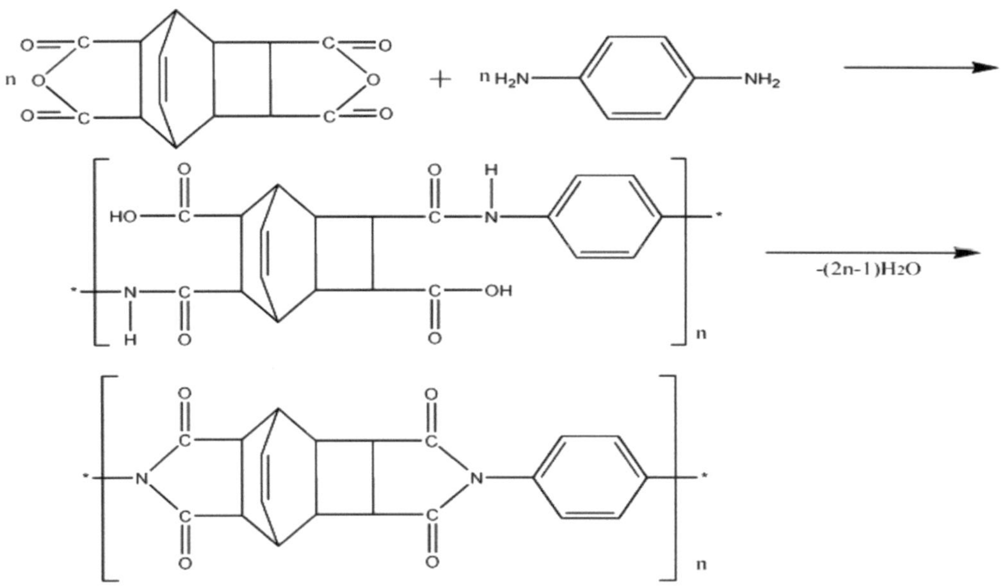

Figure 20. Polyimide based on dianhydride of dicyclodecene tetracarbonic acid and ODA

Microcapsules loaded with TMPTA and the radical photoinitiator are mixed in 5 to 15 wt. % in solution of polyimide in DMF, cast and dry, and then subjected to 100 °C for 4 hours, followed by 4 hours at 180 °C.

Environmental scanning electronic microscopy coupled with RX (ESEM/EXD) – Figure 21(a) and 21(b), show a good dispersion of the microcapsules along the polyimide film and confirm their presence.

EDX of polyimide film does not show any silicium peak - Figure 21 (a) but this peak is present in the analysis of the polyimide with silica-gel microcapsules.

4. Auto-reparation of polyimide film

Different formulation of Polyimide films have been loaded with variable concentration of microcapsules from 5 wt. % to 15 wt. %. Then the films casted are exposed to UV light (365 nm) once cracks were produced – Figures 22 & 23.

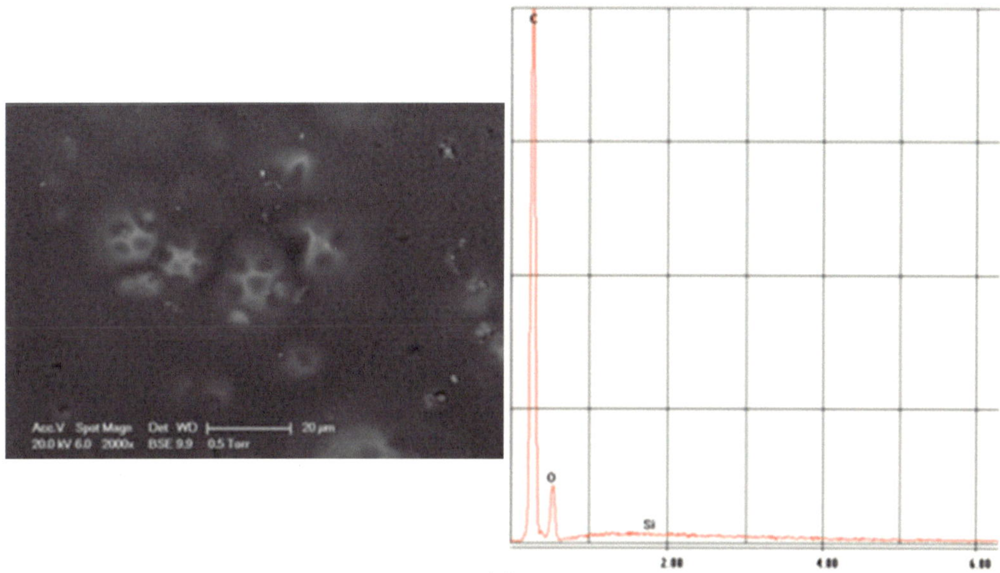

(a)

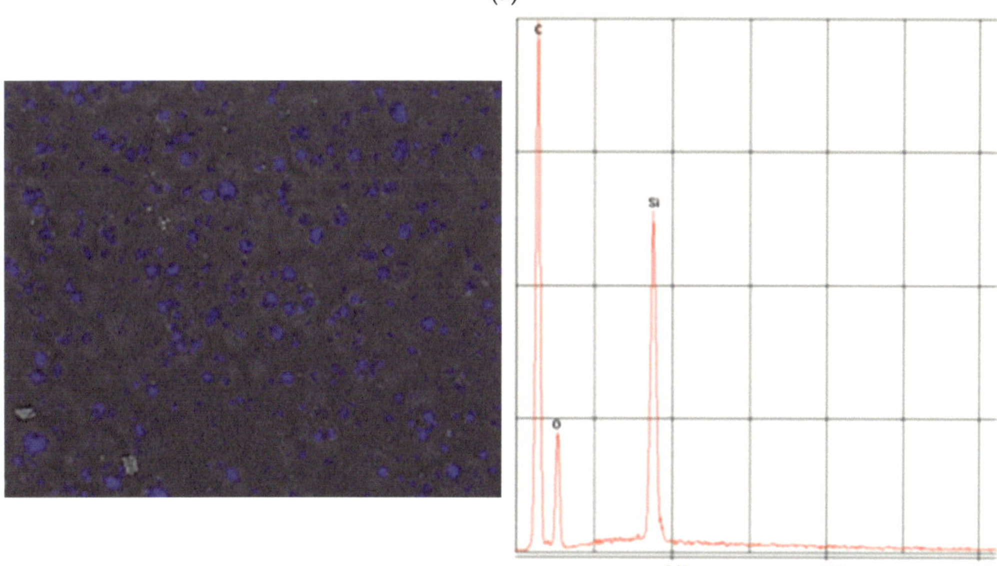

(b)

Figure 21. (a) Surface aspect of the polyimide film without microcapsules & EDX
(b) Surface aspect of the polyimide film with microcapsules & EDX

Self-healing process was followed by morphological analysis shown by environmental scanning electron microscopy ESEM.

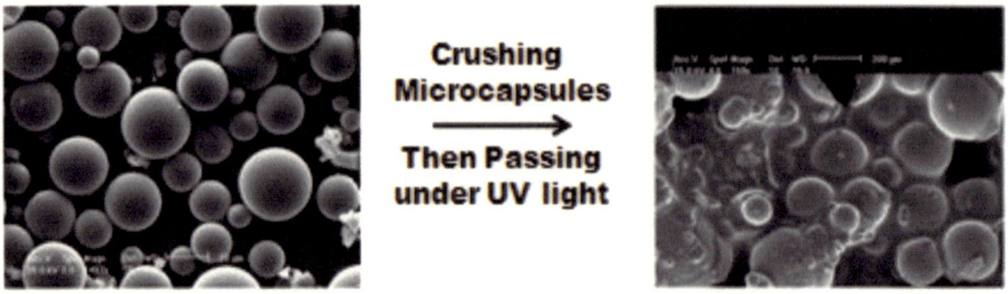

Figure 22. Auto-repairing after cracks

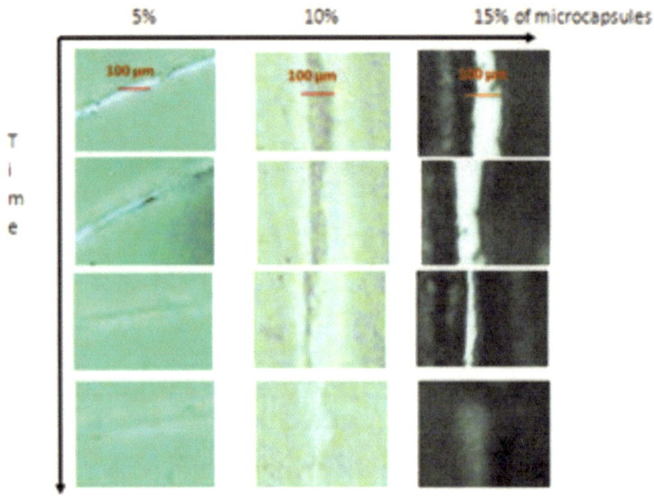

Figure 23. Sel-healing after cracks for 0, 10, 20 and 30 min of irradiation respectively

Simulation of the cosmic effect – Figures 24 (a) and (b), was provided in a simulator SES (Space Environment Simulator) at Yerevan Physics Research Institute at the following conditions (260 nm) – Table 1 below.

Experimental Conditions	Units
Vacuum	$1x10^{-5}$ Torr
Electrons Bombardment	5.0 MeV
Temperature	– 50 °C
UV Solar	2.0 kW / sq.m.
Proton Bombardment pfu	> 10 MeV

Table 1. Simulator parameters

PI films loaded with 15 wt. % of Si-microcapsules loaded with TMPTA monomer exposed 30 min are equivalent to 6 month exposure at the real space condition over geostationary orbit of the earth.

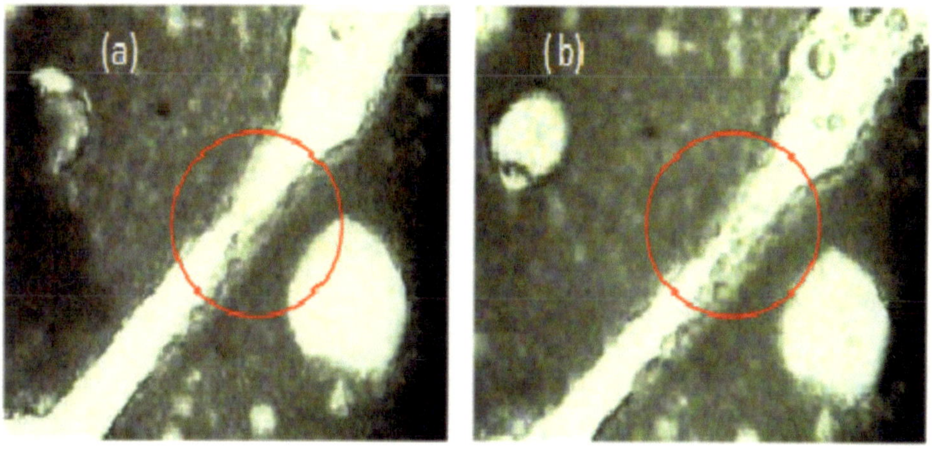

Figure 24. Cracks at time zero (a) – Cracks at exposure time 5 min under 260 nm (b)

5. Mechanical performance of the composite film

The capsules were the defects of the structure in which cracks mainly grew from. The increases in capsules concentration led to a more brittle fracture as shown in Figure 25.

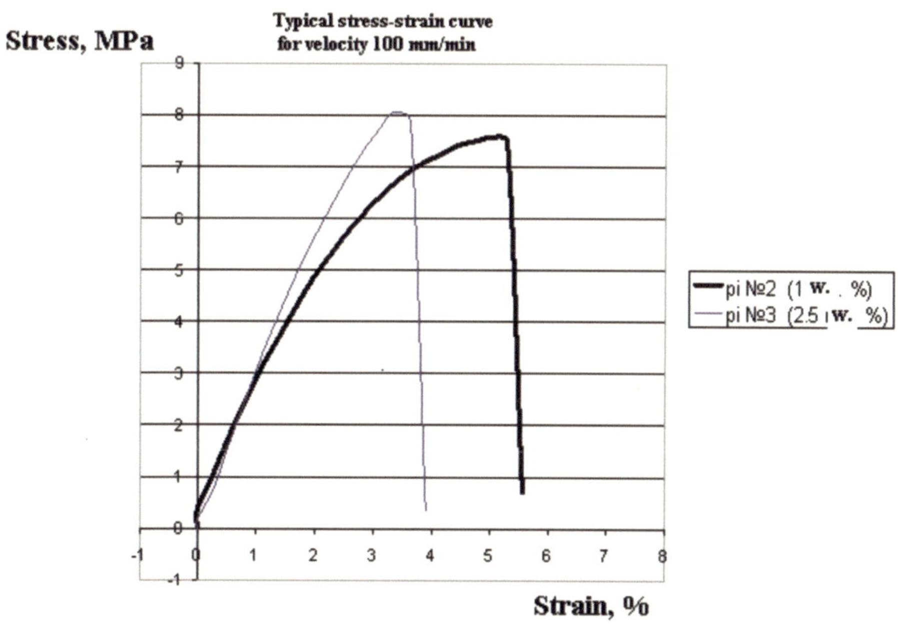

Figure 25. Effect of the wt. % loading on the strain-stress curve

Tensile curves were measured using a universal "Shimadzu" Autograph AGS machine. The test samples were strips with working size of 5x30 mm. The thickness of samples was measured by a magnetic "PosiTector 6000" instrument.

Then, a circular hole (diameter = 0.3 mm) was made in each specimen with the least (0.5 wt. %) and greatest capsules concentration (5 wt. %): Specimens pi-1 and pi-4 respectively. The holes were made by special drill press.

According to the theory of elasticity, the stress concentration coefficient near circular hole is equal to 3 ; in fact the real stress near the hole is approximately one. So, the circular hole might be considered as the standard stress concentrator.

Stress – strain curves for Specimens pi N°1 (0.5 wt. %) and pi N°4 (1.5 wt. %) with hole are represented in Figure 26.

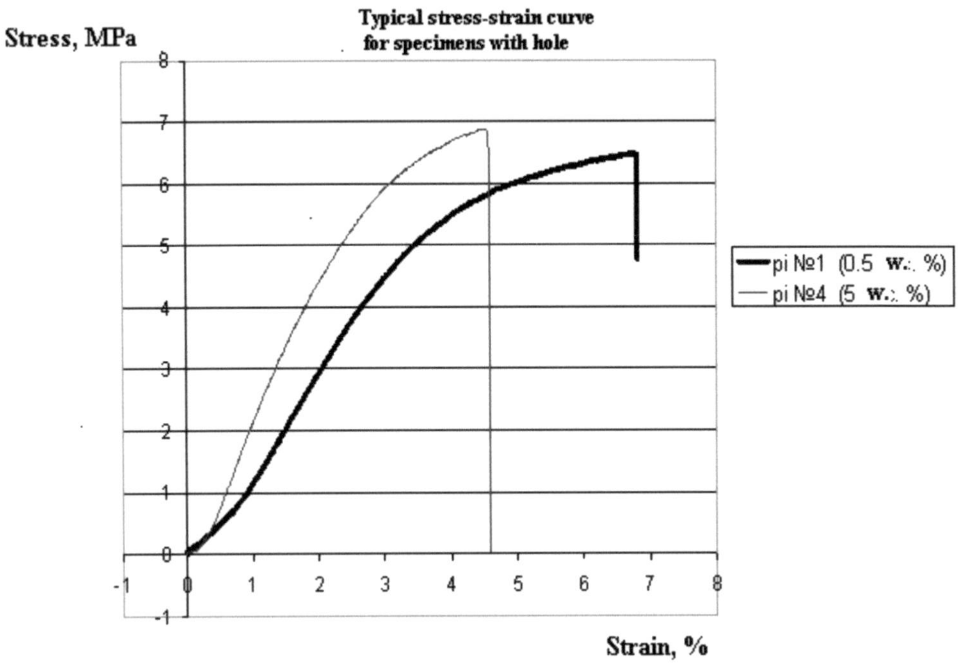

Figure 26. Effect of the wt. % loading on the strain-stress curve of the composite film with hole

The experiment showed that for Specimens pi N°1, the crack passed through the hole. So stress concentration near capsules was less than 3. But for Specimens pi N°4 the crack went through the hole. With the help of this method we can separate films with capsules into two classes: with effective stress concentration less and more than 3.

Using the polarizing microscope POLAR 312, we have detected stress fields near the capsules, even before the loading. It must be taken into account that heat input might be needed for stress relaxation to occur. On Figure 27, stress fields are represented by bright spot.

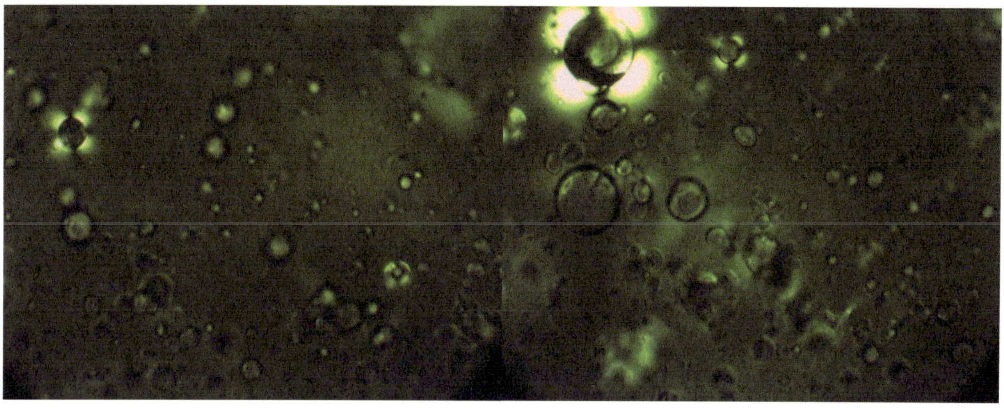

Figure 27. Stress field near hole

Around some capsules, internal strain remained even after rupture of the specimen. In the experiment the film was elongated manually under the microscope by special micromachine.

Stress fields were detected near the hole too as depicted in Figure 28.

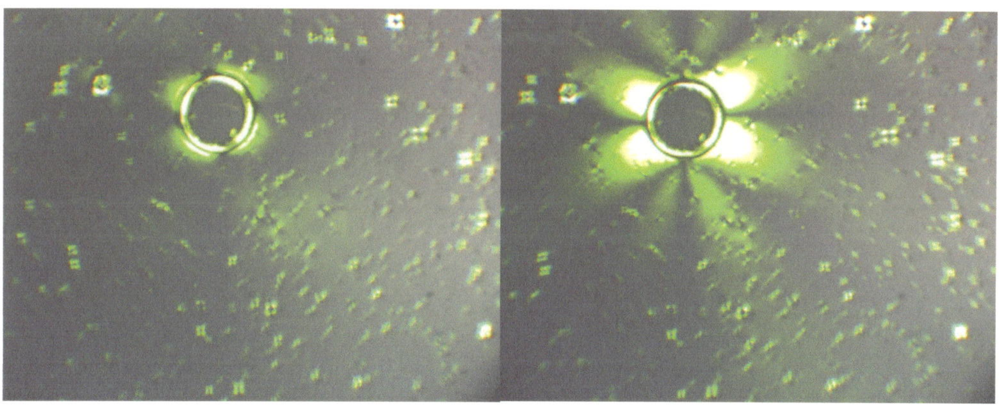

Figure 28. Stress field near hole

The increasing strain near the hole was illustrated in the photos placed in order. The cracks grew through each hole visually. For low concentration of capsules, cracks led to rupture. For high concentration, big crack arose from capsules. It meant that local concentration of stress depended on overlapping of stress fields near capsules too. Theoretically, such interaction appeared when distance between capsules was less than 5 capsule diameters – Figure 29.

These observations allowed us to conclude that stress fields near capsules are the predominant effect; capsules concentration influenced the fracture when stress fields overlapped near capsules.

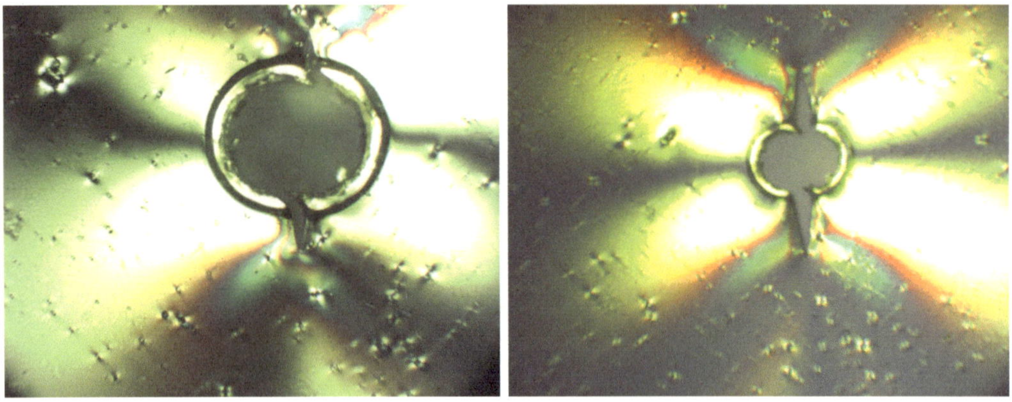

Figure 29. Stress field near hole after elongation

Test pieces were cut from polyimide films. Straight samples, 10 mm in width and 20 mm in length, were tested in tension at 2 mm/min speed with "Shimadzu" "Autograph" machine. The thickness of samples varied from sample to sample and was approximately 0.12 mm. After testing, failed samples were studied under an optical LOMO Mikmed-2 microscope. Hitachi S-520 scanning electronic microscope was used to get a high-magnification image of samples' top and side surfaces, and fracture surfaces of samples were studied with microscopes.

Figure 30 shows the typical stress-strain curves for pure PI film (without particles containing healing resin). The diagram is typical for plastic materials. Initially, the material deformed elastically, and yielding at strain more than 5 % was observed. The Young modulus of fracture deformation of samples was very high, around 130 %.

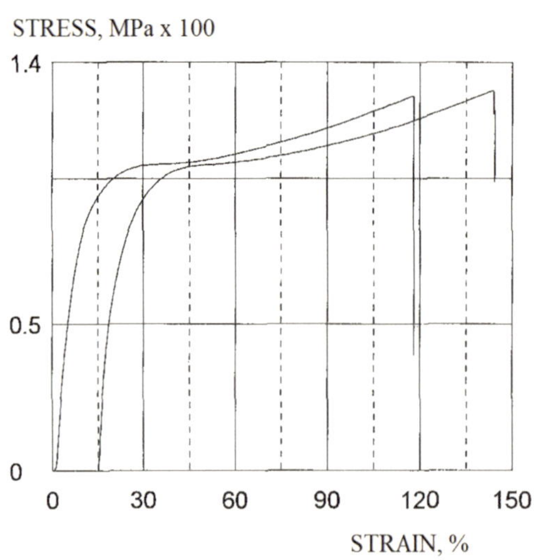

Figure 30. Typical stress-strain curve for pure PI film (without spherical particles)

Figure 31 shows typical stress-strain curves for PI containing 10 wt. % of particles with healing resin. The Young modulus of composite polymer was calculated from the slope of the stress-strain curves to be 2.6 GPa. The curves were similar to that of pure PI. However, the fracture strain was lower at approximately 50 %, due to the stress concentration near particles.

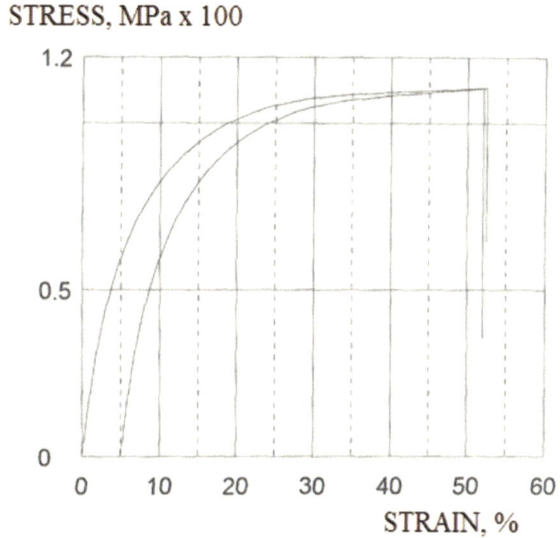

Figure 31. Typical stress-strain curves for PI containing 10 wt. % of spherical particles with healing resin

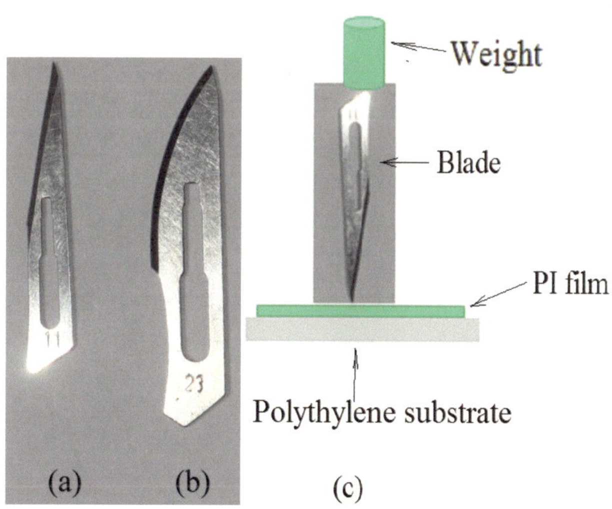

Figure 32. Changeable scalpel blades (a) and (b) used to make a crack in PI film samples. (c) – Method of making through-the-thickness crack

In each sample, two types of cracks were made. First type of cracks was through-the-thickness. These cracks were made by applying 100 g force on a changeable scalpel blade as shown in Figure 32a-c.

The resulting crack is shown in Figure 33 below where spherical particles of different size are observed.

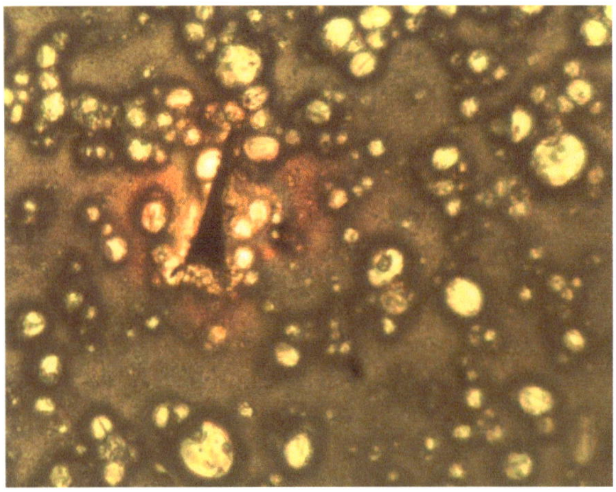

Figure 33. Throught-the-thickness triangular crack with the length of approximately 0.6 mm

Figure 34a shows a typical stress-strain curve for composite PI sample with a crack similar to that in Figure 33. The curve is similar to that of composite PI without a crack. The fracture strain was about 18 %, which was typical for ductile materials. However, fracture strain was lower than that in samples without a crack.

Figure 34b shows a stress-strain curve of a similar sample (with a crack) after healing with 200 Watts UV lamp for 30 min. The strength and fracture strain is the same as in Figure 31. Thus, this implied that the healing did not improve mechanical properties if the crack was thick. Possibly, the amount of the liquid resin was not enough to fill the entire volume of the open crack.

The fracture toughness G_{Ic} of filled PI may be estimated from the Figure 31 with the equation below:

$$G_{Ic} = \frac{\pi \sigma^2 c}{E}$$

where σ is fracture stress, c is the half of the crack length and E is the Young modulus of material. Substituting σ = 80 MPa, c = 0.6 mm and E = 2.6 GPa, the fracture toughness G_{Ic} is estimated to be 2.6 kJ/m^2. This value is typical for very tough materials.

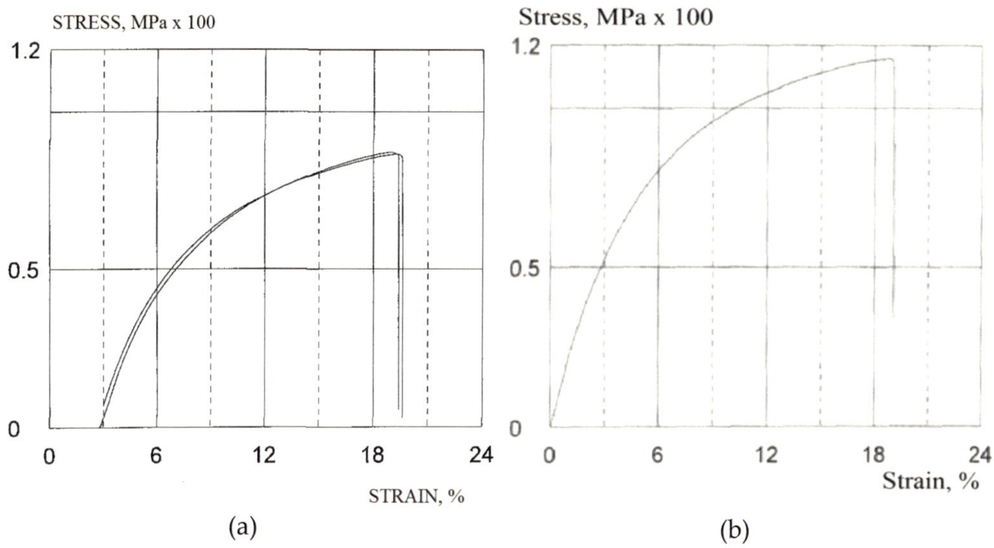

Figure 34. Typical stress-strain curve for filled PI sample with a crack. (a) – without UV irradiation; (b) – after UV irradiation

Different results were obtained from samples with crack of the second type created as shown in Figure 35.

Figure 35. Method of making surface cracks

In this case the crack was on the surface and not deep. These cracks were made by putting 100 g force on a curved scalpel blade. The crack was not wide with low opening, and could not be measured with an optical microscope.

Figure 36 shows typical stress-strain curves for samples with crack subjected to UV irradiation. In this case, two different types stress-strain curves were observed.

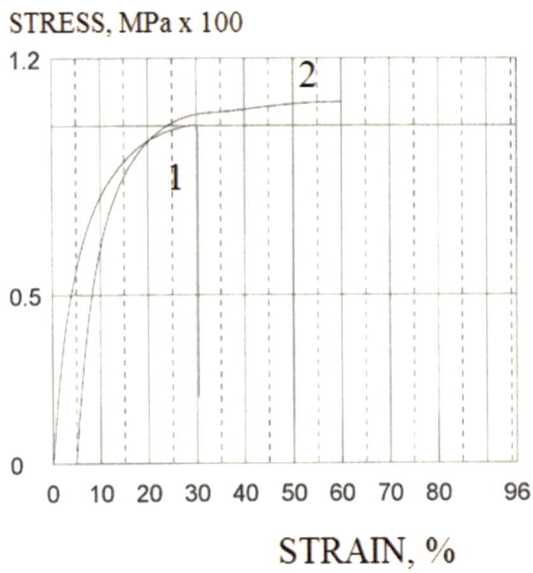

Figure 36. Two types of stress-strain curves for samples with surface cracks made as shown in Figure 35 and irradiated with UV light

The first one is shown by the curve 1 in the Figure 36. In this case the fracture strain is equal to approximately 30 %. This value is higher than that before UV irradiation (20-22 %). The stress-strain curve of the second type is shown by the curve 2 in the Figure 36. In this case the fracture strain is approximately 55 %, which was much higher than in the first case.

In the composite samples two layers were observed: the first one was pure PI and the second was PI filled with spherical particles. Samples with double-layered structure can be observed in Figure 37.

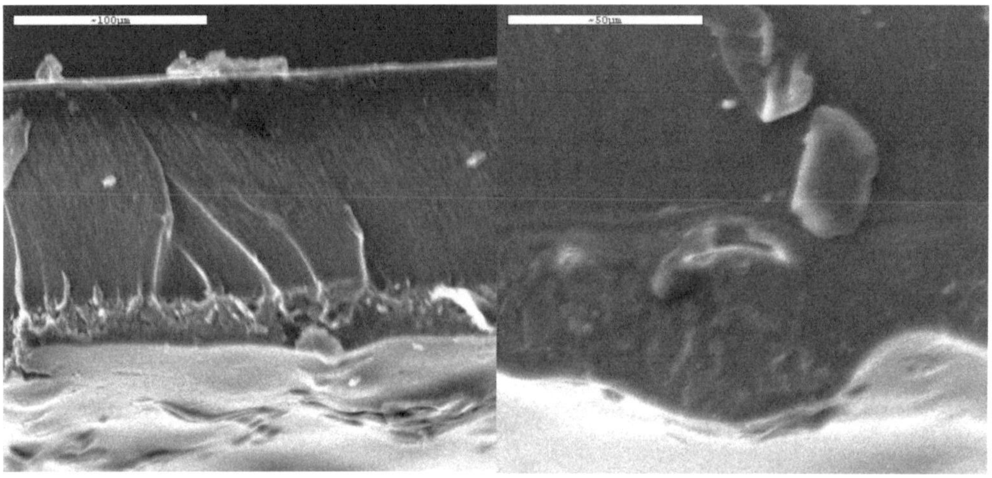

(a) (b)

Figure 37. Samples with double-layered structure frozen in liquid nitrogen, at different magnifications. In the right photo, a round particle can be noticed

Study of the fractured samples showed that the layer of the composite PI debonded from the pure PI layer, and this explained higher fracture toughness in this case. Local cracking and debonding of composite surface layer under tensile load can be observed in Figure 38.

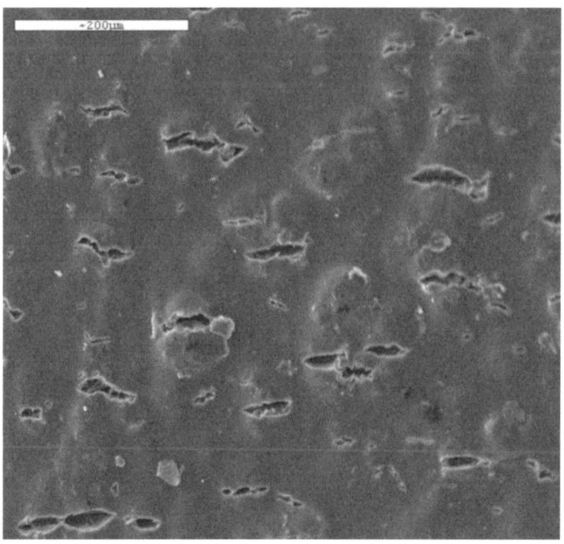

Figure 38. Local cracking of surface layer with PI filled by spherical particles

Schematic drawing of debonding of a surface filled layer near a crack is shown in Figure 39.

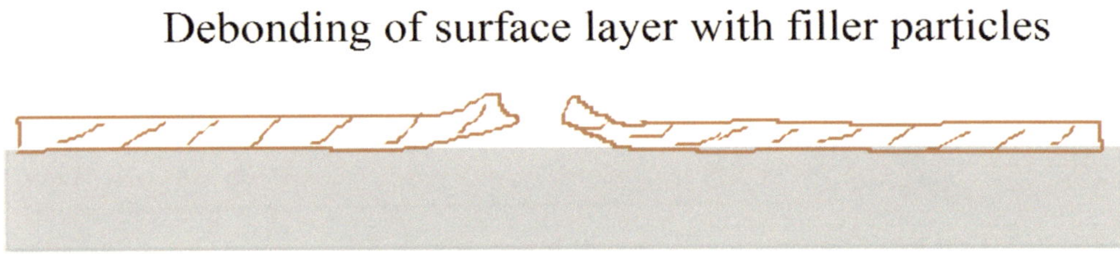

Figure 39. Schematic of debonding of composite layer at the surface near a crack

6. Formulation for low orbital spaceflight

In the condition of low orbital spaceflight (under 30 km), the presence of oxygen might reduce the photospeed of the crosslinking when radical process is used. To overcome this problem we have investigated the use of epoxies initiated by a cationic process of which kinetic is non sensitive to the presence of oxygen [10-13].

For that purpose we have considered a multifunctional epoxy formulations based on dicyclopentadiene and phenolic epoxy, such as solid Epiclon-HP720 – Figure 40.

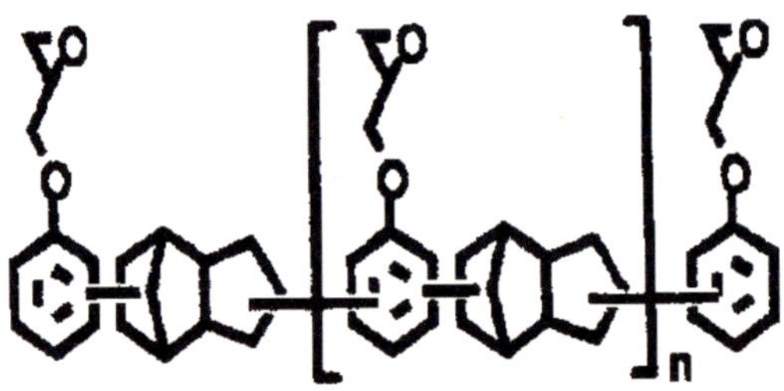

Figure 40. Structure of Epiclon-HP720

Different types of co-solvents have been studied in formulating solid epoxy resin (Epiclon-HP7200) based either:

1. on epoxy: two cycloaliphatic diepoxy monomers, both having terminal epoxy functional groups at opposite sides of the molecule, being separated by one functional carboxylate, so-called, "space" group in the case of 3,4-epoxycyclohexylmethyl-3,4-epoxycyclohexane carboxylate - Cyracure® UVR-6105 - Figure 41, and by two carboxylate "space" groups in the case of Bis-(3,4-epoxycyclohexyl) adipate - Cyracure® UVR-6128 - Figure 42, and having thus different length of chemical structure have been selected.

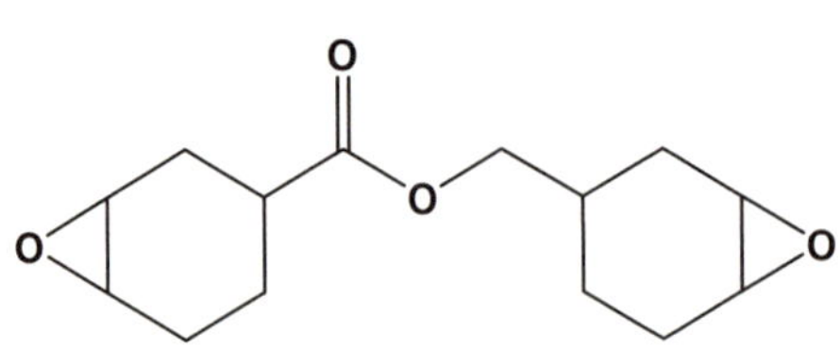

Figure 41.

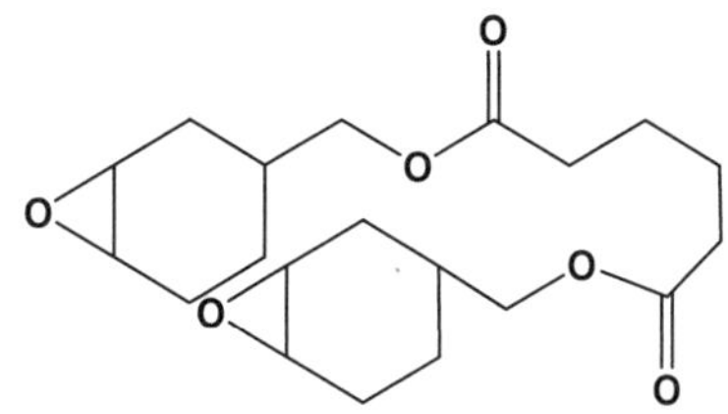

Figure 42.

2. or divinyl ether, triethylene glycoldivinyl ether - Figure 28.

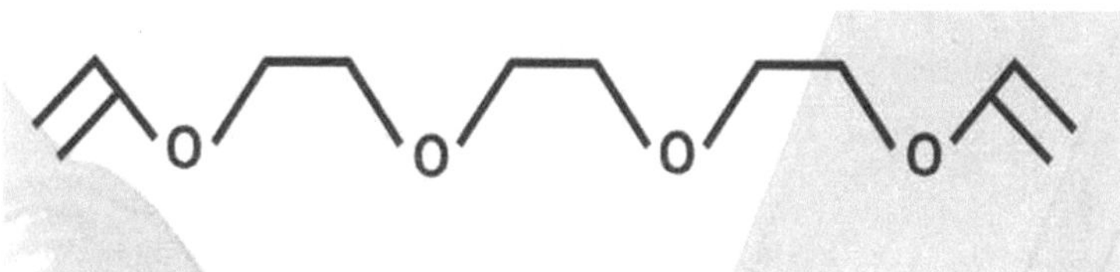

Figure 43. Chemical structure of Triethyleneglycol divinyl ether (Rapid-Cure® DVE-3)

The final objective was to formulate a system based on epoxy having the highest photoreactivity for self-healing applications.

The Photo-Differential Scanning Calorimetry (photo-DSC) [10] was used to study the cure kinetics of UV-initiated photo-polymerization of epoxy resin monomers and vinyl ether in presence of cationic photo-initiator, mainly Cyracure® UVI-6976, former Cyracure® UVI-6974, which consists of a mixture of dihexafluoroantimonate of S, S, S′, S′-tetraphenylthiobis (4,1-phenyllene) disulfonium and hexafluoroantimonate of diphenyl (4-phenylthiophenyl) sulfonium (CAS no. 89452-32-9 and 71449-78-0) at 50 wt. % in propylene carbonate - Figure 44 [11].

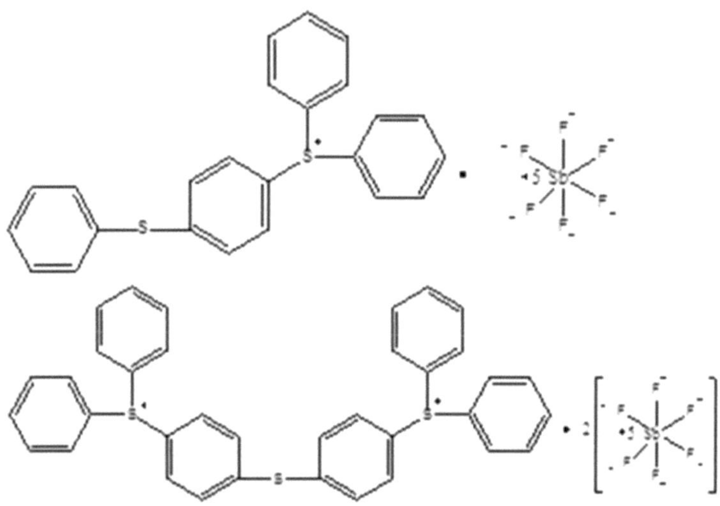

Figure 44. Cationic photoinitiator Cyracure® UVI-6976

We have studied the effect of the temperature for different solvents on the rate coefficient k according to the simplified Sestak and Berggren equation [14] and use the Arrhenius equation to calculate their activation energy Ea – Table 2.

Solvents & Epoxy	Ea (kJ/mol)
Carboxylate (Cyracure® UVR 6105)	7.02 ± 0.80
Adipate (Cyracure® UVR 6128)	4.03 ± 0.20
Vinyl ether (Rapi-Cure® DVE-3)	3.60 ± 0.42
Epiclon HP-7200/Rapi-Cure® DVE-3 (40/60 wt. %)	10.60 ± 0.75

Table 2. Activation energy for different solvents and epoxy Epiclon-HP720 solution

Photocuring kinetic of UV-initiated cationic photo-polymerization of divinyl ether shows a very low activation energy comparing to the two epoxies studied above. The Epiclon HP-7200/ Rapi-Cure DVE-3 system has been optimized by Differential Photo Calorimetry (DPC) in presence of cationic photoinitiator Cyracure® UVI-6976.

We have shown that the mixture of a multi-functional epoxy resin Epiclon HP-7200 and divinyl ether Rapid-Cure® DVE-3 for the ratio Epiclon/DVE (60 wt. % / 40 wt. %), is a valuable candidate for UV encapsulation and for further self-healing test.

7. Conclusions

Silica-oragnic microcapsules based on 50 wt. % MPTS and 50 wt. % TEOS have been selected to encapsulate the monomer trimethylol propane triacrylate TMPTA and the photoinitiator Darocur® 1173. The microcapsules prepared have diameter size of 4.00 ± 0.30 μm in optimal synthesis conditions: reaction time of 2 h and stirring rate of 450 rpm. These silica-gel macrocapsules have been chosen over urea/formaldehyde or polyurethane due to their thermal stability up to 375 °C.

We have demonstrated the feasibility of self-healing of Polyimide films as coating using a photo-radical mechanism to crosslink the TMPTA monomer.

Mechanical properties of film coatings have been investigated and found that -Stress fields near capsules are the essential factor -Large through-the-thickness and side cracks are not healed -Short surface cracks are healed and Debonding of the surface layer containing spherical particles is the most effective mechanism of healing.

For low orbital spaceflight (under 30 km) where there is the presence of small amount of oxygen, the photoreactivity of different formulations based on epoxies and vinyl ethers have been optimized by DPC, showing the feasibility of self-healing in using a photo-cationic mechanism.

Author details

A.A. Périchaud
Catalyse Ltd., Marseille, France

R.M. Iskakov
Institute of Chemical Sciences, Almaty, Kazakhstan
Kazakh British Technical University, Almaty, Kazakhstan

Andrey Kurbatov and T. Z. Akhmetov
Centre of Physico-Chemical Analyses of al-Farabi Kazakh National University, Almaty, Kazakhstan

O.Y. Prokohdko
Physics Department of al-Farabi Kazakh National University, Almaty, Kazakhstan

Irina V. Razumovskaya and Sergey L. Bazhenov
Solid State Physics Department of Federal Educational Establishment of Higher Professional Education "Moscow State Pedagogical University", Russia

P.Y. Apel
Joint Institute of Nuclear Researches, Dubna, Moscow, Russia

V. Yu. Voytekunas
Laboratory of Polymer Science & Advanced Organic Materials LEMP/MAO, CC 021, Université Montpellier II, Sciences et Techniques du Languedoc, Place Eugène Bataillon, France

M.J.M. Abadie
Laboratory of Polymer Science & Advanced Organic Materials LEMP/MAO, CC 021, Université Montpellier II, Sciences et Techniques du Languedoc, Place Eugène Bataillon, France
School of Materials Science & Engineering, Block N4.1, College of Engineering, Nanyang Technological University, Singapore

8. References

[1] M.R. Kessler – *Sefl-Healing: a New Paradigm in Material Designs.* Proc. IMechE vol. 221, Part G: J. Aerospace Engineering, 479-495 (2007)

[2] S.R. White, N.R. Sottos, Geubelle, J.S. Moore, M.R. Kessler, S.R. Sriram, E.N. Brown & S. Viswanathan - *Autonomic Healing of Polymer Composites.* Nature, 409, 794-797 (2001)

[3] S. Cho, H. M. Andersson, S.R. White, N.R. Sottos & P. Braun - *Environmentally Stable Polydimethysiloxane-Based Self-Healing of Polymers.* Advanced Materials 18 (8) 997-1000 (2006)

[4] S. Cho, H.M. Andersson, S.R. White, N.R. Sottos & P. Braun - *Environmentally Stable Polydimethysiloxane-Based Self-Healing of Polymers.* Advanced Materials 18 (8) 997-1000 (2006)

[5] A.S. Jones, J.D. Rule, J.S. Moore, N.R. Sottos & S.R. White - *Life Extension of Self-Healing Polymers with Rapidly Growing Fatigue Cracks.* Special Issue of Self-Healing Polymers and Composites, *J. Royal Society: Interface*, 4, 395-403 (2007)

[6] A.P. Esser-Kahn, S.A. Odom, N.R. Sottos, S.R. White & J.S. Moore – *Triggered Release from Polymer Capsules*. Macromolecules, 44, 5539-5553 (2011)

[7] C. Dry - U.S. patent 2004/00077844, *Multiple Function, Self-Repairing Composites with Special Adhesives.* (06/30/2006)

[8] A.A. Périchaud & M. Devassine - *Self-Repairing Composition, Self-Repairing Materials, Self-Repairing Methods and Applications.* Patent WO 2009/115671 A1

[9] R. Iskakov, A. Périchaud, A. Muzdubaeva, L. Caserta, A. Mirsakieva, B. Khudaibergenov, I. Razumovskaya & P. Apel - *Trimethylolpropanetriacrylate Loaded Nanonpores Pc Films with Self-Healing,* Potential Proceeding 3rd International Conference on Self-Healing Materials Bath, UK. 27-29 June 2011

[10] M.J.M. Abadie & V. Yu. Voytekunas - *New Trends in UV Curing,* Eur.Chem.Tech.Journal, 6 67-77 (2004)

[11] M.J.M. Abadie, N.K. Chia & F.Y.C. Boey - *Cure Kinetics for the Ultraviolet Cationic Polymerization of Cycloliphatic and Diglycidyl Ether of Bisphenol-A (DGEBA) Epoxy Systems with Sulfonium Salt Using an Auto Catalytic Model.* J. Appl. Polym. Sc . 7, 86-97 (2002)

[12] L.L. Ionescu-Vasii & M.J.M. Abadie - *Kinetic Model of Photoinduced Polymerisation of Phenyl Glycidyl Ether Monomer.* Polym. International 1998; 47, 221-225 (1998)

[13] A. Lebel , J. Couve & M.J.M. Abadie - *Study of the Photolysis of Sulfonium Salts. Cationic Photoinitiators.* C.R. Acad Sci. Paris, Série II, 1, 201-207 (1998)

[14] J. Sestak & G. Berggren - *Study of the Kinetics of the Mechanism of Solid-State Reactions at Increasing Temperatures.* Thermochimica Acta 1971; 3: 1-11.